## The Nature of Mathematical Modeling

This is a book about the nature of mathematical modeling, and about the kinds of techniques that are useful for modeling (both natural and otherwise). It is oriented towards simple efficient implementations on computers.

The text has three parts. The first covers exact and approximate analytical techniques (ordinary differential and difference equations, partial differential equations, variational principles, stochastic processes); the second, numerical methods (finite differences for ODEs and PDEs, finite elements, cellular automata); and the third, model inference based on observations (function fitting, data transforms, network architectures, search techniques, density estimation, filtering and state estimation, linear and nonlinear time series).

Each of these essential topics would be the worthy subject of a dedicated text, but such a narrow treatment obscures the connections among old and new approaches to modeling. By covering so much material so compactly, this book helps bring it to a much broader audience. Each chapter presents a concise summary of the core results in an area, providing an accessible introduction to what they can (and cannot) do, enough background to use them to solve typical problems, and then pointers into the specialized research literature. The text is complemented by a Website and extensive worked problems that introduce extensions and applications. This essential book will be of great value to anyone seeking to develop quantitative and qualitative descriptions of complex phenomena, from physics to finance.

Professor Neil Gershenfeld leads the Physics and Media Group at the MIT Media Lab, and codirects the Things That Think research consortium. His laboratory investigates the relationship between the content of information and its physical representation, from building molecular quantum computers to building musical instruments for collaborations ranging from Yo-Yo Ma to Penn & Teller. He has a BA in Physics from Swarthmore College, was a technician at Bell Labs, received a PhD in Applied Physics from Cornell University, and he was a Junior Fellow of the Harvard Society of Fellows.

# The Nature of Mathematical Modeling

Neil Gershenfeld

PUBLISHED BY THE PRESS SYNDICATE OF THE UNIVERSITY OF CAMBRIDGE
The Pitt Building, Trumpington Street, Cambridge, United Kingdom

CAMBRIDGE UNIVERSITY PRESS
The Edinburgh Building, Cambridge CB2 2RU, UK    www.cup.cam.ac.uk
40 West 20th Street, New York, NY 10011-4211, USA    www.cup.org
10 Stamford Road, Oakleigh, Melbourne 3166, Australia
Ruiz de Alarcón 13, 28014 Madrid, Spain

© Cambridge University Press 1999

This book is in copyright. Subject to statutory exception
and to the provisions of relevant collective licensing agreements,
no reproduction of any part may take place without
the written permission of Cambridge University Press.

First published 1999
Reprinted (with corrections) 1999

Printed in the United Kingdom at the University Press, Cambridge

*Typeface* Monotype Ehrhardt $10\frac{1}{2}/13$pt    *System* LaTeX $2_\varepsilon$    [EPC]

*A catalogue record of this book is available from the British Library*

*Library of Congress Cataloguing in Publication data*
Gershenfeld, Neil A.
    The nature of mathematical modeling / Neil Gershenfeld.
       p. cm.
    Includes bibliographical references and index.
    ISBN 0-521-57095-6
    1. Mathematical models.    I. Title.
QA401.G47    1998
511'.8–dc21     98-22029    CIP

ISBN 0 521 57095 6 hardback

# Contents

|  |  |
|---|---|
|  | *page* x |
| *Preface* | xi |
| **1 Introduction** | 1 |
|   1.1 Selected References | 3 |
| **Part One: Analytical Models** | 5 |
| **2 Ordinary Differential and Difference Equations** | 9 |
|   2.1 Linear Differential Equations | 9 |
|   2.2 Systems of Differential Equations and Normal Modes | 12 |
|   2.3 Laplace Transforms | 13 |
|   2.4 Perturbation Expansions | 17 |
|   2.5 Discrete Time Equations | 18 |
|   2.6 $z$-Transforms | 19 |
|   2.7 Selected References | 21 |
|   2.8 Problems | 22 |
| **3 Partial Differential Equations** | 24 |
|   3.1 The Origin of Partial Differential Equations | 24 |
|   3.2 Linear Partial Differential Equations | 26 |
|   3.3 Separation of Variables | 27 |
|     3.3.1 Rectangular Coordinates | 28 |
|     3.3.2 Cylindrical Coordinates | 29 |
|     3.3.3 Spherical Coordinates | 31 |
|   3.4 Transform Techniques | 32 |
|   3.5 Selected References | 33 |
|   3.6 Problems | 33 |
| **4 Variational Principles** | 34 |
|   4.1 Variational Calculus | 34 |
|     4.1.1 Euler's Equation | 34 |
|     4.1.2 Integrals and Missing Variables | 36 |
|     4.1.3 Constraints and Lagrange Multipliers | 37 |

4.2 Variational Problems . . . . . . . . . . . . . . . . . . . . . . 38
    4.2.1 Optics: Fermat's Principle . . . . . . . . . . . . . . . 38
    4.2.2 Analytical Mechanics: Hamilton's Principle . . . . . . . . 38
    4.2.3 Symmetry: Noether's Theorem . . . . . . . . . . . . . 39
4.3 Rigid Body Motion . . . . . . . . . . . . . . . . . . . . . . 40
4.4 Selected References . . . . . . . . . . . . . . . . . . . . . 43
4.5 Problems . . . . . . . . . . . . . . . . . . . . . . . . . . 43

## 5 Random Systems     44
5.1 Random Variables . . . . . . . . . . . . . . . . . . . . . . 44
    5.1.1 Joint Distributions . . . . . . . . . . . . . . . . . . . 46
    5.1.2 Characteristic Functions . . . . . . . . . . . . . . . . 48
5.2 Stochastic Processes . . . . . . . . . . . . . . . . . . . . . 50
    5.2.1 Distribution Evolution Equations . . . . . . . . . . . . 51
    5.2.2 Stochastic Differential Equations . . . . . . . . . . . . 55
5.3 Random Number Generators . . . . . . . . . . . . . . . . . 56
    5.3.1 Linear Congruential . . . . . . . . . . . . . . . . . . 57
    5.3.2 Linear Feedback . . . . . . . . . . . . . . . . . . . . 59
5.4 Selected References . . . . . . . . . . . . . . . . . . . . . 60
5.5 Problems . . . . . . . . . . . . . . . . . . . . . . . . . . 60

# Part Two: Numerical Models     63

## 6 Finite Differences: Ordinary Differential Equations     67
6.1 Numerical Approximations . . . . . . . . . . . . . . . . . . 67
6.2 Runge–Kutta Methods . . . . . . . . . . . . . . . . . . . . 70
6.3 Beyond Runge–Kutta . . . . . . . . . . . . . . . . . . . . . 72
6.4 Selected References . . . . . . . . . . . . . . . . . . . . . 76
6.5 Problems . . . . . . . . . . . . . . . . . . . . . . . . . . 76

## 7 Finite Differences: Partial Differential Equations     78
7.1 Hyperbolic Equations: Waves . . . . . . . . . . . . . . . . . 79
7.2 Parabolic Equations: Diffusion . . . . . . . . . . . . . . . . 81
7.3 Elliptic Equations: Boundary Values . . . . . . . . . . . . . . 84
7.4 Selected References . . . . . . . . . . . . . . . . . . . . . 91
7.5 Problems . . . . . . . . . . . . . . . . . . . . . . . . . . 91

## 8 Finite Elements     93
8.1 Weighted Residuals . . . . . . . . . . . . . . . . . . . . . 93
8.2 Rayleigh–Ritz Variational Methods . . . . . . . . . . . . . . 99
8.3 Selected References . . . . . . . . . . . . . . . . . . . . . 100
8.4 Problems . . . . . . . . . . . . . . . . . . . . . . . . . . 101

## 9 Cellular Automata and Lattice Gases     102
9.1 Lattice Gases and Fluids . . . . . . . . . . . . . . . . . . . 103
9.2 Cellular Automata and Computing . . . . . . . . . . . . . . 107

|       |                                                      |      |
|------:|------------------------------------------------------|-----:|
| 9.3   | Selected References                                  | 109  |
| 9.4   | Problems                                             | 110  |

## Part Three: Observational Models     111

## 10 Function Fitting     115
- 10.1 Model Estimation . . . . . 116
- 10.2 Least Squares . . . . . 117
- 10.3 Linear Least Squares . . . . . 118
  - 10.3.1 Singular Value Decomposition . . . . . 119
- 10.4 Nonlinear Least Squares . . . . . 122
  - 10.4.1 Levenberg–Marquardt Method . . . . . 124
- 10.5 Estimation, Fisher Information, and the Cramér–Rao Inequality . . . . 125
- 10.6 Selected References . . . . . 127
- 10.7 Problems . . . . . 127

## 11 Transforms     128
- 11.1 Orthogonal Transforms . . . . . 128
- 11.2 Fourier Transforms . . . . . 129
- 11.3 Wavelets . . . . . 131
- 11.4 Principal Components . . . . . 136
- 11.5 Selected References . . . . . 138
- 11.6 Problems . . . . . 138

## 12 Architectures     139
- 12.1 Polynomials . . . . . 139
  - 12.1.1 Padé Approximants . . . . . 139
  - 12.1.2 Splines . . . . . 141
- 12.2 Orthogonal Functions . . . . . 142
- 12.3 Radial Basis Functions . . . . . 145
- 12.4 Overfitting . . . . . 147
- 12.5 Curse of Dimensionality . . . . . 148
- 12.6 Neural Networks . . . . . 150
  - 12.6.1 Back Propagation . . . . . 152
- 12.7 Regularization . . . . . 153
- 12.8 Selected References . . . . . 155
- 12.9 Problems . . . . . 155

## 13 Optimization and Search     156
- 13.1 Multidimensional Search . . . . . 157
- 13.2 Local Minima . . . . . 161
- 13.3 Simulated Annealing . . . . . 162
- 13.4 Genetic Algorithms . . . . . 164
- 13.5 The Blessing of Dimensionality . . . . . 166

|       | 13.6 Selected References | 167 |
|---|---|---|
|       | 13.7 Problems | 168 |

## 14  Clustering and Density Estimation — 169
- 14.1 Histogramming, Sorting, and Trees ... 169
- 14.2 Fitting Densities ... 172
- 14.3 Mixture Density Estimation and Expectation-Maximization ... 174
- 14.4 Cluster-Weighted Modeling ... 178
- 14.5 Selected References ... 185
- 14.6 Problems ... 185

## 15  Filtering and State Estimation — 186
- 15.1 Matched Filters ... 186
- 15.2 Wiener Filters ... 187
- 15.3 Kalman Filters ... 189
- 15.4 Nonlinearity and Entrainment ... 195
- 15.5 Hidden Markov Models ... 197
- 15.6 Selected References ... 203
- 15.7 Problems ... 203

## 16  Linear and Nonlinear Time Series — 204
- 16.1 Linear Time Series ... 205
- 16.2 The Breakdown of Linear Systems Theory ... 207
- 16.3 State-Space Reconstruction ... 208
- 16.4 Characterization ... 213
  - 16.4.1 Dimensions ... 214
  - 16.4.2 Lyapunov Exponents ... 216
  - 16.4.3 Entropies ... 217
- 16.5 Forecasting ... 220
- 16.6 Selected References ... 224
- 16.7 Problems ... 224

## *Appendix 1* Graphical and Mathematical Software — 225
- A1.1 Math Packages ... 226
  - A1.1.1 Programming Environments ... 226
  - A1.1.2 Interactive Environments ... 228
- A1.2 Graphics Tools ... 230
  - A1.2.1 Postscript ... 230
  - A1.2.2 X Windows ... 234
  - A1.2.3 OpenGL ... 240
  - A1.2.4 Java ... 244
- A1.3 Problems ... 249

## *Appendix 2* Network Programming — 250
- A2.1 OSI, TCP/IP, and All That ... 250

A2.2 Socket I/O . . . . . . . . . . . . . . . . . . . . . . . . 251
  A2.3 Parallel Programming . . . . . . . . . . . . . . . . . . 254

*Appendix 3* **Benchmarking**     257

*Appendix 4* **Problem Solutions**     259
  A4.1 Introduction . . . . . . . . . . . . . . . . . . . . . . . 259
  A4.2 Ordinary Differential and Difference Equations . . . . . . . . . . . 259
  A4.3 Partial Differential Equations . . . . . . . . . . . . . . . . . 266
  A4.4 Variational Principles . . . . . . . . . . . . . . . . . . . . 269
  A4.5 Random Systems . . . . . . . . . . . . . . . . . . . . . 271
  A4.6 Finite Differences: Ordinary Differential Equations . . . . . . . . . 276
  A4.7 Finite Differences: Partial Differential Equations . . . . . . . . . . 281
  A4.8 Finite Elements . . . . . . . . . . . . . . . . . . . . . . 289
  A4.9 Cellular Automata and Lattice Gases . . . . . . . . . . . . . . 292
  A4.10 Function Fitting . . . . . . . . . . . . . . . . . . . . . . 302
  A4.11 Transforms . . . . . . . . . . . . . . . . . . . . . . . . 305
  A4.12 Architectures . . . . . . . . . . . . . . . . . . . . . . . 309
  A4.13 Optimization and Search . . . . . . . . . . . . . . . . . . 315
  A4.14 Clustering and Density Estimation . . . . . . . . . . . . . . . 319
  A4.15 Filtering and State Estimation . . . . . . . . . . . . . . . . 323
  A4.16 Linear and Nonlinear Time Series . . . . . . . . . . . . . . . 325

*Bibliography*     330
*Index*     340

For GLADYS AND WALTER
who taught me how to build models

# Preface

This is a book about the nature of mathematical modeling, and about the kinds of techniques that are useful for modeling systems (both natural and otherwise). It is oriented towards simple efficient implementations on computers. Digital systems are routinely used for modeling purposes ranging from characterizing and transmitting realities to experimenting with possibilities to realizing fantasies; whether the goal is to reproduce our world or to create new worlds, there is a recurring need for compact, rich descriptions of how systems behave.

This text, like its companion *The Physics of Information Technology* [Gershenfeld, 1999a], grew out of many questions from students that indicated that the pressure for specialization within disciplines often leads to a lack of awareness of alternative approaches to a problem. Just as a typical physicist learns about variational principles, but not about how they can be used to derive finite element algorithms, an engineer who learns about finite elements might not know about techniques for exact or approximate analytical solutions of variational problems. Many people learn about how continuum equations arise from microscopic dynamics, and about how finite difference approximations of them can be used for numerical solutions, but not about how it is possible to stop before the continuum description by using lattice gases. And few outside of specialized research communities encounter emerging ideas such as state-space reconstruction for nonlinear systems, or methods for managing complexity in machine learning.

Each of these topics can (and should) be the subject of a semester-long course, and so to cover this range I necessarily have had to sacrifice some depth. Although this brisk pace risks missing important points, the alternative for many people may be no exposure at all to this material. I've tried to strike a balance by introducing the basic ideas, giving simple but complete and useful algorithms, and then providing pointers into the literature. Other than assuming some calculus, linear algebra, and programming skills, the book is self-contained. My hope is that the inexperienced reader will find this collection of topics to be a useful introduction to what is known, readers with more experience will be able to find usable answers to specific questions, and that advanced readers will find this to be a helpful platform from which to see further [Merton, 1993].

Included here are a number of "old-fashioned" subjects such as ordinary differential equations, and "new-fashioned" ones such as neural networks. One of the major goals is to make clear how the latter relate to the former. The modern profusion of fashionable techniques has led to equally exaggerated claims of success and failure. There really are some important underlying new insights that were not part of traditional practice, but these are best viewed in that richer context. The study of neural networks, for example, has helped lead to more generally applicable lessons about functional approximation and search in high-dimensional spaces. I hope that by demystifying the hype surrounding some of these ideas I can help bring out their real magic.

The study of modeling is inseparable from the practice of modeling. I've tried to keep this text concise by focusing on the important underlying ideas, and left the refinements and applications to the problem sets as much as possible. I've included the problem solutions because I use problem sets not as a test to grade correct answers, but rather as a vehicle to teach problem solving skills and supporting material. My students are cautioned that although the solutions are freely available, their grade is based on how they approach the problems in class rather than on what answers they get, and so reciting my solutions back is both obvious and a waste of everyone's time. This organization helps them, and me, focus attention where it should be: on developing solutions instead of looking them up.

In this text I usually start with the assumption that the governing equations or experimental observations to be modeled are already known, and leave the introduction of specific applications to the relevant literature (see, for example, [Gershenfeld, 1999a] for details, and [Gershenfeld, 1999b] for context). Therefore, an important component of the corresponding course at MIT is a semester modeling project that provides a broader chance for the students to develop practical modeling skills in particular domains. This is the best way I've found to communicate the matters of taste and judgement that are needed to turn the raw material in this text into useful results. These projects are always a pleasure to supervise, ranging from studying the spread of gossip to the spread of fire. Some of the most valuable lessons have come from students sharing their experience building models. To help extend that interchange to all of the readers of this text there is a Web page at http://www.cup.cam.ac.uk/online/nmm that provides pointers to explore, comment on, and contribute modeling projects.

I am a great admirer of real mathematicians, who I define to be people who always are aware of the difference between what they know and what they think they know; I certainly am not one. I hope that my presumption in reducing whole disciplines to ten or so pages apiece of essential ideas is exceeded by the value of such a compact presentation. For my inevitable sins of omission, comission, and everything in between, I welcome your feedback at nmm@media.mit.edu.

It has been a great pleasure to watch my sketchy lecture notes grow up into this text under the guidance of the many students and colleagues who have helped shape it by their thoughtful, challenging, exasperating, inspiring questions and comments (with a particular thanks to F. Joseph Pompei). And I am grateful to the Media Lab and its community of sponsors and collaborators who have created an environment that lets dreams be chased wherever they lead.

Cambridge, MA                                                                                                                  Neil Gershenfeld
April, 1998

# 1 Introduction

How would you describe

- The flickering of a flame?
- The texture of an oil painting?
- Highway traffic during a rush hour?
- Twinkling stars?
- Breaking glass?
- A bowling ball hitting pins?
- Melting ice?
- The flight of a paper airplane?
- The sound of a violin?

These questions do not have simple answers: many are active research areas. There cannot be a single recipe that covers this whole menu. There are many possible levels of description; choosing among them depends on your goals and on the available tools. This text is a tour through those spaces. For example, if you seek to make a mathematical model of a violin, you could use a numerical model based on a first-principles description. This lets you match your model parameters to measurements on a real instrument, and change parameters between a Stradivarius and a Guarneri. However, running it in real time will require a supercomputer, and the effort to find good parameters for the model is almost as much work as building a real violin. Alternatively, you could try to use an analytical (pencil-and-paper) solution to the governing equations; in return for some large approximations you may be able to find a useful explicit solution, but it might not sound very good. Finally, you could forget about the underlying governing equations entirely and experimentally try to find an effective description of how the player's actions are related to the sound made by the instrument (which is a reasonable thing to do because dissipation and symmetries in a system reduce the effective number of degrees of freedom [Temam, 1988]). These three approaches (analytical, numerical, and observational) comprise the three parts of this book.

To build a model there are many decisions that must be made, either explicitly or more often implicitly. Some of these are shown in Figure 1.1. Each of these is a continuum rather than a discrete choice. This list is not exhaustive, but it's important to keep

Figure 1.1. Some levels of description for mathematical model building.

returning to it: many efforts fail because of an unintentional attempt to decribe either too much or too little.

These are *meta-modeling* questions. There are no rigorous ways to make these choices, but once they've been decided there are rigorous ways to use them. There's no single definition of a "best" model, although quasi-religious wars are fought over the question. One good attempt is the *Minimum Description Length* principle [Rissanen, 1986], essentially Occam's Razor: the best model is the one that is the smallest (including the information to specify both the form of the model and the values of the parameters). Unfortunately, this has two serious problems: finding the minimum description length for a given problem is an uncomputable task, and it says nothing about the error metric that will be used to judge the model. A stock trader, civil engineer, cardiologist, and video game designer have very different standards for success. They differ in the prior information they have about their problem, and the posterior criteria that they will use to evaluate and update their model. Ultimately, the strongest useful statement is that the best model is the one that works best for you.

Surprisingly little ambition is needed to exceed the performance of almost any available computer, and conversely computer hardware speeds have been racing ahead of the development of software tools to use them effectively. Where computational speed is most important, the examples in this book will use efficient portable low-level tools (such as C and X Windows). On the other hand, where algorithm clarity is most important, high-level environments will be used (such as Matlab). The appendices provide brief introductions to these environments.

No single reference text covers the range of subjects in this book. To help access the literature, each chapter ends with a list of relevant general sources, and then cites the more specialized literature as needed throughout. Where important ideas are introduced without any references they are either so well known that they need no further citation, or are my own results that I have not published elsewhere (the context should make this distinction clear). And I've used *URL*s (*World Wide Web Uniform Resource Locators*) where possible to provide pointers to information on the Internet.

## 1.1 SELECTED REFERENCES

[Press *et al.*, 1992] Press, William H., Teukolsky, Saul A., Vetterling, William T., & Flannery, Brian P. (1992). *Numerical Recipes in C: The Art of Scientific Computing*. 2nd edn. New York, NY: Cambridge University Press.

> This is warmly recommended for almost any numerical problem. The numerical analysis literature is full of rigorous results that have little bearing on solving practical problems; *Numerical Recipes* gracefully merges theoretical insights with practical tricks for most useful algorithms. It's one of those rare books that's immediately useful by a beginner but that continues to hold new insights for an expert.

[Pearson, 1990] Pearson, Carl E. (1990). *Handbook of Applied Mathematics: Selected Results and Methods*. 2nd edn. New York: Van Nostrand Reinhold.

> This is a good example of one of a number of such large reference volumes that survey applied mathematics.

*Part One*
Analytical Models

The first part of the book looks at *analytical models*. These are models that you can at least in theory write down with nothing more than a pencil and a piece of paper, hopefully arriving at an explicit *closed-form* solution. Analytical modeling is often, but not always, done with *analytic functions* [Saff & Snider, 1993], and so we will usually assume that the functions we encounter can be expanded in a power series. Analytical models have been, and continue to be, of great importance because of their power: where they are applicable, it can be possible to deduce everything there is to know about a system. The cost for this power is limited applicability – much of the world is simply too complicated to describe this way.

Analytical models are still important in approximate techniques that do require computers. This includes numerical methods, which can use pieces of analytical solutions to make the numerical steps much more effective, and symbolic methods that can quantitatively expand the effective size of your piece of paper with significant qualitative implications (such as the ability to do much higher-order perturbation theory).

The first chapter covers *ordinary differential equations*, where a collection of variables change as a function of one independent variable (such as time). The orbits of the planets are a classic example. The next chapter adds more independent degrees of freedom (such as space) to arrive at *partial differential equations* to describe, for example, the ripples on the surface of a lake. There is an intimate connection between local differential equations and the global properties of a system, introduced in the following chapter on *variational methods*. The last chapter looks at solutions for *stochastic systems*. While being exact about something random might appear to be paradoxical, there are many powerful techniques for exactly describing the *distribution* of a random variable without saying anything about the particular value of the variable.

# 2 Ordinary Differential and Difference Equations

## 2.1 LINEAR DIFFERENTIAL EQUATIONS

Change is the most interesting aspect of most systems, hence the central importance across disciplines of differential equations. An *ordinary differential equation* (*ODE*) is an equation (or system of equations) written in terms of an unknown function and its derivatives with respect to a single independent variable (such as time). Examples include the familiar equations of classical mechanics and electrical circuits. In the next chapter we will consider *partial differential equations* (*PDEs*), which have multiple independent variables (such as space, for example in fluid flow or electrodynamics). The subject of differential equations can appear to be quite tedious. In part it is: it is like learning spelling and grammar as a necessary prelude to the study of Shakespeare. And in part it isn't: there can be beautiful structure lurking behind what appear to be very simple differential equations. This chapter will concentrate on the canon of linear (or nearly linear) differential equations; after detouring through many other supporting topics the book will return to consider nonlinear differential equations in the closing chapter on time series.

The simplest differential equation can immediately be solved by integration

$$\frac{dy}{dt} = f(t) \Rightarrow dy = f(t)\, dt$$

$$\Rightarrow y(t_1) - y(t_0) = \int_{t_0}^{t_1} f(t)\, dt \tag{2.1}$$

(a point that is surprisingly often forgotten). The *order* of a differential equation is the highest derivative that occurs, and so the preceeding example is a first-order equation. If every term involves either the unknown function or its derivatives the equation is said to be *homogeneous*; if there is a term that depends on the independent variable alone (i.e., a forcing term) then the equation is *inhomogeneous*. If the unknown function does not appear within powers or more complicated functions, then the differential equation is linear, and can be written in terms of a linear operator $L_N(y)$ defined by

$$L_N(y) \equiv \frac{d^N y}{dt^N} + A_1(t)\frac{d^{N-1} y}{dt^{N-1}} + \cdots + A_{N-1}(t)\frac{dy}{dt} + A_N(t) y \ . \tag{2.2}$$

There is no need for an $A_0$ coefficient because it can be eliminated by dividing all the other terms by it. $L_N(y) = f(t)$ is an inhomogeneous equation, and $L_N(y) = 0$ is the associated homogeneous equation.

Linear differential equations are particularly important, in part because they occur so often (particularly in systems that are not strongly driven), and in part because general techniques exist for solving them (whether or not they really apply to what might be a nonlinear problem). Although this can be a bit like the proverbial drunk looking for lost change under a street lamp, it is sensible if it is the only illumination available.

The solution of an $N$th-order linear differential equation will contain $N$ unknown constants that are determined by *boundary conditions*. If it is an *initial-value problem*, the initial values of $N$ independent functions of the variable and its derivatives are given (usually, $y(0)$, $dy/dt(0)$, ..., $d^{N-1}y/dt^{N-1}(0)$). For a *boundary-value problem*, boundary conditions are given at both the beginning and the end of an interval.

An $N$-order homogeneous equation $L_N(y) = 0$ will have $N$ linearly independent solutions $u_1(t)$, $u_2(t)$, ..., $u_N(t)$. By superposition, an arbitrary linear combination of them will also be a solution:

$$y_g(t) = \sum_{n=1}^{N} C_n u_n(t) \ . \tag{2.3}$$

This is the *general solution*; any solution of the homogeneous equation can be represented by an appropriate choice of the $C_n$'s. If a *particular solution* of the inhomogeneous problem can be found ($L_N(y_p) = f(t)$), then the *complete solution* is $y(t) = y_g(t) + y_p(t)$. The general solution represents the transient response of the system to the boundary conditions, and the particular solution is the result of the forcing of the system by the inhomogeneous term.

The simplest linear differential equation has constant coefficients:

$$\frac{d^N y}{dt^N} + A_1 \frac{d^{N-1} y}{dt^{N-1}} + \cdots + A_{N-1} \frac{dy}{dt} + A_N y = f(t) \ . \tag{2.4}$$

An important technique for solving differential equations is to guess the functional form of a solution (called an *ansatz*, or trial answer), substitute it in, and then see if the free parameters can be adjusted to make the solution work. Because the solution of a differential equation is unique as long as the functions defining it are reasonably smooth and bounded [Coddington & Levinson, 1984], if you find *a* solution then that is *the* solution. If we try the guess $y = e^{rt}$ for the solution of the homogeneous part of equation (2.4), the result of substituting it in is the *characteristic equation*

$$r^N + A_1 r^{N-1} + \cdots + A_{N-1} r + A_N = 0 \ . \tag{2.5}$$

This $N$th-order polynomial has $N$ roots. The real part of the roots represent exponentially growing or decaying solutions, and the complex part oscillatory behavior. If all of the roots are distinct:

$$r^N + A_1 r^{N-1} + \cdots + A_{N-1} r + A_N = (r - r_1)(r - r_2) \cdots (r - r_N) \tag{2.6}$$

then the general solution is

$$y_g = \sum_{n=1}^{N} C_n e^{r_n t} \ . \tag{2.7}$$

This gives the $N$ linearly independent solutions required for a general solution. However, if a root has a higher multiplicity

$$r^N + A_1 r^{N-1} + \cdots + A_{N-1} r + A_N = (r - r_1)^M (r - r_{M+1}) \cdots (r - r_N) \tag{2.8}$$

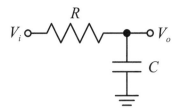

Figure 2.1. An RC circuit.

then this will provide fewer than $N$ solutions. The missing solutions are found by recognizing that if $L_N(e^{rt}) = 0$ then

$$\frac{\partial}{\partial r} L_N(e^{rt}) = L_N\left(\frac{\partial e^{rt}}{\partial r}\right) = L_N(te^{rt}) = 0 \ ,$$

$$\frac{\partial^2}{\partial r^2} L_N(e^{rt}) = L_N\left(\frac{\partial^2}{\partial r^2} e^{rt}\right) = L_N(t^2 e^{rt}) = 0 \ , \quad (2.9)$$

and so forth. Therefore, the $M$ functions

$$(C_1 + C_2 t + C_3 t^2 + \cdots + C_M t^{M-1}) e^{r_1 t} \quad (2.10)$$

are linearly independent solutions to $L_N(y) = 0$, and so these can be used as the $M$ solutions associated with the $M$-fold root. It might appear that this trick can be used to generate arbitrarily many solutions by continuing to differentiate, but this is not so: a derivative of an order higher than the multiplicity of equation (2.8) will give the useless equation $0 = 0$.

As a simple example of a linear constant-coefficient differential equation, consider a circuit consisting of a resistor and a capacitor (Figure 2.1). The current into the node from the resistor is $(V_i - V_o)/R$, and the current out of the node into the capacitor is $C\dot{V}_o$, and so the governing equation for this circuit is

$$C\dot{V}_o = \frac{V_i - V_o}{R} \quad (2.11)$$

or

$$RC\dot{V}_o + V_o = V_i \ . \quad (2.12)$$

The characteristic equation gives

$$RC \cdot r + 1 = 0 \Rightarrow r = \frac{-1}{RC} \Rightarrow V_o = Ae^{-t/RC} \ . \quad (2.13)$$

The undriven response of the circuit is to exponentially discharge the capacitor. Now, let's assume periodic forcing $V_i = \exp(i\omega t)$ and look for a particular solution at this frequency. The voltage in the circuit is of course a real number; by representing it as a complex number we can simultaneously keep track of both phase components (sin and cos). Plugging in the ansatz $V_o = A\exp(i\omega t)$ gives

$$RCAi\omega + A = 1 \Rightarrow A = \frac{1}{1 + i\omega RC} \ . \quad (2.14)$$

At low frequencies the output is equal to the input; at high frequencies it rolls off as $1/\omega$ (it is a low-pass filter) and is out of phase by 90°. Problem 2.1 covers the important example of a damped, driven harmonic oscillator.

This completes (more-or-less) everything that there is to know about solving linear differential equations. The theory is simple and useful. The situation is very different for nonlinear differential equations, where amidst a sea of insoluble problems live special tricks for some tractable equations, approximation methods based on some nearby exactly soluble problems, and qualitative insights into the behavior of classes of solutions. Because of this, the study of nonlinear differential equations requires either a lot of specialized attention or else numerical methods.

Another extension of this basic theory is to coupled systems of equations. Once again, little general can be said about nonlinear systems, but for the case of linear couplings it is possible to find exact solutions. The next section looks at this for the important case of coupled oscillators.

## 2.2 SYSTEMS OF DIFFERENTIAL EQUATIONS AND NORMAL MODES

The $N$th-order linear differential equation (2.4) can be written as a first-order equation for an $N$-dimensional vector

$$\frac{d}{dt} \begin{pmatrix} y_0 \\ y_1 \\ \vdots \\ y_{N-2} \\ y_{N-1} \end{pmatrix} = \quad (2.15)$$

$$\begin{pmatrix} 0 & 1 & 0 & \cdots & 0 \\ 0 & 0 & 1 & \cdots & 0 \\ \vdots & \vdots & \vdots & \ddots & \vdots \\ 0 & 0 & 0 & \cdots & 1 \\ -A_N & -A_{N-1} & -A_{N-2} & \cdots & -A_1 \end{pmatrix} \begin{pmatrix} y_0 \\ y_1 \\ \vdots \\ y_{N-2} \\ y_{N-1} \end{pmatrix} + \begin{pmatrix} 0 \\ 0 \\ \vdots \\ 0 \\ f(t) \end{pmatrix} .$$

This transformation does not make the problem any simpler (it can be solved by diagonalizing the matrix, which requires solving exactly the same characteristic equation), but it can be convenient to simplify notation by using a vector first-order equation [Gershenfeld et al., 1983].

This is a simple example of a system of differential equations. Such systems also arise whenever there are interactions; an important special case is an unforced, undamped system of masses with coordinates $(y_1, y_2, \ldots, y_N) \equiv \vec{y}$ that have a restoring force that is an arbitrary linear combination of their positions. The corresponding vector equation is

$$\frac{d^2 \vec{y}}{dt^2} + \mathbf{A} \cdot \vec{y} = 0 \ . \quad (2.16)$$

If the coupling matrix **A** is diagonal ($A_{ij} = 0$ for $i \neq j$) then the oscillators will be independent, but if it isn't then they won't. Let's look for a new set of variables $\vec{z} \equiv \mathbf{M}^{-1} \cdot \vec{y}$, defined by an unknown transformation **M**, for which these equations decouple:

$$\frac{d^2 \vec{z}}{dt^2} + \mathbf{D} \cdot \vec{z} = 0 \ , \qquad (2.17)$$

where **D** is a diagonal matrix. The required transformation **M** can be found by changing variables:

$$\frac{d^2 \vec{y}}{dt^2} + \mathbf{A} \cdot \vec{y} = 0$$

$$\mathbf{M} \cdot \frac{d^2 \vec{z}}{dt^2} + \mathbf{A} \cdot \mathbf{M} \cdot \vec{z} = 0$$

$$\frac{d^2 \vec{z}}{dt^2} + \mathbf{M}^{-1} \cdot \mathbf{A} \cdot \mathbf{M} \cdot \vec{z} = 0$$

$$\frac{d^2 \vec{z}}{dt^2} + \mathbf{D} \cdot \vec{z} = 0 \qquad (2.18)$$

and so $\mathbf{M}^{-1} \cdot \mathbf{A} \cdot \mathbf{M} = \mathbf{D}$ or $\mathbf{A} \cdot \mathbf{M} = \mathbf{M} \cdot \mathbf{D}$. This will be the case (remember that **D** is diagonal) if the columns of **M** are the eigenvectors of **A** (the diagonal elements of **D** will then be the eigenvalues of **M**). This procedure is called *diagonalizing*. The new variables here are called *normal modes* [Goldstein, 1980, Scheck, 1990] and behave exactly like independent oscillators. There will be as many normal modes as there are degrees of freedom, unless there are fewer distinct eigenvectors because of degenerate eigenvalues. Problem 2.2 finds the normal modes for a simple system.

## 2.3 LAPLACE TRANSFORMS

Using the characteristic equation to solve a differential equation requires separate steps to find the general solution, search for a particular solution, and solve for the coefficients to match the boundary conditions. *Laplace transforms* provide a convenient alternative, turning many differential equations into an algebraic problem and giving the complete solution in a single step.

The *one-sided Laplace transform* of a function $f(t)$ is defined by

$$\mathcal{L}\{f(t)\} \equiv F(s) = \int_0^\infty e^{-st} f(t) \, dt \ . \qquad (2.19)$$

If the integral extended from $-\infty$ to $\infty$ this would be the *two-sided* Laplace transform. The one-sided transform explicitly includes the initial conditions of the system at $t = 0$, and for this reason we will use it; the two-sided transform is used for steady-state problems for which the initial conditions do not matter.

The Laplace transform is a generalization of the Fourier transform to an arbitrary complex argument. Its usefulness for differential equations comes from recognizing that differentiation just multiplies that Laplace transfrom by $s$:

$$\mathcal{L}\left\{\frac{df(t)}{dt}\right\} = \int_0^\infty e^{-st}\frac{df(t)}{dt}\,dt$$

$$= e^{-st}f(t)\bigg|_0^\infty + s\int_0^\infty e^{-st}f(t)\,dt$$

$$= sF(s) - f(0) \;, \tag{2.20}$$

where the second step follows by integrating by parts

$$\int_A^B u\,dv = uv\bigg|_A^B - \int_A^B v\,du \;. \tag{2.21}$$

Similarly, for second derivatives

$$\mathcal{L}\left\{\frac{d^2f(t)}{dt^2}\right\} = \int_0^\infty e^{-st}\frac{d^2f(t)}{dt^2}\,dt$$

$$= e^{-st}\frac{df(t)}{dt}\bigg|_0^\infty + s\int_0^\infty e^{-st}\frac{df(t)}{dt}\,dt$$

$$= s^2 F(s) - sf(0) - \frac{df(0)}{dt} \;, \tag{2.22}$$

and by induction for higher derivatives

$$\mathcal{L}\left\{\frac{d^N f(t)}{dt^N}\right\} = s^N F(s) - s^{N-1}f(0) - s^{N-2}\frac{df(0)}{dt}$$

$$- s^{N-3}\frac{d^2 f(0)}{dt^2} - \cdots - \frac{d^{N-1}f(0)}{dt^{N-1}} \tag{2.23}$$

(see [Hildebrand, 1976] for the continuity requirements assumed by these results). There is a corresponding relationship for integrals:

$$\mathcal{L}\left\{\int_0^t f(u)\,du\right\} = \int_0^\infty e^{-st}\int_0^t f(u)\,du\,dt$$

$$= -\frac{e^{-st}}{s}\int_0^t f(u)\,du\bigg|_0^\infty + \frac{1}{s}\int_0^\infty e^{-st}f(t)\,dt$$

$$= \frac{1}{s}F(s) \;. \tag{2.24}$$

The Laplace transform turns differential and integral equations into algebraic equations. Without proof, Table 2.1 gives a number of other Laplace transform pairs (much longer tables are available in any mathematics handbook). Where needed it is assumed that the arguments vanish for $t < 0$.

In order to solve a differential equation by using Laplace transforms, the steps are

1. Laplace transform the differential equation into an algebraic equation, including the initial conditions.
2. Solve this new equation for the unknown function in terms of the transform variable $s$.
3. Find the inverse transform.

There is not an automatic way to invert a Laplace transform. The easiest approach, when it works, is to look up the inverse in a table of Laplace transforms. Many more problems can be handled by first doing a *partial fraction expansion* to simplify them.

Table 2.1. *Selected Laplace transforms.*

$$\mathcal{L}\{t^N f(t)\} = (-1)^N \frac{d^N F(s)}{ds^N}$$

$$\mathcal{L}\{e^{at} f(t)\} = F(s-a)$$

$$\mathcal{L}\left\{\int_0^t f(t-u)g(u)\,du\right\} \equiv \mathcal{L}\{f*g\} = \text{convolution} = F(s)G(s)$$

$$\mathcal{L}\{1\} = \frac{1}{s}$$

$$\mathcal{L}\{e^{-at}\} = \frac{1}{s+a}$$

$$\mathcal{L}\{\sin at\} = \frac{a}{s^2 + a^2}$$

$$\mathcal{L}\{\cos at\} = \frac{s}{s^2 + a^2}$$

$$\mathcal{L}\{\sinh at\} = \frac{a}{s^2 - a^2}$$

$$\mathcal{L}\{\cosh at\} = \frac{s}{s^2 - a^2}$$

$$\mathcal{L}\left\{\frac{1}{a-b}(-e^{-at} + e^{-bt})\right\} = \frac{1}{(s+a)(s+b)}$$

$$\mathcal{L}\left\{\frac{1}{a-b}(ae^{-at} - be^{-bt})\right\} = \frac{s}{(s+a)(s+b)}$$

$$\mathcal{L}\left\{\frac{e^{-at}}{(a-b)(a-c)} - \frac{e^{-bt}}{(a-b)(b-c)} - \frac{e^{-ct}}{(a-c)(c-b)}\right\} = \frac{1}{(s+a)(s+b)(s+c)}$$

$$\mathcal{L}\left\{-\frac{ae^{-at}}{(a-b)(a-c)} + \frac{be^{-bt}}{(a-b)(b-c)} + \frac{ce^{-ct}}{(a-c)(c-b)}\right\} = \frac{s}{(s+a)(s+b)(s+c)}$$

$$\mathcal{L}\{\delta(t-t_0)\} = e^{-st_0}$$

$$\mathcal{L}\{t^N\} = \frac{N!}{s^{N+1}}$$

$$\mathcal{L}\left\{\frac{t^{N-1}e^{-at}}{(N-1)!}\right\} = \frac{1}{(s+a)^N}$$

The transform $F(s)$ of a constant-coefficient linear differential equation will be the ratio of two polynomials

$$F(s) = \frac{N(s)}{D(s)}, \qquad (2.25)$$

called a *rational function*. The roots of the numerator $N(s)$ are called *zeros*, and the roots of the denominator $D(s)$ are called *poles*. Let the poles be $\{p_i\}_{i=1}^N$. The *partial fraction expansion* of an $F(s)$ with distinct poles and a numerator of lower degree than the denominator is [Pearson, 1990]

$$F(s) = \sum_{i=1}^{N} \frac{A_i}{s - p_i} \qquad (2.26)$$

where

$$A_i = \lim_{s \to p_i} F(s)(s - p_i) \quad . \qquad (2.27)$$

The definition of the coefficients can be verified by substitution:

$$F(s)(s - p_i)|_{s=p_i} = A_i + \sum_{\substack{j \neq i}}^{N} A_j \frac{s - p_i}{s - p_j}\bigg|_{s=p_i} = A_i \quad . \qquad (2.28)$$

For repeated poles $p_n$ with a multiplicity $n$, the terms in the partial fraction expansion are of the form

$$F(s) = \frac{A_n}{(s - p_n)^n} + \frac{A_{n-1}}{(s - p_n)^{n-1}} + \cdots + \frac{A_1}{s - p_n} \quad , \qquad (2.29)$$

with

$$A_n = \lim_{s \to p_n} F(s)(s - p_n)^n \quad . \qquad (2.30)$$

Finally, some problems can be solved by doing a complex contour integral to invert equation (2.19) [Saff & Snider, 1993].

Let's return to the simple example of an RC circuit:

$$RC\dot{V}_o(t) + V_o(t) = V_i(t)$$

$$RC[sV_o(s) - V_o(0)] + V_o(s) = V_i(s)$$

$$V_o(s) = \frac{V_i(s)}{1 + sRC} + \frac{RCV_o(0)}{1 + sRC}$$

$$V_o(t) = (RC)^{-1} e^{-t/RC} * V_i(t) + V_o(0) e^{-t/RC} \qquad (2.31)$$

(from Table 2.1). We see immediately that the output is the sum of the transient decay of the initial state and the convolution of the exponential decay with the input. If as before we take the forcing to be $\exp(i\omega t)$,

$$V_o(s) = \frac{1}{1 + sRC} \frac{1}{s - i\omega} + \frac{RCV_o(0)}{1 + sRC}$$

$$V_o(t) = \frac{1}{1 + i\omega RC} \left( e^{i\omega t} - e^{-t/RC} \right) + V_o(0) e^{-t/RC} \qquad (2.32)$$

(using Table 2.1). We did not see the second part of the first term before because we did not worry about satisfying the initial conditions; here it arises naturally from the use of the Laplace transform.

The output $y(t)$ from an arbitrary causal linear time-invariant system (one in which the future is uniquely determined by the past, and the equations do not change over time, such as a constant-coefficient linear differential equation) is related to the input $x(t)$ by convolution with respect to the impulse response $h(t)$ (the response of the system to a delta-function input)

$$y(t) = \int_0^t h(t')x(t-t')\,dt' \quad . \tag{2.33}$$

Since convolution in the time domain equals multiplication of Laplace transforms,

$$H(s) = \frac{Y(s)}{X(s)} \quad . \tag{2.34}$$

$H(s)$, the Laplace transform of the impulse response, is called the system *transfer function* and is equal to the ratio of the Laplace transforms of the input and the output.

A general transfer function for a finite-dimensional linear system will be the ratio of two polynomials. Since the polynomials can be constructed from knowledge of the roots, the location of the poles and zeros completely characterizes the response of the system. A great deal therefore can be learned about a system from the placement of its poles and zeros. Most importantly, for the system to be globally stable, all of the poles must lie in the left half–plane $\text{Re}(p_i) < 0$ (recall that $\mathcal{L}\{\exp(at)\} = 1/(s-a)$, and so if $\text{Re}(a) > 0$ then the solution will diverge). Similarly, poles off of the real axis are associated with oscillatory solutions.

If the input is $x = \exp(i\omega t)$ then the Laplace transform of the output is

$$Y(s) = \frac{H(s)}{s - i\omega} \tag{2.35}$$

(from Table 2.1). If $H(s)$ is a rational function, the output can be found by a partial fraction expansion

$$Y(s) = R(s) + \frac{A}{s - i\omega} \quad , \tag{2.36}$$

where $R(s)$ is another rational function, and

$$A = \lim_{s \to i\omega} Y(s)(s - i\omega) = H(i\omega) \quad . \tag{2.37}$$

$R(s)$ is a transient that does not depend on the input, and hence can be ignored in finding the driven response. Therefore the steady-state output is

$$y(t) = H(i\omega)e^{i\omega t} \quad . \tag{2.38}$$

The amplitude and phase of the output at a frequency $\omega$ are given by the transfer function evaluated at $i\omega$.

## 2.4 PERTURBATION EXPANSIONS

One is much more likely to encounter a differential equation that is not analytically soluble than one that is. If the problem is related to one that does have an analytical solution, it can be possible to find an approximate solution by an expansion around the known one. The key is to recognize a small parameter $\epsilon$ in the problem that in the limit $\epsilon \to 0$ reduces to the known problem. Equally useful is a very large parameter, since its inverse can be used to develop an approximation.

For example, consider a harmonic oscillator with a weak nonlinear dissipation

$$\ddot{x} + x + \epsilon \dot{x}^2 = 0 \quad . \tag{2.39}$$

Clearly, when $\epsilon$ is small the solution will not be too far from simple harmonic motion. As an ansatz we will expand the solution in powers of the small parameter:

$$x(t) = x_0(t) + \epsilon x_1(t) + \epsilon^2 x_2(t) + \mathcal{O}(\epsilon^3) \quad . \tag{2.40}$$

The notation $\mathcal{O}(\epsilon^3)$ stands for terms of order $\epsilon^3$ and higher. Now, plug this in and collect terms based on their order. All of the terms with the same order of $\epsilon$ must satisfy the differential equation independently of the others, because their relationship will change if $\epsilon$ is arbitrarily varied:

$$\ddot{x}_0 + x_0 + \epsilon \dot{x}_0^2 + \epsilon \ddot{x}_1 + \epsilon x_1 + \epsilon^2 \dot{x}_1^2 + \cdots = 0$$

$$\mathcal{O}(\epsilon^0) : \ddot{x}_0 + x_0 = 0$$

$$\mathcal{O}(\epsilon^1) : \ddot{x}_1 + x_1 + \dot{x}_0^2 = 0 \tag{2.41}$$

and so forth. The lowest-order equation is just that for the unperturbed oscillator, and the higher-order equations give corrections. This is a hierarchy that can be solved in order, first finding $x_0$, then using it to find $x_1$, and on up to the desired order of approximation. If the initial conditions are chosen so that $x_0 = \exp(it)$, then this gives for $x_1$

$$\ddot{x}_1 + x_1 = e^{i2t} \quad . \tag{2.42}$$

The homogeneous solution is $A \exp(it) + B \exp(-it)$, and a particular solution can be found by plugging in $x_1 = C \exp(i2t)$, which gives $C = -1/3$. The nonlinearity has to first order introduced a harmonic into the system, a familiar feature of strongly driven systems.

Perturbation approximations are useful if there is a natural notion of a small deviation from a soluble problem. Another important kind of approximation is an *asymptotic expansion*, which is a series expansion that applies for very large arguments (such as $t \to \infty$). See, for example, [Pearson, 1990].

## 2.5 DISCRETE TIME EQUATIONS

It is a common and usually reasonable approximation to consider physical systems to be continuous, and so differential equations apply. But this is not the case for digital systems, which usually are discretized in time (there is a system clock). Conveniently, the theory of discrete time equations is essentially identical to that for differential equations.

Take $y(k)$ to be a series defined only at integer values of the time $k$ (replace $k$ with $k\Delta t$ throughout this section if the time step is not an integer). An $N$-th order linear constant coefficient difference equation for $y(k)$ is

$$y(k) + A_1 y(k-1) + \cdots + A_{N-1} y(k-N+1) + A_N y(k-N) = f(k) \ . \tag{2.43}$$

Substituting the ansatz $y(k) = r^k$ into the homogeneous equation ($f(k) = 0$) and cancelling the $r^k$ terms gives the same characteristic equation we saw for differential equations,

$$r^N + A_1 r^{N-1} + \cdots + A_{N-1} r + A_N = 0 \ . \tag{2.44}$$

Once again, if all of the roots of this $N$-th order polynomial are distinct

$$r^N + A_1 r^{N-1} + \cdots + A_{N-1} r + A_N = (r - r_1)(r - r_2) \cdots (r - r_N) \tag{2.45}$$

then the general solution is

$$y_g = \sum_{n=1}^{N} C_n r_n^k \ , \tag{2.46}$$

where the $C_n$'s are unknown coefficients determined by the boundary equations. A complete solution is made up of this general solution plus a particular solution to the inhomogeneous problem.

If some of the roots of the characteristic polynomial are repeated,

$$r^N + A_1 r^{N-1} + \cdots + A_{N-1} r + A_N = (r - r_1)^M (r - r_{M+1}) \cdots (r - r_N) \tag{2.47}$$

then recognizing that the operations of differentiation and shifting in time can be interchanged, and repeating the argument associated with equation (2.9), we see that the extra solutions associated with the repeated root are

$$\frac{\partial}{\partial r} L(r^k) = L\left(\frac{\partial r^k}{\partial r}\right) = L(k r^{k-1}) = 0 \ ,$$

$$\frac{\partial^2}{\partial r^2} L(r^k) = L\left(\frac{\partial^2}{\partial r^2} r^k\right) = L(k(k-1) r^{k-2}) = 0 \ , \tag{2.48}$$

and so forth.

## 2.6 $z$-TRANSFORMS

For a series $y(k)$, the *one-sided z-transform* is defined by

$$\mathcal{Z}\{y(k)\} \equiv Y(z) = \sum_{k=0}^{\infty} y(k) z^{-k} \ . \tag{2.49}$$

As with the Laplace transform, a two-sided transform can also be defined for signals that start at $-\infty$, extending the discrete Fourier transform to the complex plane.

Just as differentiation is the most important property of Laplace transforms, time-shifting is the most important property for $z$-transforms. Consider a series delayed by

Table 2.2. *Selected z-transforms.*

$$\mathcal{Z}\{kf(k)\} = -z\frac{dF(z)}{dz}$$

$$\mathcal{Z}\{a^k f(k)\} = F(z/a)$$

$$\mathcal{Z}\left\{\sum_{n=0}^{k} f(k-n)g(n)\right\} \equiv \mathcal{Z}\{f * g\} = F(z)G(z)$$

$$\mathcal{Z}\{1\} = \frac{z}{z-1}$$

$$\mathcal{Z}\{a^k\} = \frac{z}{z-a}$$

$$\mathcal{Z}\{\delta(k-n)\} = z^{-n}$$

one time step:

$$\mathcal{Z}\{y(k-1)\} = \sum_{k=0}^{\infty} y(k-1)z^{-k}$$

$$= \sum_{k'=-1}^{\infty} y(k')z^{-k'}z^{-1} \quad (k' = k-1)$$

$$= z^{-1}\sum_{k'=0}^{\infty} y(k')z^{-k'} + z^{-1}y(-1)z$$

$$= z^{-1}Y(z) + y(-1) \ . \tag{2.50}$$

Similarly, for a delay of two,

$$\mathcal{Z}\{y(k-2)\} = \sum_{k=0}^{\infty} y(k-2)z^{-k}$$

$$= \sum_{k'=-2}^{\infty} y(k')z^{-k'}z^{-2} \quad (k' = k-2)$$

$$= z^{-2}\sum_{k'=0}^{\infty} y(k')z^{-k'} + z^{-2}y(-1)z + z^{-2}y(-2)z^2$$

$$= z^{-2}Y(z) + z^{-1}y(-1) + y(-2) \ , \tag{2.51}$$

and so forth for longer delays. The $z$-transform turns a difference equation into an algebraic equation. Without proof, Table 2.2 gives a few other $z$-transforms (where $f(k) = 0$ for $k < 0$); see the references for many more.

Many of the properties of the Laplace transform carry over to the $z$-transform. If the input forcing to a constant-coefficient linear difference equation is a unit impulse $x(k) = \delta(k)$ ($\delta(0) = 1$, $\delta(k \neq 0) = 0$), then the solution $y(k) = h(k)$ is defined to be the impulse response of the system. The output for an arbitrary input is given by the convolution with the impulse response

$$y(k) = \sum_{n=0}^{k} h(n)x(k-n) \tag{2.52}$$

and in the $z$ domain the transfer function (the $z$-transform of the impulse response) is the ratio of the transforms of the input and the output

$$H(z) = \frac{Y(z)}{X(z)} \quad . \tag{2.53}$$

The stability of a system is determined by the location of the poles of the transfer function; for it to be stable for any bounded input signal the poles must have a complex magnitude of less than 1 (recall that $\mathcal{Z}\{a^k\} = z/(z-a)$, which has a pole at $a$, so $|a|$ must be less than 1 for $a^k$ to remain bounded).

The frequency response can be found with a calculation similar to the continuous time case. If the input is $x(k) = \exp(i\omega\delta_t k)$, and $H(z)$ is the ratio of two polynomials $N(z)/D(z)$, then

$$\begin{aligned} Y(z) &= H(z)\frac{z}{z - e^{i\omega\delta_t}} \\ &= \frac{N(z)}{D(z)}\frac{z}{z - e^{i\omega\delta_t}} \\ &= \frac{A_1}{D(z)} + \frac{A_2}{z - e^{i\omega\delta_t}} \end{aligned} \tag{2.54}$$

with

$$A_2 = Y(z)(z - e^{i\omega\delta_t})\big|_{z=e^{i\omega\delta_t}} = H(e^{i\omega\delta_t})e^{i\omega\delta_t} \quad . \tag{2.55}$$

Therefore the output is

$$y(k) = H(e^{i\omega\delta_t})e^{i\omega\delta_t}e^{i\omega\delta_t k} = H(e^{i\omega\delta_t})e^{i\omega\delta_t(k+1)} \quad . \tag{2.56}$$

The solution is delayed by the one time step required to be advanced by the transfer function, and the amplitude and phase are determined by the transfer function evaluated on the unit circle at $H(\exp(i\omega\delta_t))$. Problem 2.3 looks at the solution for, and frequency response of, a simple digital filter.

## 2.7 SELECTED REFERENCES

[Strang, 1986] Strang, Gilbert (1986). *Introduction to Applied Mathematics*. Wellesley, MA: Wellesley-Cambridge Press.

A very readable introduction to applied mathematics, including linear algebra and differential equations.

[Zwillinger, 1992] Zwillinger, Daniel (1992). *Handbook of Differential Equations*. 2nd edn. New York, NY: Academic Press.

A good meta-index to most known techniques (analytical and numerical) for studying differential equations.

[Kamen, 1990] Kamen, Edward W. (1990). *Introduction to Signals and Systems*. 2nd edn. New York, NY: Macmillan.

Because of its importance in engineering practice, there are many books with a title similar to this that cover continuous and discrete time systems (Laplace and $z$-transforms). Kamen's book provides a nice balance between rigor and insight.

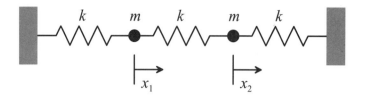

Figure 2.2. Two coupled harmonic oscillators.

[Oppenheim & Schafer, 1989] Oppenheim, A.V., & Schafer, R.W. (1989). *Discrete-Time Signal Processing*. Englewood Cliffs, NJ: Prentice Hall.

A definitive reference for one of the most important applications of the theory of discrete time systems, digital signal processing.

## 2.8 PROBLEMS

(2.1) Consider the motion of a damped, driven harmonic oscillator (such as a mass on a spring, a ball in a well, or a pendulum making small motions):

$$m\ddot{x} + \gamma\dot{x} + kx = e^{i\omega t} \quad . \tag{2.57}$$

(a) Under what conditions will the governing equations for small displacements of a particle around an arbitrary 1D potential minimum be simple undamped harmonic motion?

(b) Find the solution to the homogeneous equation, and comment on the possible cases. How does the amplitude depend on the frequency?

(c) Find a particular solution to the inhomogeneous problem by assuming a response at the driving frequency, and plot its magnitude and phase as a function of the driving frequency for $m = k = 1, \gamma = 0.1$.

(d) For a driven oscillator the $Q$ or *Quality factor* is defined as the ratio of the center frequency to the width of the curve of the average energy (kinetic + potential) in the oscillator versus the driving frequency (the width is defined by the places where the curve falls to half its maximum value). For an undriven oscillator the $Q$ is defined to be the ratio of the energy in the oscillator to the energy lost per radian (one cycle is $2\pi$ radians). Show that these two definitions are equal, assuming that the damping is small. How long does it take the amplitude of a 100 Hz oscillator with a $Q$ of $10^9$ to decay by $1/e$?

(e) Now find the solution to equation (2.57) by using Laplace transforms. Take the initial condition as $x(0) = \dot{x}(0) = 0$.

(f) For an arbitrary potential minimum, work out the form of the lowest-order correction to simple undamped unforced harmonic motion.

(2.2) Explicitly solve (and try to simplify) the system of differential equations for two coupled harmonic oscillators (see Figure 2.2; don't worry about the initial transient), and then find the normal modes by matrix diagonalization.

(2.3) A common simple digital filter used for smoothing a signal is
$$y(k) = \alpha y(k-1) + (1-\alpha)x(k) \quad , \tag{2.58}$$
where $\alpha$ is a parameter that determines the response of the filter. Use $z$-transforms to solve for $y(k)$ as a function of $x(k)$ (assume $y(k < 0) = 0$). What is the amplitude of the frequency response?

# 3 Partial Differential Equations

Partial differential equations (PDEs) are equations that involve rates of change with respect to continuous variables. The configuration of a rigid body is specified by six numbers, but the configuration of a fluid is given by the continuous distribution of the temperature, pressure, and so forth. The dynamics for the rigid body take place in a finite-dimensional configuration space; the dynamics for the fluid occur in an infinite-dimensional configuration space. This distinction usually makes PDEs much harder to solve than ODEs, but here again there will be simple solutions for linear problems. Classic domains where PDEs are used include acoustics, fluid flow, electrodynamics, and heat transfer.

## 3.1 THE ORIGIN OF PARTIAL DIFFERENTIAL EQUATIONS

In the preceeding chapter we saw how the solution for two coupled harmonic oscillators simplifies into two independent normal modes. What does the solution look like if there are 10 oscillators? $10^{10}$? Are there any simplifications? Not surprisingly, the answer is yes.

Consider an infinite chain of oscillators (Figure 3.1). The governing equation for the $n$th mass is

$$m\ddot{y}_n = -k(y_n - y_{n+1}) - k(y_n - y_{n-1})$$
$$\ddot{y}_n = \frac{k}{m}(y_{n+1} - 2y_n + y_{n-1})$$
$$= \underbrace{k\,\delta x}_{\tau}\;\underbrace{\frac{\delta x}{m}}_{1/\rho}\;\frac{y_{n+1} - 2y_n + y_{n-1}}{\delta x^2} \quad . \tag{3.1}$$

The two prefactors are the average spring constant $\tau$ and mass density $\rho$ (remember that springs add inversely proportionally), and the final term is just an approximation to the second spatial derivative:

$$\frac{\partial y}{\partial x} \approx \frac{y_{n-1} - y_n}{\delta x}$$
$$\frac{\partial^2 y}{\partial x^2} \approx \frac{1}{\delta x}\left[\frac{y_{n+1} - y_n}{\delta x} - \frac{y_n - y_{n-1}}{\delta x}\right]$$
$$= \frac{y_{n+1} - 2y_n + y_{n-1}}{\delta x^2} \quad . \tag{3.2}$$

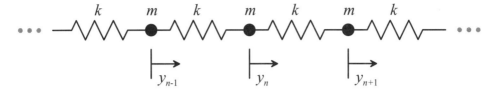

Figure 3.1. A chain of harmonic oscillators.

Therefore in the limit of a small spacing between the springs, the system of ordinary differential equations for a chain of harmonic oscillators reduces to a single partial differential equation

$$\frac{\partial^2 y}{\partial t^2} = \frac{\tau}{\rho}\frac{\partial^2 y}{\partial x^2} . \tag{3.3}$$

This equation is solved by a travelling wave. To see this, substitute a general solution $y = f(x + ct)$:

$$c^2 f'' = \frac{\tau}{\rho} f''$$

$$c = \pm\sqrt{\frac{\tau}{\rho}} . \tag{3.4}$$

This represents an arbitrary disturbance travelling to the right and left with a velocity $c$: the location of the origin (for example) of $f$ is determined by $x + ct = 0 \Rightarrow x/t = -c$. If there are nonlinearities the velocity will no longer be independent of the shape of the pulse: different wavelengths will travel at different speeds, a phenomenon called *dispersion*. Note that unlike the case for ODEs, the general solution involves an undetermined function and not just undetermined constants.

This same equation can be found directly by considering the transverse motion of an infinitesimal element of a continuous string that has a density of $\rho$ and a tension $\tau$ (Figure 3.2). The governing equation for the transverse displacement $y$ of this element is

$$ma = F$$
$$\rho\, dx\, \frac{\partial^2 y}{\partial t^2} = \tau \sin\theta \Big|_{x+dx} - \tau \sin\theta \Big|_{x}$$
$$\approx \tau \tan\theta \Big|_{x+dx} - \tau \tan\theta \Big|_{x}$$
$$\approx \tau \frac{\partial y}{\partial x}\Big|_{x+dx} - \tau \frac{\partial y}{\partial x}\Big|_{x}$$
$$\frac{\partial^2 y}{\partial t^2} \approx \frac{\tau}{\rho}\frac{\partial^2 y}{\partial x^2} . \tag{3.5}$$

As a final example of the origin of partial differential equations, consider a highway for which position is measured by $x$, the density of cars by $\rho(x)$, and the rate at which cars pass a point by $I(x)$. In a time interval $dt$, the difference in the number of cars that enter and leave an interval $dx$ is $[I(x) - I(x+dx)]dt$. This must be equal to the change

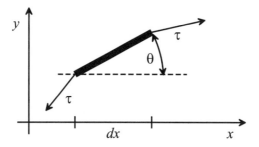

Figure 3.2. An infinitesimal piece of a string.

in the number of cars at that interval $[\rho(t+dt) - \rho(t)]dx$ because the total number of cars is (usually) conserved. Equating these gives

$$[\rho(t+dt) - \rho(t)]dx = [I(x) - I(x+dx)]dt$$

$$\frac{\rho(t+dt) - \rho(t)}{dt} + \frac{I(x+dx) - I(x)}{dx} = 0$$

$$\frac{\partial \rho}{\partial t} + \frac{\partial I}{\partial x} = 0 \quad . \tag{3.6}$$

Now, assume that a driver's response can be modeled by a relationship between the traffic flow and density $I(x) = f(\rho(x), x)$. Multiplying both sides of equation (3.6) by $\partial I/\partial \rho$ gives

$$\frac{\partial \rho}{\partial t}\frac{\partial I}{\partial \rho} + \frac{\partial I}{\partial x}\frac{\partial I}{\partial \rho} = 0$$

$$\frac{\partial I}{\partial t} + \frac{\partial f(\rho, x)}{\partial \rho}\frac{\partial I}{\partial x} = 0 \quad . \tag{3.7}$$

Modeling traffic with a PDE can be a very good approximation, and can explain many observed traffic phenomena such as shock fronts and stationary disturbances [Whitham, 1974].

## 3.2 LINEAR PARTIAL DIFFERENTIAL EQUATIONS

As with ordinary differential equations, we will immediately specialize to linear partial differential equations, both because they occur so frequently and because they are amenable to analytical solution. A general linear second-order PDE for a field $\varphi(x, y)$ is

$$A\frac{\partial^2 \varphi}{\partial x^2} + B\frac{\partial^2 \varphi}{\partial x \partial y} + C\frac{\partial^2 \varphi}{\partial y^2} + D\frac{\partial \varphi}{\partial x} + E\frac{\partial \varphi}{\partial y} + F\varphi = G \quad , \tag{3.8}$$

where $G(x, y)$ is specified in some portion of the $(x, y)$ plane and the solution must be determined in another portion.

A *characteristic* of a PDE is a surface across which there can be a discontinuity in the value or derivative of the solution. These define the domains which can be influenced by parts of the boundary conditions, much like the concept of a light cone in the space-time plane of special relativity [Taylor & Wheeler, 1992]. The characteristics of equation (3.8) are determined by the roots of a quadratic polynomial and accordingly can have the form of a hyperbola, a parabola, or an ellipse based on the sign of the discriminant $B^2 - 4AC$ [Hildebrand, 1976, Pearson, 1990]. The standard forms of these three cases define the most common PDEs that we will study:

- $B^2 - 4AC > 0$ (*hyperbolic*)

$$\nabla^2 \varphi = \frac{1}{c^2} \frac{\partial^2 \varphi}{\partial t^2} \quad ; \tag{3.9}$$

- $B^2 - 4AC = 0$ (*parabolic*)

$$\nabla^2 \varphi = \frac{1}{D} \frac{\partial \varphi}{\partial t} \quad ; \tag{3.10}$$

- $B^2 - 4AC < 0$ (*elliptic*)

$$\nabla^2 \varphi = \rho \quad . \tag{3.11}$$

The first of these is a wave equation (like we found for the coupled harmonic oscillators), the second is a diffusion equation (for example, for heat or for ink), and the third is *Poisson's equation* (or *Laplace's equation* if the source term $\rho = 0$) and arises in boundary value problems (for example, for electric fields or for fluid flow).

## 3.3 SEPARATION OF VARIABLES

These three important partial differential equations can be reduced to systems of ordinary differential equations by the important technique of *separation of variables*. The logic of this technique may be confusing upon first aquaintance, but it rests on the uniqueness of solutions to differential equations: as with ODEs, if you can find *any* solution that solves the equation and satisfies the boundary conditions, then it is *the* solution. We will assume as an ansatz that the dependence of the solution on space and time can be written as a product of terms that each depend on a single coordinate, and then see if and how this can be made to solve the problem.

To start, the time dependence can be separated by assuming a solution of the form $\varphi(\vec{x}, t) = \psi(\vec{x})T(t)$. There is no time dependence for Laplace's equation; trying this in the diffusion equation gives

$$T(t)\nabla^2 \psi(\vec{x}) = \frac{1}{D} \psi(\vec{x}) \frac{\partial T(t)}{\partial t} \quad . \tag{3.12}$$

Dividing both sides by $\psi T$ results in no $t$ dependence on the left hand side and no $\vec{x}$ dependence on the right hand side, so both sides must be equal to some constant because the space and time variables can be varied arbitrarily. By convention, taking this constant to be $-k^2$ gives

$$\frac{1}{\psi(\vec{x})} \nabla^2 \psi(\vec{x}) = \frac{1}{D} \frac{1}{T(t)} \frac{dT}{dt} = -k^2 \quad . \tag{3.13}$$

The $t$ equation can immediately be integrated to find

$$T(t) = Ae^{-k^2 Dt} \quad , \tag{3.14}$$

and the $\vec{x}$ equation is *Helmholtz's equation*

$$\nabla^2 \psi(\vec{x}) + k^2 \psi(\vec{x}) = 0 \quad . \tag{3.15}$$

Similarly, for the wave equation this separation gives

$$\frac{1}{\psi(\vec{x})} \nabla^2 \psi(\vec{x}) = \frac{1}{c^2} \frac{1}{T} \frac{d^2 T}{dt^2} = -k^2 \quad . \tag{3.16}$$

The time equation is solved by

$$T(t) = A \sin(kct) + B \cos(kct) \quad , \tag{3.17}$$

and the space equation is Helmholtz's equation again.

Solving Helmholtz's equation will depend on the coordinate system used for the problem. There are three common ones used in 3D, based on the symmetry of the problem: rectangular, cylindrical, and spherical. Writing the derivative operators in each of these systems is a straightforward exercise in applying the chain rule to the coordinate definitions.

### 3.3.1 Rectangular Coordinates

Writing the Laplacian $\nabla^2$ in rectangular coordinates leads to Helmholtz's equation as

$$\nabla^2 \psi + k^2 \psi = \frac{\partial^2 \psi}{\partial x^2} + \frac{\partial^2 \psi}{\partial y^2} + \frac{\partial^2 \psi}{\partial z^2} + k^2 \psi = 0 \quad . \tag{3.18}$$

Assume that $\psi(\vec{x}) = X(x) Y(y) Z(z)$, substitute this in, and divide by it:

$$\frac{1}{X(x)} \frac{d^2 X}{dx^2} + \frac{1}{Y(y)} \frac{d^2 Y}{dy^2} + \frac{1}{Z(z)} \frac{d^2 Z}{dz^2} + k^2 = 0 \quad . \tag{3.19}$$

Since each term depends only on $x$, $y$, or $z$, the only way that this equation can hold is if each has a constant value (determined by the boundary conditions)

$$\frac{1}{X(x)} \frac{d^2 X}{dx^2} = -k_1^2, \quad \frac{1}{Y(y)} \frac{d^2 Y}{dy^2} = -k_2^2, \quad \frac{1}{Z(z)} \frac{d^2 Z}{dz^2} = -k_3^2 \tag{3.20}$$

with $k_1^2 + k_2^2 + k_3^2 = k^2$. Each of these can be integrated to find

$$\begin{aligned} X &= A_1 e^{i k_1 x} + B_1 e^{-i k_1 x} \\ Y &= A_2 e^{i k_2 y} + B_2 e^{-i k_2 y} \\ Z &= A_3 e^{i k_3 z} + B_3 e^{-i k_3 z} \quad . \end{aligned} \tag{3.21}$$

Multiplying these back together, the spatial solution has the form

$$\psi(\vec{x}) = A e^{i \vec{k} \cdot \vec{x}} \tag{3.22}$$

with $\vec{k} \cdot \vec{k} = k^2$.

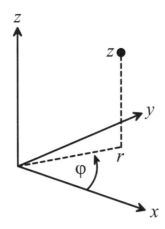

Figure 3.3. Cylindrical coordinate system.

As an example, let's return to the 1D wave equation that we found from a chain of harmonic oscillators

$$\frac{\partial^2 y}{\partial x^2} = \frac{1}{c^2} \frac{\partial^2 y}{\partial t^2} \quad . \tag{3.23}$$

With the separation $y(x,t) = X(x)T(t)$ this becomes

$$\frac{d^2 T}{dt^2} + c^2 k^2 T = 0 \qquad \frac{d^2 X}{dx^2} + k^2 X = 0 \quad , \tag{3.24}$$

solved by

$$T = A \sin ckt + B \cos ckt \qquad X = C \sin kx + D \cos kx \quad . \tag{3.25}$$

We know that the chain must be fixed at the ends ($X(0) = X(L) = 0$). This implies that $D = 0$, and that allowable values of the separation constant $k$ are $k_n = n\pi/L$ for integer $n$. Therefore the general solution is

$$y(x,t) = \sum_n \sin\left(\frac{n\pi}{L} x\right) \left[A_n \sin\left(c\frac{n\pi}{L} t\right) + B_n \cos\left(c\frac{n\pi}{L} t\right)\right] \quad . \tag{3.26}$$

These are the normal modes of a string, with the oscillation frequency of each mode proportional to the number of cycles across the string.

### 3.3.2 Cylindrical Coordinates

In cylindrical coordinates (Figure 3.3), the Helmholtz equation is

$$\nabla^2 \psi + k^2 \psi = \frac{\partial^2 \psi}{\partial r^2} + \frac{1}{r} \frac{\partial \psi}{\partial r} + \frac{1}{r^2} \frac{\partial^2 \psi}{\partial \varphi^2} + \frac{\partial^2 \psi}{\partial z^2} + k^2 \psi = 0 \quad . \tag{3.27}$$

Once again, try separating by substituting in $\psi = R(r)\Phi(\varphi)Z(z)$ and dividing by it:

$$\frac{1}{R}\left[\frac{d^2 R}{dr^2} + \frac{1}{r}\frac{dR}{dr}\right] + \frac{1}{r^2}\frac{d^2 \Phi}{d\varphi^2} + \frac{1}{Z}\frac{d^2 Z}{dz^2} + k^2 = 0 \quad . \tag{3.28}$$

The terms will cancel if

$$\frac{1}{\Phi}\frac{d^2\Phi}{d\varphi^2} = -m^2$$

$$\frac{1}{Z}\frac{d^2Z}{dz^2} = \alpha^2 - k^2$$

$$\frac{1}{R}\left[\frac{d^2R}{dr^2} + \frac{1}{r}\frac{dR}{dr}\right] - \frac{m^2}{r^2} + \alpha^2 = 0 \tag{3.29}$$

for constants $\alpha$ and $m$ (these definitions are conventional). The first equation is easily solved:

$$\Phi = A\sin m\varphi + B\cos m\varphi \quad . \tag{3.30}$$

For the solution to be single valued $\Phi(\varphi + 2\pi) = \Phi(\varphi)$, and so $m$ must be an integer. The second equation is similarly solved:

$$Z = Ce^{z\sqrt{\alpha^2-k^2}} + De^{-z\sqrt{\alpha^2-k^2}} \quad . \tag{3.31}$$

Rewriting the radial equation in terms of $r = \rho/\alpha$,

$$\frac{d^2R}{d\rho^2} + \frac{1}{\rho}\frac{dR}{d\rho} + \left(1 - \frac{m^2}{\rho^2}\right)R = 0 \quad . \tag{3.32}$$

This is *Bessel's equation*; its solution is given by Bessel functions

$$R = EJ_m(\alpha r) + FN_m(\alpha r) \quad . \tag{3.33}$$

$N_m$ is singular as $r \to 0$ while $J_m$ is not, so if the solution is finite at the origin $F = 0$. If the radial solution must vanish for some $r$ value it is necessary to know where the zeros of $J_m$ occur; these are tabulated in many sources (such as [Abramowitz & Stegun, 1965]). The lowest ones are

$$J_0(x) = 0 \Rightarrow x \approx 2.405, 5.520, 8.654, \ldots$$
$$J_1(x) = 0 \Rightarrow x \approx 3.832, 7.016, 10.173, \ldots$$
$$J_2(x) = 0 \Rightarrow x \approx 5.136, 8.417, 11.620, \ldots \quad . \tag{3.34}$$

If $\alpha = 0$, the radial equation becomes

$$\frac{d^2R}{dr^2} + \frac{1}{r}\frac{dR}{dr} - \frac{m^2}{r^2}R = 0 \quad , \tag{3.35}$$

which is solved by

$$R(r) = \begin{cases} Gr^m + Hr^{-m} & (m \neq 0) \\ G + H\ln r & (m = 0) \end{cases} \quad . \tag{3.36}$$

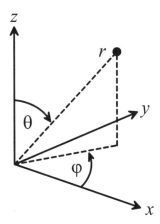

Figure 3.4. Spherical coordinate system.

### 3.3.3 Spherical Coordinates

Finally, in spherical coordinates (Figure 3.4) we want to separate

$$\nabla^2 \psi + k^2 \psi = 0$$

$$\frac{1}{r}\frac{\partial^2}{\partial r^2}(r\psi) + \frac{1}{r^2 \sin\theta}\left[\frac{\partial}{\partial\theta}\left(\sin\theta\frac{\partial\psi}{\partial\theta}\right) + \frac{1}{\sin\theta}\frac{\partial^2\psi}{\partial\varphi^2}\right] + k^2\psi = 0 \ . \tag{3.37}$$

Let's start with the radial part: $\psi = R(r)Y(\theta,\varphi)$

$$\frac{1}{R}\frac{1}{r}\frac{d^2}{dr^2}(rR) + \frac{1}{r^2}\frac{1}{Y\sin\theta}\left[\frac{\partial}{\partial\theta}\left(\sin\theta\frac{\partial Y}{\partial\theta}\right) + \frac{1}{\sin\theta}\frac{\partial^2 Y}{\partial\varphi^2}\right] + k^2 = 0$$

$$\Rightarrow \frac{1}{Y\sin\theta}\left[\frac{\partial}{\partial\theta}\left(\sin\theta\frac{\partial Y}{\partial\theta}\right) + \frac{1}{\sin\theta}\frac{\partial^2 Y}{\partial\varphi^2}\right] = -\lambda \ ,$$

$$\frac{1}{R}\frac{1}{r}\frac{d^2}{dr^2}(rR) + k^2 - \frac{\lambda}{r^2} = 0 \tag{3.38}$$

for a constant $\lambda$. If $k^2 \neq 0$, substituting $r = \rho/k$ and then $R = S/\sqrt{\rho}$ gives Bessel's equation again:

$$\frac{d^2 S}{d\rho^2} + \frac{1}{\rho}\frac{dS}{d\rho} + \left(1 - \frac{\lambda + 1/4}{\rho^2}\right)S = 0 \ , \tag{3.39}$$

with the solution

$$R = A\frac{1}{\sqrt{kr}}J_{\sqrt{\lambda+1/4}}(kr) + B\frac{1}{\sqrt{kr}}N_{\sqrt{\lambda+1/4}}(kr) \ . \tag{3.40}$$

If $k^2 = 0$, the radial equation simplifies to

$$\frac{1}{r}\frac{d^2}{dr^2}(rR) - \frac{\lambda}{r^2}R = 0 \ , \tag{3.41}$$

solved by

$$R = Ar^{(-1+\sqrt{1+4\lambda})/2} + Br^{(-1-\sqrt{1+4\lambda})/2} \ . \tag{3.42}$$

Now separate the angular parts with $Y = \Theta(\theta)\Phi(\varphi)$:

$$\frac{1}{\Theta}\frac{1}{\sin\theta}\frac{d}{d\theta}\left(\sin\theta\frac{d\Theta}{d\theta}\right) + \frac{1}{\sin^2\theta}\frac{1}{\Phi}\frac{d^2\Phi}{d\varphi^2} + \lambda = 0$$

$$\Rightarrow \frac{1}{\Phi}\frac{d^2\Phi}{d\varphi^2} = -m^2 \quad,$$

$$\frac{1}{\sin\theta}\frac{d}{d\theta}\left(\sin\theta\frac{d\Theta}{d\theta}\right) + \left(\lambda - \frac{m^2}{\sin^2\theta}\right)\Theta = 0 \quad. \tag{3.43}$$

The solution of the $\varphi$ equation is

$$\Phi = Ae^{im\varphi} + Be^{-im\varphi} \quad. \tag{3.44}$$

For the $\theta$ equation, substitute $x = \cos\theta$:

$$\frac{d^2\Theta}{dx^2} - \frac{2x}{1-x^2}\frac{d\Theta}{dx} + \frac{1}{1-x^2}\left[\lambda - \frac{m^2}{1-x^2}\right]\Theta = 0 \quad. \tag{3.45}$$

This is *Legendre's equation*, solved by the Legendre functions

$$\Theta = C\Theta_l^m(x) + DQ_l^m(x) \tag{3.46}$$

with $l(l+1) = \lambda$.

## 3.4 TRANSFORM TECHNIQUES

In the previous section we saw that separation of variables can turn partial differential equations into ordinary differential equations; this can also sometimes be done by taking the Fourier transform. For example, start with the PDE

$$\nabla^2\varphi + A\frac{\partial^2\varphi}{\partial t^2} + B\frac{\partial\varphi}{\partial t} = 0 \quad. \tag{3.47}$$

$\varphi(\vec{x}, t)$ is related to its Fourier transform $\Phi(\vec{k}, t)$ by

$$\varphi(\vec{x}, t) = \int_{-\infty}^{\infty} \Phi(\vec{k}, t)e^{i\vec{k}\cdot\vec{x}}\, d\vec{k}$$

$$\Phi(\vec{k}, t) = \frac{1}{2\pi}\int_{-\infty}^{\infty} \varphi(\vec{x}, t)e^{-i\vec{k}\cdot\vec{x}}\, d\vec{x} \quad. \tag{3.48}$$

Substituting in the transform for $\varphi$ in equation (3.47), exchanging the order of differentiation and integration, and grouping terms,

$$\int_{-\infty}^{\infty}\left[-k^2\Phi(\vec{k}, t) + A\frac{\partial^2}{\partial t^2}\Phi(\vec{k}, t) + B\frac{\partial}{\partial t}\Phi(\vec{k}, t)\right]e^{i\vec{k}\cdot\vec{x}}\, d\vec{k} = 0 \quad. \tag{3.49}$$

The only way that the integral can equal zero for all $x$ is if the integrand vanishes. The integrand now depends only on $\Phi$ and its time derivatives; if a solution can be found to this ordinary differential equation in $t$

$$-k^2\Phi(\vec{k}, t) + A\frac{\partial^2}{\partial t^2}\Phi(\vec{k}, t) + B\frac{\partial}{\partial t}\Phi(\vec{k}, t) = 0 \tag{3.50}$$

then it will solve the integral equation. A solution to this differential equation can be multiplied by an arbitrary function of $\vec{k}$ and still be a solution; this function is determined from initial conditions by the transform

$$\Phi(\vec{k}, 0) = \frac{1}{2\pi} \int_{-\infty}^{\infty} \varphi(\vec{x}, 0) e^{-i\vec{k} \cdot \vec{x}} \, d\vec{x} \quad . \tag{3.51}$$

## 3.5 SELECTED REFERENCES

[Hildebrand, 1976] Hildebrand, Francis B. (1976). *Advanced Calculus for Applications*. 2nd edn. Englewood Cliffs, NJ: Prentice–Hall.

This basic text provides good background for PDEs (and ODEs).

[Wyld, 1976] Wyld, Henry W. (1976). *Mathematical Methods for Physics*. Reading, MA: W.A. Benjamin.

The mathematical methods used in physics are full of partial differential equations; Wyld's book is a very readable but comprehensive introduction.

[Whitham, 1974] Whitham, Gerald B. (1974). *Linear and Nonlinear Waves*. New York, NY: Wiley-Interscience.

Everything you ever wanted to know about wave equations.

## 3.6 PROBLEMS

(3.1) Consider a round drumhead of radius $L$. For small displacements its motion is described by a linear wave equation. Find the frequencies of the six lowest oscillation modes, and plot the shape of the modes.

(3.2) Solve a 1D diffusion equation with Fourier transforms.

(3.3) Assume a crowded room full of generous children who have varying amounts of candy. Let $\varphi_{n,m}(t_i)$ be the amount of candy held by the $n, m$th child at time $t_i$. Because of the crowding, the children are approximately close-packed on a square grid. The children want to equalize the amount of candy, but it is so noisy that they can only talk to their nearest neighbors (although they are wearing watches). Find a simple strategy for them to use that results in the candy being evenly distributed, and in the continuum limit find a familiar PDE that is equivalent to this strategy.

# 4 Variational Principles

So far, we have discussed a variety of clever ways to solve differential equations, but have given less attention to where these differential equations come from. In this chapter we will look at a very powerful general approach to finding governing equations for a broad class of systems: variational principles. These replace local rules with global constraints which can be much easier to understand, and which can then be used to derive the local equations. The foundations of many disciplines can be written either in local or global forms; historically this has been the subject of intense religious debates (figuratively and literally).

## 4.1 VARIATIONAL CALCULUS

### 4.1.1 Euler's Equation

A variational principle is one that states a problem in terms of an unknown function that makes an integral take on an extremum (a maximum or a minimum; frequently a problem is constrained so that it is not necessary to distinguish between these). For example, let's say that we seek the function $y(x)$ that minimizes the distance between two points in a plane (not a very ambitious problem). If $ds$ is an element of the path length along this curve, the total length of the curve is

$$\mathcal{I} = \int_{x_1}^{x_2} ds = \int_{x_1}^{x_2} \sqrt{dx^2 + dy^2} = \int_{x_1}^{x_2} \sqrt{1 + \left(\frac{dy}{dx}\right)^2} \, dx \quad (4.1)$$

(the integral is broken into segments if the curve is not single-valued). More generally (in 2D), an integral defining a variational constraint may be written as

$$\mathcal{I} = \int_{x_1}^{x_2} f[y(x), \dot{y}(x), x] \, dx \quad (4.2)$$

($\dot{y} = dy/dx$). We are writing $y$ and $\dot{y}$ independently because the function may involve either or both; they will soon become related through a differential equation. To find the solution to this problem, let's start as we often do by assuming that we've already solved it. Let $y(x)$ be that solution, and let

$$y(x, \alpha) = y(x) + \alpha \eta(x) \quad (4.3)$$

represent the solution with an arbitrary curve $\eta$ added that has the property $\eta(x_1) =$

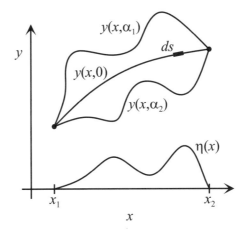

Figure 4.1. Variational paths.

$\eta(x_2) = 0$ (Figure 4.1). Therefore, $y(x, 0) = y(x)$ and by definition $I$ is extremal for $\alpha = 0$:

$$\frac{dI}{d\alpha} = 0 \quad . \tag{4.4}$$

This is sometimes written $\delta I = 0$. To solve this equation, first differentiate under the integral and apply the chain rule:

$$\frac{dI}{d\alpha} = \int_{x_1}^{x_2} \left[ \frac{\partial f}{\partial y} \frac{\partial y}{\partial \alpha} + \frac{\partial f}{\partial \dot{y}} \frac{\partial \dot{y}}{\partial \alpha} \right] dx = 0 \quad . \tag{4.5}$$

The second term can be integrated by parts:

$$\int_{x_1}^{x_2} \frac{\partial f}{\partial \dot{y}} \frac{\partial \dot{y}}{\partial \alpha} = \int_{x_1}^{x_2} \frac{\partial f}{\partial \dot{y}} \frac{d}{dx} \frac{\partial y}{\partial \alpha} \, dx = \underbrace{\frac{\partial f}{\partial \dot{y}} \frac{\partial y}{\partial \alpha} \bigg|_{x_1}^{x_2}}_{0} - \int_{x_1}^{x_2} \frac{d}{dx} \left( \frac{\partial f}{\partial \dot{y}} \right) \underbrace{\frac{\partial y}{\partial \alpha}}_{\eta(x)} dx \quad . \tag{4.6}$$

Therefore

$$\frac{dI}{d\alpha} = \int_{x_1}^{x_2} \left[ \frac{\partial f}{\partial y} - \frac{d}{dx} \left( \frac{\partial f}{\partial \dot{y}} \right) \right] \eta(x) \, dx = 0 \quad . \tag{4.7}$$

Since this must hold for all choices of $\eta$, the expression in the square brackets must vanish:

$$\frac{\partial f}{\partial y} - \frac{d}{dx} \left( \frac{\partial f}{\partial \dot{y}} \right) = 0 \quad . \tag{4.8}$$

This is *Euler's equation* for a variational extremum. Repeating this derivation for a function that depends on $D$ variables $f(y_1(x), y_2(x), \ldots, y_D(x), \dot{y}_1(x), \dot{y}_2(x), \ldots, \dot{y}_D(x), x)$, Euler's equation becomes a set of $D$ equations

$$\frac{\partial f}{\partial y_i} - \frac{d}{dx} \left( \frac{\partial f}{\partial \dot{y}_i} \right) = 0 \quad . \tag{4.9}$$

This is a common, correct, and confusing notation. The partial derivatives are taken with respect to the symbols $y$ and $\dot{y}$ as if they are unrelated, then their usual relationship is restored to solve the resulting differential equation. This procedure keeps track of the chain rule for how they appear in the function, and is straightforward to perform in practice. But the notation is strained because it is really expressing an algorithm rather than a functional relationship. An intriguing alternative made possible by computer symbolic manipulation is to give up on conventional mathematical notation entirely and do variational problems entirely in an algorithmic language [Abelson et al., 1997].

Returning to equation (4.1), we can now see that

$$f = \sqrt{1 + \dot{y}^2} \Rightarrow \frac{\partial f}{\partial y} = 0, \; \frac{\partial f}{\partial \dot{y}} = \frac{\dot{y}}{\sqrt{1 + \dot{y}^2}} \tag{4.10}$$

and so Euler's equation is

$$\frac{d}{dx}\left(\frac{\dot{y}}{\sqrt{1+\dot{y}^2}}\right) = 0 \quad . \tag{4.11}$$

Integrating,

$$\frac{\dot{y}}{\sqrt{1+\dot{y}^2}} = C \Rightarrow \dot{y}^2 = \frac{C^2}{1-C^2} \tag{4.12}$$

for any $C$. The slope is constant; we have just made the significant discovery that the shortest distance between two points is a straight line.

### 4.1.2 Integrals and Missing Variables

If some of the variables are missing in the function $f$, first integrals of Euler's equation exist and can be used to help find solutions. This is an example of the deep connection between symmetry and conserved quantities, to be developed in Section 4.2.3.

If $f$ does not depend on $y$, then

$$\frac{d}{dx}\left(\frac{\partial f}{\partial \dot{y}}\right) = 0 \Rightarrow \frac{\partial f}{\partial \dot{y}} = C \tag{4.13}$$

where $C$ is an integration constant determined by the boundary conditions. Similarly, if $f$ does not depend on $x$, Euler's equations becomes a first-order ordinary differential equation

$$f - \dot{y}\frac{\partial f}{\partial \dot{y}} = C \quad . \tag{4.14}$$

To see this, differentiate with respect to $x$:

$$\frac{d}{dx}\left(f - \dot{y}\frac{\partial f}{\partial \dot{y}}\right) = \frac{\partial f}{\partial y}\dot{y} + \frac{\partial f}{\partial \dot{y}}\ddot{y} - \frac{\partial f}{\partial \dot{y}}\ddot{y} - \frac{\partial^2 f}{\partial y \partial \dot{y}}\dot{y}^2 - \frac{\partial^2 f}{\partial \dot{y}^2}\dot{y}\ddot{y}$$

$$= \dot{y}\left(\frac{\partial f}{\partial y} - \frac{\partial^2 f}{\partial y \partial \dot{y}}\dot{y} - \frac{\partial^2 f}{\partial \dot{y}^2}\ddot{y}\right) \quad . \tag{4.15}$$

Equation (4.14) then follows from this because the right hand side vanishes, since for the case $\partial f/\partial x = 0$ Euler's equation becomes

$$\frac{\partial f}{\partial y} - \frac{d}{dx}\left(\frac{\partial f}{\partial \dot{y}}\right) = \frac{\partial f}{\partial y} - \frac{\partial^2 f}{\partial y \partial \dot{y}}\dot{y} - \frac{\partial^2 f}{\partial \dot{y}^2}\ddot{y} - \underbrace{\frac{\partial^2 f}{\partial x \partial \dot{y}}}_{0} = 0 \quad . \tag{4.16}$$

Finally, if $f$ does not depend on $\dot{y}$ then Euler's equation reduces to the algebraic equation

$$f(x, y) = C \quad . \tag{4.17}$$

The integrals frequently involve a path length $\sqrt{1 + \dot{y}^2}$ in the argument, which can lead to *hyperbolic functions* in the solution. These are defined by analogy with trigonometric functions

$$\cos(x) = \frac{e^{ix} + e^{-ix}}{2} \qquad \sin(x) = \frac{e^{ix} - e^{-ix}}{2i}$$

$$\cosh(x) \equiv \frac{e^x + e^{-x}}{2} \qquad \sinh(x) \equiv \frac{e^x - e^{-x}}{2} \tag{4.18}$$

(pronounced "cosh" and "sinch"). They arise because the derivatives of their inverses have the simple form

$$\frac{d}{dx}\cosh^{-1}\left(\frac{x}{a}\right) = \frac{a}{\sqrt{x^2 - a^2}} \qquad \frac{d}{dx}\sinh^{-1}\left(\frac{x}{a}\right) = \frac{a}{\sqrt{x^2 + a^2}} \quad .$$

### 4.1.3 Constraints and Lagrange Multipliers

Often, a variational problem $\delta \int_1^2 f\, dx = 0$ comes with an ancillary integral constraint

$$\int_{x_1}^{x_2} g[y(x), \dot{y}(x), x]\, dx = C \tag{4.19}$$

for some constant $C$. For example, the problem may be to find a minimal energy curve or surface with a given length or area (Problem 4.1). This is handled by recognizing that if $f$ solves the variational problem and $g$ satisfies the constraint equation, then $h = f + \lambda g$ will also satisfy Euler's equation for any $\lambda$. This is because

$$\int_{x_1}^{x_2} h\, dx = \int_{x_1}^{x_2} (f + \lambda g)\, dx = \int_{x_1}^{x_2} f\, dx + \lambda \int_{x_1}^{x_2} g\, dx = \int_{x_1}^{x_2} f\, dx + \lambda C \tag{4.20}$$

and so if $f$ is extremal then $h$ will also be (the other term is a constant). Solving Euler's equation for $f + \lambda g$ introduces the new variable, $\lambda$, called a *Lagrange multiplier*, into the solution. It parameterizes a family of solutions all of which are extremal with a value of the constraint integral that depends on the choice of $\lambda$, which is then determined by the boundary conditions on the problem. If there is more than one constraint,

$$\int_{x_1}^{x_2} g_1[y(x), \dot{y}(x), x]\, dx = C_1$$

$$\int_{x_1}^{x_2} g_2[y(x), \dot{y}(x), x]\, dx = C_2 \quad , \tag{4.21}$$

then there will be a Lagrange multiplier for each constraint equation $h = f + \lambda_1 g_1 + \lambda_2 g_2$.

The logic of using a Lagrange multiplier is hard to follow when first encountered, but straightforward to implement and is an extremely important trick. For example, statistical mechanics is derived by finding the population distribution that maximizes the entropy of a system with a set of constraints; the Lagrange multiplier associated with a fixed average energy gives the temperature of the system, and the one associated with a fixed average number of particles gives the chemical potential of a particle.

## 4.2 VARIATIONAL PROBLEMS

### 4.2.1 Optics: Fermat's Principle

Many physical laws can be derived from variational principles; depending on one's taste this may be viewed as simply convenient or deeply significant. An important example is provided by optics. The velocity $v$ of a photon in a medium is related to its velocity in a vacuum $c$ by the *index of refraction* $n = c/v$. *Fermat's principle* states that a light ray will choose the path that minimizes the total transmit time. This means that the following variation vanishes between the start and the end of the path:

$$\delta \int_1^2 n \, ds = \delta c \int_1^2 \frac{ds}{v} = \delta c \int_1^2 dt = 0 \quad , \tag{4.22}$$

and so $f = n \, ds$ satisfies Euler's equation. In 2D this leads to the equation

$$\begin{aligned} 0 &= \frac{\partial}{\partial y}\left(n(y)\sqrt{1+\dot{y}^2}\right) - \frac{d}{dx}\left[\frac{\partial}{\partial \dot{y}}\left(n(y)\sqrt{1+\dot{y}^2}\right)\right] \\ &= \frac{\partial n(y)}{\partial y}\sqrt{1+\dot{y}^2} - \frac{d}{dx}\frac{n(y)\,\dot{y}}{\sqrt{1+\dot{y}^2}} \quad . \end{aligned} \tag{4.23}$$

Problem 4.2 solves this at the interface between two media.

### 4.2.2 Analytical Mechanics: Hamilton's Principle

For a classical conservative system (one in which all the forces can be derived from potentials, which means that there is no dissipation), the *Lagrangian* is equal to the difference between the kinetic energy $U$ and the potential energy $V$

$$\mathcal{L}(q_1(t), \dot{q}_1(t), \ldots, t) = U - V \quad . \tag{4.24}$$

The *generalized coordinates* $q_i(t)$ are degrees of freedom that can be varied independently and that are chosen by their convenience for specifying the potential and kinetic energy. The great virtue of a Lagrangian formulation of a problem is that by using generalized coordinates it is possible to avoid explicitly writing the constraint forces that appear in a conventional coordinate system.

The integral over a path of the Lagrangian is called the *action*

$$\mathcal{I} = \int_1^2 \mathcal{L}(q_1(t), \dot{q}_1(t), \ldots, t) \, dt \quad . \tag{4.25}$$

According to Hamilton's variational principle, the system's trajectory in its configuration space will be the one that makes the action extremal ($\delta \mathcal{I} = 0$). Therefore, the Lagrangian satisfies Euler's equations:

$$\frac{\partial \mathcal{L}}{\partial q_i} - \frac{d}{dt}\left(\frac{\partial \mathcal{L}}{\partial \dot{q}_i}\right) = 0 \quad . \tag{4.26}$$

In this context these are called the Euler–Lagrange equations. They allow the governing equations to be found directly from knowledge of the energy in a system. As a simple example, take a particle in a quadratic potential so that the Lagrangian is

$$\mathcal{L} = \frac{1}{2}m\dot{y}^2 - \frac{1}{2}ky^2$$

$$\frac{\partial \mathcal{L}}{\partial y} = -ky$$

$$\frac{d}{dt}\frac{\partial \mathcal{L}}{\partial \dot{y}} = \frac{d}{dt}m\dot{y} = m\ddot{y}$$

$$m\ddot{y} + ky = 0 \quad . \tag{4.27}$$

This of course is just a simple harmonic oscillator. Problem 4.3 covers an example for which the equations are much harder to find without the use of a Lagrangian.

### 4.2.3 Symmetry: Noether's Theorem

Symmetries of systems are coordinate transformations that leave the governing equations unchanged; in this section we will see that for each symmetry there is a conserved quantity (an *integral of the motion*) that is invariant along the system's trajectory in its configuration space. These invariants are so valuable because each one decreases by 1 the effective dimensionality of the system of differential equations that must be solved.

Let $f_s$ be a coordinate transformation operator that is parameterized by $s$ (for example, a shift $f_s$ that replaces $q$ with $q + s$) and that does not change the Lagrangian, with $f_{s=0}$ equal to the identity transformation that makes no change to the system. If $\vec{q}(t) = (q_1(t), q_2(t), \ldots)$ is a solution of the Euler–Lagrange equations then by assumption $\vec{q}(s,t) = f_s[\vec{q}(t)]$ is also a solution:

$$\frac{d}{dt}\frac{\partial \mathcal{L}}{\partial \dot{q}_i}[\vec{q}(s,t), \dot{\vec{q}}(s,t)] = \frac{\partial \mathcal{L}}{\partial q_i}[\vec{q}(s,t), \dot{\vec{q}}(s,t)] \quad . \tag{4.28}$$

Also, since the Lagrangian is invariant under the transformation

$$0 = \frac{d}{ds}\mathcal{L}[\vec{q}(s,t), \dot{\vec{q}}(s,t)] = \sum_i \left[\frac{\partial \mathcal{L}}{\partial q_i}\frac{dq_i(s,t)}{ds} + \frac{\partial \mathcal{L}}{\partial \dot{q}_i}\frac{d\dot{q}_i(s,t)}{ds}\right] \quad . \tag{4.29}$$

Combining these two equations gives

$$0 = \sum_i \left[\frac{d}{dt}\left(\frac{\partial \mathcal{L}}{\partial \dot{q}_i}\right)\frac{dq_i}{ds} + \frac{\partial \mathcal{L}}{\partial \dot{q}_i}\frac{d}{dt}\left(\frac{dq_i}{ds}\right)\right]$$

$$\equiv \frac{d\mathcal{N}}{dt} \quad , \tag{4.30}$$

where

$$\mathcal{N} = \sum_i \frac{\partial \mathcal{L}}{\partial \dot{q}_i}\frac{d}{ds}f_s(q_i) \quad . \tag{4.31}$$

This is called *Noether's theorem* (after the mathematician Emmy Noether). It is true for all $s$, but in particular it is true at $s = 0$. Therefore, the integral invariants can be found by considering an infinitesimal neighborhood of the identity transformation.

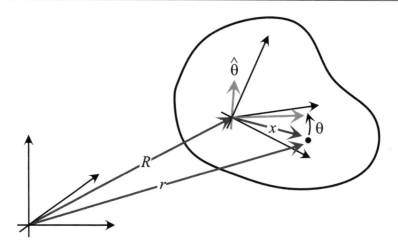

Figure 4.2. Axes for the motion of a rigid body.

The simplest example is translational invariance $f_s(q_i) = q_i + s$. The corresponding integral is

$$\mathcal{N} = \frac{\partial \mathcal{L}}{\partial \dot{q}_i} \frac{d}{ds}(q_i + s) = \frac{\partial \mathcal{L}}{\partial \dot{q}_i} \quad . \tag{4.32}$$

If $\dot{q}_i$ enters into $\mathcal{L}$ as $m\dot{q}_i^2/2$, then this integral is just the momentum $m\dot{q}_i$ (which could be linear or angular depending on the generalized coordinate $q_i$). Conservation of momentum is really a statement of the translational invariance of a system.

## 4.3 RIGID BODY MOTION

An important application of Lagrange's equations is to the motion of *rigid bodies*, such as a spinning top. Developing this connection will require a small detour to describe the configuration of a rigid body.

Let the distribution of mass be $\rho(\vec{x})$, and let the origin for $\vec{x}$ be taken to be the center of mass so that

$$\int \rho(\vec{x}) \vec{x} \, d\vec{x} = 0 \quad . \tag{4.33}$$

This discussion is for a continuous mass distribution, but it also applies to a discrete distribution if the integrals are replaced with sums.

Consider the motion of a point $\vec{x}$ in the rigid body. Its position can also be measured from a fixed external coordinate system ($\vec{r}$), and in this frame let the position of the center of mass be $\vec{R}$. In addition to translation the body can also rotate; let the rotation be by an angle $\theta$ around an axis $\hat{\theta}$ (Figure 4.2). If both the position of the center of mass and the angle change, the position of the point as seen in the fixed frame changes by

$$d\vec{r} = d\vec{R} + d\hat{\theta} \times \vec{x} \quad . \tag{4.34}$$

Dividing both sides by $dt$ shows that the velocity of the point is the sum of the velocity of the center of mass of the rigid body and a term that depends on the rotation:

$$\vec{v} = \vec{V} + \vec{\omega} \times \vec{x} \qquad (4.35)$$

($\vec{\omega} \equiv d\hat{\theta}/dt$). Therefore in a rotating frame the time derivative operator gains an extra cross-product term

$$\left(\frac{d\vec{f}}{dt}\right)_{\text{rotating}} \mapsto \left(\frac{d\vec{f}}{dt}\right)_{\text{fixed}} + \omega \times \vec{f} \quad . \qquad (4.36)$$

Now let's calculate the kinetic energy of the body so that we can find the Lagrangian:

$$\begin{aligned}
U &= \frac{1}{2} \int \rho(\vec{x}) |\vec{v}(\vec{x})|^2 \, d\vec{x} \\
&= \frac{1}{2} \int \rho(\vec{x}) |\vec{V} + \vec{\omega} \times \vec{x}|^2 \, d\vec{x} \\
&= \frac{1}{2} |\vec{V}|^2 \int \rho(\vec{x}) \, d\vec{x} + \vec{V} \cdot \int \rho(\vec{x}) \, \vec{\omega} \times \vec{x} \, d\vec{x} + \frac{1}{2} \int \rho(\vec{x}) |\vec{\omega} \times \vec{x}|^2 \, d\vec{x} \\
&= \frac{1}{2} M |\vec{V}|^2 + (\vec{V} \times \vec{\omega}) \cdot \underbrace{\int \rho(\vec{x}) \vec{x} \, d\vec{x}}_{0} + \frac{1}{2} \int \rho(\vec{x}) |\vec{\omega} \times \vec{x}|^2 \, d\vec{x}
\end{aligned} \qquad (4.37)$$

(the last line uses the identity $\vec{A} \cdot (\vec{B} \times \vec{C}) = (\vec{A} \times \vec{B}) \cdot \vec{C}$). The kinetic energy is the sum of two terms. The first one just depends on the translational motion of the center of mass, and the second depends only on the internal rotation. The second term can be simplified by writing it in terms of a dot product. Letting the angle between $\vec{\omega}$ and $\vec{x}$ be $\alpha$,

$$\begin{aligned}
|\vec{\omega} \times \vec{x}|^2 &= |\vec{\omega}|^2 |\vec{x}|^2 \sin^2(\alpha) \\
&= |\vec{\omega}|^2 |\vec{x}^2| (1 - \cos^2(\alpha)) \\
&= |\vec{\omega}^2| |\vec{x}|^2 - (\vec{\omega} \cdot \vec{x})^2 \\
&= \sum_i \sum_j \omega_i (|\vec{x}|^2 \delta_{ij} - x_i x_j) \omega_j \quad .
\end{aligned} \qquad (4.38)$$

Plugging this back into equation (4.37),

$$U = \frac{1}{2} M |\vec{V}|^2 + \frac{1}{2} \sum_i \sum_j \omega_i \underbrace{\int \rho(\vec{x}) [|\vec{x}|^2 \delta_{ij} - x_i x_j] \, d\vec{x}}_{\equiv J_{ij}} \omega_j$$

$$= \frac{1}{2} M |\vec{V}|^2 + \frac{1}{2} \vec{\omega}^T \cdot \mathbf{J} \cdot \vec{\omega} \qquad (4.39)$$

(where $\vec{\omega}^T$ is the transpose of the vector $\vec{\omega}$). J is called the *inertia tensor*. If the coordinate system in the rigid body is transformed to diagonalize the inertia tensor (as we did for normal modes), the eigenvalues $(I_1, I_2, I_3)$ are called the *moments of inertia* around the eigenvectors called the *principal axes*, and the kinetic energy reduces to

$$U = \frac{1}{2} M |\vec{V}|^2 + \frac{1}{2} I_1 \omega_1^2 + \frac{1}{2} I_2 \omega_2^2 + \frac{1}{2} I_3 \omega_3^2 \quad . \qquad (4.40)$$

The inertia tensor is diagonal with respect to the principal axes, but it is not diagonal

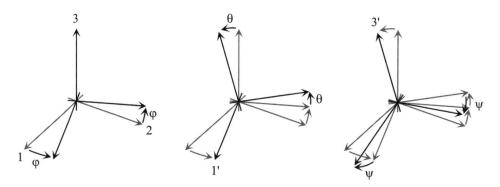

Figure 4.3. Definition of Euler angles.

when viewed from a fixed external frame. Therefore, to calculate the kinetic energy it will be convenient to transform from the fixed frame into the moving principal axes frame. This transformation can be parameterized by taking for generalized coordinates the *Euler angles* which can be varied independently (Figure 4.3). In the standard definition used for rigid body motion, there is a rotation of $\varphi$ around the third axis, a rotation of $\theta$ around the new position of the first axis, and then a rotation of $\psi$ around the new position of the third axis. As the body moves, these angles are measured between the principal axes and a coordinate system that moves with the center of mass but remains aligned with the external frame.

The angular velocities around the body's principal axes can be related to the derivatives of the Euler angles through simple trigonometry in Figure 4.3:

$$\begin{aligned} \omega_1 &= \dot{\theta} \cos \psi + \dot{\varphi} \sin \theta \sin \psi \\ \omega_2 &= -\dot{\theta} \sin \psi + \dot{\varphi} \sin \theta \cos \psi \\ \omega_3 &= \dot{\varphi} \cos \theta + \dot{\psi} \quad . \end{aligned} \tag{4.41}$$

Therefore, the rotational kinetic energy is

$$\begin{aligned} U_{rot} &= \frac{1}{2} \sum I_i \omega_i^2 . \\ &= \frac{1}{2} I_1 (\dot{\theta} \cos \psi + \dot{\varphi} \sin \theta \sin \psi)^2 + \frac{1}{2} I_2 (-\dot{\theta} \sin \psi + \dot{\varphi} \sin \theta \cos \psi)^2 \\ &\quad + \frac{1}{2} I_3 (\dot{\varphi} \cos \theta + \dot{\psi})^2 \quad . \end{aligned} \tag{4.42}$$

Applying Lagrange's equation to this gives the *Euler equations* for rigid body motion. Ignoring the translational energy, and assuming there are no external forces, these take a particularly simple form:

$$\begin{aligned} I_1 \dot{\omega}_1 &= (I_2 - I_3) \omega_2 \omega_3 \\ I_2 \dot{\omega}_2 &= (I_3 - I_1) \omega_3 \omega_1 \\ I_3 \dot{\omega}_3 &= (I_1 - I_2) \omega_1 \omega_2 \quad . \end{aligned} \tag{4.43}$$

If the principal axes are ordered so that $I_3 > I_2 > I_1$, the middle equation has a positive coefficient while the first and third have negative coefficients. Small perturbations about

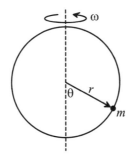

Figure 4.4. Mass on a hoop.

the middle axis will grow exponentially. To see this experimentally, take an object such as an eraser or a book and try to throw it up while spinning it around each of the principal axes in turn.

## 4.4 SELECTED REFERENCES

[Goldstein, 1980] Goldstein, Herbert (1980). *Classical Mechanics*. 2nd edn. Reading, MA: Addison-Wesley.

[Scheck, 1990] Scheck, Florian (1990). *Mechanics : From Newton's Laws to Deterministic Chaos*. New York, NY: Springer-Verlag.

The development of variational methods is closely tied to classical mechanics. Goldstein's book is a classic text, and Scheck provides a nice treatment of modern techniques.

## 4.5 PROBLEMS

(4.1) Consider a chain of length $L$ and density $\rho$ that is hanging between two posts. Find the general form for the shape that minimizes the potential energy. Remember that the potential energy of a segment $ds$ is $\rho \, ds \, g \, y$.

(4.2) Consider a light ray travelling in a medium that has index of refraction $n_1$ for $x < 0$ and $n_2$ for $x > 0$. As it crosses the line $x = 0$ its angle with respect to the $x$ axis will change. Solve Euler's equation to find the simple relationship between the two angles.

(4.3) Find the Lagrangian for a mass on a spinning hoop (Figure 4.4), and use it to find the equation of motion. What is its form for small oscillations?

# 5 Random Systems

So far we have been studying deterministic systems. But the world around us is not very deterministic; there are fluctuations in everything from a cup of coffee to the global economy. In principle, these could be described from first principles, accounting for the actions of every molecule in the cup or every person on the planet. This is hopeless of course, for even if we could determine the initial condition of the whole system, any tiny errors we made would rapidly grow so that our model would no longer be exact. Even if that wasn't a problem, such an enormous model would offer little insight beyond observing the original system.

Fortunately, random (or *stochastic*) systems can be as simple to understand as deterministic systems, if we're careful to ask the right questions. It's a mistake to try to guess the value of the toss of a fair coin, because it is completely uncertain. But we can quite precisely answer questions about the likelihood of events such as seeing a particular string of heads and tails, and easily model a typical sequence of tosses. In fact, simple stochastic systems are as straightforward as simple deterministic ones. Just as deterministic systems get harder to understand as they become more complex, stochastic systems become harder to understand as they become less ideally random, and approximation methods are needed to describe this deviation.

Different problems lie at different ends of this continuum. For a bridge builder, the bridge had better remain standing with near-certain probability, so the uncertainty is $\approx 1 - \epsilon$, where $\epsilon$ is very small. For a financial trader, markets are nearly random, and so any (legal) trading strategy must start with the assuption of near-random uncertainty. There, the probabilities are $\approx 0.5 + \epsilon$. The representations appropriate to $1 - \epsilon$ are different from the ones aimed at $0.5 + \epsilon$, but they do share the central role of variables that are random rather than deterministic.

Our first step will be to look at how to describe the properties of a random variable, and the inter-relationships among a collection of them. Then we will turn to stochastic processes, which are random systems that evolve in time. The chapter closes by looking at algorithms for generating random numbers.

## 5.1 RANDOM VARIABLES

The most important concept in all of stochastic systems is the idea of a *random variable*. This is a fluctuating quantity, such as the hiss from an amplifier, that is described by governing equations in terms of probability distributions. The crucial distinction between random variables and ordinary deterministic variables is that it is not possible to predict the value of a random variable, but it is possible to predict the probability for seeing a given event in terms of the random variable (such as the likelihood of observing a value in a particular range, or the functional form of the power spectrum). An ensemble of identical stochastic processes will generate different values for the random variables, but each member of the ensemble (or *realization*) will obey these same distribution laws. It is important to distinguish between the random variables that appear in distribution functions and the particular values that are obtained in a single realization of a stochastic process by drawing from the distribution.

The simplest random system consists of values $x$ taken from a distribution $p(x)$. For example, in a coin toss $x$ can be heads or tails, and $p(\text{heads}) = p(\text{tails}) = 1/2$. In this case $x$ takes on discrete values; it is also possible for a random variable to come from a continuous distribution. For a continuous variable, $p(x)\,dx$ is the probability to observe a value between $x$ and $x + dx$, and more generally

$$\int_a^b p(x)\,dx \tag{5.1}$$

is the probability to observe $x$ between $a$ and $b$. For a continuous variable, remember that $p(x)$ must always appear with a $dx$ if you want to make a statement about the likelihood of an observable event – a common mistake is to try to evaluate the likelihood of an event by using the value of the density $p(x)$ alone without integrating it over an interval. That is meaningless; among many problems it can easily lead to the impossible result of probabilities greater than 1.

Now consider a string $(x_1, x_2, \ldots, x_N)$ of $N$ $x$'s drawn from a distribution $p(x)$ (such as many coin tosses). The average value (or *expectation value*) of a function $f(x)$ is defined by

$$\langle f(x) \rangle = \lim_{N \to \infty} \frac{1}{N} \sum_{i=1}^{N} f(x_i) \tag{5.2}$$

$$= \int_{-\infty}^{\infty} f(x)\,p(x)\,dx \quad .$$

(in the statistical literature, the expectation is usually written $E[f(x)]$ instead of $\langle f(x) \rangle$). The equivalence of these two equations can be taken as an empirical definition of the probability distribution $p(x)$. This is for a continuous variable; here and throughout this chapter, for a discrete variable the integral is replaced by a sum over the allowed values. If $p(x)$ is not defined from $-\infty$ to $\infty$ then the integral extends over the values for which it is defined (this is called the *support*). A trivial example of an expectation value is

$$\langle 1 \rangle = \int_{-\infty}^{\infty} p(x)\,dx = 1 \tag{5.3}$$

(since probability distributions must be normalized). An important expectation is the *mean value*

$$\bar{x} \equiv \langle x \rangle = \int_{-\infty}^{\infty} x\,p(x)\,dx \quad . \tag{5.4}$$

The mean value might never actually occur (for example $p(x)$ might be a *bimodal* distribution with two peaks that vanish at $p(\bar{x})$). Therefore, another useful quantity is the value of $x$ at which $p(x)$ is a maximum, the *maximum likelihood* value.

The mean value tells us nothing about how big the fluctuations in $x$ are around it. A convenient way to measure this is by the *variance* $\sigma_x^2$, defined to be the average value of the square of the deviation around the mean

$$\begin{aligned}
\sigma_x^2 &\equiv \langle (x - \bar{x})^2 \rangle \\
&= \langle x^2 - 2x\bar{x} + \bar{x}^2 \rangle \\
&= \langle x^2 \rangle - 2\langle x \rangle \bar{x} + \bar{x}^2 \\
&= \langle x^2 \rangle - 2\bar{x}^2 + \bar{x}^2 \\
&= \langle x^2 \rangle - \langle x \rangle^2
\end{aligned} \quad (5.5)$$

(remember that $\bar{x} = \langle x \rangle$ is just a constant). The square root of the variance is called the *standard deviation* $\sigma_x$. To calculate the variance, we need to know the mean value, and the *second moment*

$$\langle x^2 \rangle = \int_{-\infty}^{\infty} x^2 \, p(x) \, dx \quad . \quad (5.6)$$

It is similarly possible to define the higher-order moments $\langle x^n \rangle$.

### 5.1.1 Joint Distributions

Now let's consider two random variables $x$ and $y$, such as the result from throwing a pair of dice, that are specified by a joint density $p(x, y)$. The expected value of a function that depends on both $x$ and $y$ is

$$\langle f(x, y) \rangle = \int_{-\infty}^{\infty} \int_{-\infty}^{\infty} f(x, y) \, p(x, y) \, dx \, dy \quad . \quad (5.7)$$

$p(x, y)$ must be normalized, so that

$$\int_{-\infty}^{\infty} \int_{-\infty}^{\infty} p(x, y) \, dx \, dy = 1 \quad . \quad (5.8)$$

It must also be normalized with respect to each variable, so that

$$p(x) = \int_{-\infty}^{\infty} p(x, y) \, dy \quad (5.9)$$

and

$$p(y) = \int_{-\infty}^{\infty} p(x, y) \, dx \quad . \quad (5.10)$$

Integrating a variable out of a joint distribution is called *marginalizing* over the variable.

For joint random variables a very important quantity is $p(x|y)$ ("the probability of $x$ given $y$"). This is the probability of seeing a particular value of $x$ if we already know the value of $y$, and is defined by *Bayes' rule*

$$p(x|y) = \frac{p(x, y)}{p(y)} \quad , \quad (5.11)$$

which takes the joint probability and divides out from it the known scalar probability. This is easily extended to combinations of more variables,

$$\begin{aligned} p(x,y,z) &= p(x|y,z)\, p(y,z) \\ &= p(x|y,z)\, p(y|z)\, p(z) \\ &= p(x,y|z)\, p(z) \quad , \end{aligned} \quad (5.12)$$

manipulations that will recur in later chapters.

If $x$ and $y$ are *independent*, $p(x,y) = p(x)\,p(y)$. The probability of seeing two independent events is the product of the probabilities of seeing each alone. For independent variables, the conditional distribution will then depend on just one variable, $p(x|y) = p(x)$. This provides a convenient way to remember the form of Bayes' rule, because for independent variables $p(x|y) = p(x)p(y)/p(y) = p(x)$. For *uncorrelated* variables, $\langle xy \rangle = \langle x \rangle \langle y \rangle$. Independent variables are always uncorrelated, but the converse need not be true (although it often is).

Bayes' rule has an almost mystical importance to the (frequently vocal) community of *Bayesians*, who object to any distribution that is not conditional. Equation (5.2) represents a *frequentist* viewpoint, in which probabilites are defined by observed fractions of events. If you are told that the probability of flipping a coin and seeing a head is 50%, you can perform a series of trials to check that. But, if you're told that the probability of rain tomorrow is 50%, you can't check that in the same way. The day will happen only once, so it doesn't make sense to discuss an ensemble of tomorrows. This kind of probability is really a conclusion drawn from a collection of models and prior beliefs. Since you almost always believe something, when you announce to the world $p$(observation), you are really saying $p$(observation|prior beliefs). From Bayes' rule, we know that these are related by

$$p(\text{observation}|\text{prior beliefs}) = \frac{p(\text{observation, prior beliefs})}{p(\text{prior beliefs})}$$

$$= \frac{p(\text{prior beliefs}|\text{observation})\, p(\text{observation})}{p(\text{prior beliefs})} \quad . \quad (5.13)$$

A term $p$(prior beliefs) is called a *prior*, and no well-equipped Bayesian would be caught without one. While this insistence can be dogmatic, it has helped guide many people to recognize and clearly state their procedure for handling advance knowledge about a system and how it gets updated based on experience. We'll have much more to say about this in Chapter 10.

To work with random variables we must understand how their distributions change when they are transformed in mathematical expressions. First, if we know a distribution $p(x)$, and are given a change of coordinates $y(x)$, then what is the distribution $p(y)$? This is easily found by remembering that probabilities are defined in terms of the value of the distribution at a point times a differential element, and equating these:

$$\begin{aligned} p(x)\,|dx| &= p(y)\,|dy| \\ \Rightarrow p(x) &= p(y)\left|\frac{dy}{dx}\right| \quad . \end{aligned} \quad (5.14)$$

In higher dimensions, the transformation of a distribution is done by multiplying it by the *Jacobian*, the determinant of the matrix of partial derivatives of the change of coordinates.

If we take $p(y) = 1$ in the unit interval and integrate both sides of equation (5.14), this relationship has a useful implication

$$\int_0^x p(x')\,dx' = \int_0^x 1 \frac{dy(x')}{dx}\,dx'$$

$$P(x) = y(x) \ . \tag{5.15}$$

This means that if we choose the transformation $y(x)$ to be the integral $P(x)$ of a given distribution $p(x)$ ($P$ is called the *cumulative distribution*), and pick random values of $y$ from a uniform distribution, then the corresponding values of $x$ will be distributed according to $p$.

The next question is how to combine random variables. The simplest operation is adding two independent random variables, $x_1$ drawn from $p_1$, and $x_2$ drawn from $p_2$, to create a new random variable $y = x_1 + x_2$. To find the probability distribution $p(y)$ for $y$, we must add up the probabilities for each of the different ways that $x_1$ and $x_2$ can add up to that value of $y$. The probability of seeing a particular pair of $x_1$ and $x_2$ is given by the product of their probabilities, and so integrating we see that

$$\begin{aligned} p(y) &= \int_{-\infty}^{\infty} p_1(x) p_2(y-x)\,dx \\ &= p_1(x) * p_2(x) \ . \end{aligned} \tag{5.16}$$

The probability distribution for the sum of two random variables is the convolution of the individual distributions. Now consider the average of $N$ variables

$$y_N = \frac{x_1 + x_2 + \cdots + x_N}{N} \tag{5.17}$$

that are independent and identically distributed (often abbreviated as *iid*), and let's look at what happens as $N \to \infty$. The distribution of $y$ is equal to the distribution of $x$ convolved with itself $N$ times, and since taking a Fourier transform turns convolution into multiplication, the Fourier transform of the distribution of $y$ is equal to the product of the $N$ transforms of the distribution of $x$. This suggests an important role for Fourier transforms in studying random processes.

### 5.1.2 Characteristic Functions

The Fourier transform of a probability distribution is called the *characteristic function*, and is equal to the expectation value of the complex exponential

$$\langle e^{ikx} \rangle = \int_{-\infty}^{\infty} e^{ikx} p(x)\,dx \ . \tag{5.18}$$

For a multivariate distribution, the characteristic function is

$$\langle e^{i\vec{k}\cdot\vec{x}} \rangle = \int_{-\infty}^{\infty} \cdots \int_{-\infty}^{\infty} e^{i\vec{k}\cdot\vec{x}} p(\vec{x})\,d\vec{x} \ . \tag{5.19}$$

Now let's look at the characteristic function for the deviation of $y_N$, the average of $N$ iid random numbers $x_1, \ldots, x_N$, from the average value $\bar{x}$:

$$\begin{aligned}
\langle e^{ik(y_N-\bar{x})} \rangle &= \langle e^{ik(x_1+x_2+\cdots+x_N-N\bar{x})/N} \rangle \\
&= \langle e^{ik[(x_1-\bar{x})+\cdots+(x_N-\bar{x})]/N} \rangle \\
&= \langle e^{ik(x-\bar{x})/N} \rangle^N \\
&= \left\langle 1 + \frac{ik}{N}(x-\bar{x}) - \frac{k^2}{2N^2}(x-\bar{x})^2 + \mathcal{O}\left(\frac{k^3}{N^3}\right) \right\rangle^N \\
&= \left[ 1 + 0 - \frac{k^2\sigma_x^2}{2N^2} + \mathcal{O}\left(\frac{k^3}{N^3}\right) \right]^N \\
&\approx e^{-k^2\sigma_x^2/2N} \quad .
\end{aligned} \qquad (5.20)$$

In deriving this we have used the fact that the $x_i$ are independent and identically distributed, and done a Taylor expansion around the average value. The last line follows because

$$\lim_{N\to\infty} \left[1 + \frac{x}{N}\right]^N = e^x \qquad (5.21)$$

(which can be verified by comparing the Taylor series of both sides), and we can neglect higher-order terms in the limit $N \to \infty$. To find the probability distribution for $y$ we now take the inverse transform

$$\begin{aligned}
p(y_N - \bar{x}) &= \frac{1}{2\pi} \int_{-\infty}^{\infty} e^{-k^2\sigma_x^2/2N} e^{-ik(y-\bar{x})} \, dk \\
&= \sqrt{\frac{N}{2\pi\sigma_x^2}} e^{-N(y_N-\bar{x})^2/2\sigma_x^2} \quad .
\end{aligned} \qquad (5.22)$$

Something remarkable has happened: in the limit $N \to \infty$, the sum of $N$ variables approaches a *Gaussian* distribution, independent of the distribution of the variables! This is called the *Central Limit Theorem*, and explains why Gaussians are so important in studying random processes. The Gaussian distribution is also called the *normal* distribution, because it is so, well, normal. Since the standard deviation is $\sigma_x/\sqrt{N}$, which vanishes as $N \to \infty$, equation (5.22) also contains the *Law of Large Numbers*: the average of a large number of random variables approaches the mean, independent of the distribution.

The characteristic function takes on an interesting form if the complex exponential is expanded in a power series:

$$\langle e^{i\vec{k}\cdot\vec{x}} \rangle = \int e^{i\vec{k}\cdot\vec{x}} p(\vec{x}) \, d\vec{x}$$

$$= \sum_{n_1=0}^{\infty} \sum_{n_2=0}^{\infty} \cdots \sum_{n_d=0}^{\infty} \frac{(ik_1)^{n_1}}{n_1!} \frac{(ik_2)^{n_2}}{n_2!} \cdots \frac{(ik_N)^{n_N}}{n_N!} \langle x_1^{n_1} x_2^{n_2} \cdots x_N^{n_N} \rangle \quad . \qquad (5.23)$$

This provides a relationship between the moments of a distribution and the distribution itself (via its Fourier transform). For this reason, the characteristic function is also called the *moment generating function*. If the characteristic function is known, the moments can be found by taking derivatives of the appropriate order and evaluating at $\vec{k} = 0$ (only one term will be nonzero).

Another important object is the logarithm of the characteristic function. If we choose to write this as a power series in $k$ of the form (for the 1D case)

$$\log\langle e^{ikx}\rangle = \sum_{n=1}^{\infty} \frac{(ik)^n}{n!} C_n \quad , \tag{5.24}$$

this defines the *cumulants* $C_n$ (note that the sum starts at $n = 1$ because $\log 1 = 0$ and so there is no constant term). The cumulants can be found by comparing this to the power series expansion of the characteristic function,

$$\exp\left(\sum_{n=1}^{\infty} \frac{(ik)^n}{n!} C_n\right) = \sum_{n=0}^{\infty} \frac{(ik)^n}{n!} \langle x^n \rangle \quad , \tag{5.25}$$

expanding the exponential as

$$e^x = 1 + x + \frac{x^2}{2} + \frac{x^3}{6} + \cdots \quad , \tag{5.26}$$

and grouping terms by order of $k$. The cumulants have an interesting connections to Gaussianity (Problem 5.1).

## 5.2 STOCHASTIC PROCESSES

It is now time for time to appear in our discussion of random systems. When it does, this becomes the study of *stochastic processes*. We will look at two ways to bring in time: the evolution of probability distributions for variables correlated in time, and stochastic differential equations.

If $x(t)$ is a time-dependent random variable, its Fourier transform

$$X(\nu) = \lim_{T\to\infty} \int_{-T/2}^{T/2} e^{i2\pi\nu t} x(t)\, dt \tag{5.27}$$

is also a random variable but its *power spectral density* $S(\nu)$ is not:

$$S(\nu) = \langle |X(\nu)|^2\rangle = \langle X(\nu) X^*(\nu)\rangle \tag{5.28}$$

$$= \lim_{T\to\infty} \frac{1}{T} \int_{-T/2}^{T/2} e^{i2\pi\nu t} x(t)\, dt \int_{-T/2}^{T/2} e^{-i2\pi\nu t'} x(t')\, dt'$$

(where $X^*$ is the complex conjugate of $X$, replacing $i$ with $-i$). The inverse Fourier transform of the power spectral density has an interesting form,

$$\int_{-\infty}^{\infty} S(\nu) e^{-i2\pi\nu\tau}\, d\nu$$

$$= \int_{-\infty}^{\infty} \langle X(\nu) X^*(\nu)\rangle e^{-i2\pi\nu\tau}\, d\nu$$

$$= \lim_{T\to\infty} \frac{1}{T} \int_{-\infty}^{\infty} \int_{-T/2}^{T/2} e^{i2\pi\nu t} x(t)\, dt \int_{-T/2}^{T/2} e^{-i2\pi\nu t'} x(t')\, dt'\, e^{-i2\pi\nu\tau}\, d\nu$$

$$= \lim_{T\to\infty} \frac{1}{T} \int_{-\infty}^{\infty} \int_{-T/2}^{T/2} \int_{-T/2}^{T/2} e^{i2\pi\nu(t-t'-\tau)} \, d\nu \, x(t)x(t') \, dt \, dt'$$

$$= \lim_{T\to\infty} \frac{1}{T} \int_{-T/2}^{T/2} \int_{-T/2}^{T/2} \delta(t-t'-\tau) x(t)x(t') \, dt \, dt'$$

$$= \lim_{T\to\infty} \frac{1}{T} \int_{-T/2}^{T/2} x(t)x(t-\tau) \, dt$$

$$= \langle x(t)x(t-\tau) \rangle \quad , \tag{5.29}$$

found by using the Fourier transform of a *delta function*

$$\int_{-\infty}^{\infty} e^{-i2\pi\nu t} \delta(t) \, dt = 1 \quad \Rightarrow \quad \delta(t) = \int_{-\infty}^{\infty} e^{i2\pi\nu t} \, dt \quad , \tag{5.30}$$

where the delta function is defined by

$$\int_{-\infty}^{\infty} f(x)\delta(x-x_0) \, dx = f(x_0) \quad . \tag{5.31}$$

This is the *Wiener–Khinchin* theorem. It relates the spectrum of a random process to its *autocovariance function*, or, if it is normalized by the variance, the *autocorrelation function* (which features prominently in time series analysis, Chapter 16).

### 5.2.1 Distribution Evolution Equations

A natural way to describe a stochastic process is in terms of the probability to see a sample value $x$ at a time $t$ (written $x_t$) given a history of earlier values

$$p(x_t|x_{t_1}, x_{t_2}, \ldots) \quad . \tag{5.32}$$

Given starting values for $x$ this determines the probability distribution of the future values. If the distribution depends only on the time differences and not on the absolute time,

$$p(x_t|x_{t-\tau_1}, x_{t-\tau_2}, \ldots) \quad , \tag{5.33}$$

then the process is said to be *stationary* (sometimes qualified by calling it *narrow-sense* or *strict* stationarity). A more modest definition asks only that the means and covariances of a process be independent of time, in which case the process is said to be *weak* or *wide-sense* stationary.

If the conditional distribution is limited to a finite history

$$p(x_t|x_{t-\tau_1}, x_{t-\tau_2}, \ldots, x_{t-\tau_N}) \tag{5.34}$$

this is said to be an $N$th-order *Markov process*. If it depends on just the previous value

$$p(x_t|x_{t-\tau}) \tag{5.35}$$

it is simply called a Markov process, and if $x$ and $t$ are discrete variables

$$p(x_t|x_{t-1}) \tag{5.36}$$

it becomes a *Markov Chain*. As with ODEs, an $N$th-order Markov process for a scalar variable can always be converted to a first-order Markov process in an $N$-dimensional variable.

Given the conditional distribution for one time step $p(x_t|x_{t-\tau})$ we can find the distribution two time steps ahead $p(x_{t+\tau}|x_{t-\tau})$ by adding up the probabilities for all of the ways to get from $x_{t-\tau}$ to $x_{t+\tau}$ through the intermediate value $x_t$. The probability for each possible path is the product of the probability to get from $x_{t-\tau}$ to $x_t$ times the probability to go from $x_t$ to $x_{t+\tau}$. For a discrete system this is a sum, and for a continuous system it is the integral

$$p(x_{t+\tau}|x_{t-\tau}) = \int_{-\infty}^{\infty} p(x_{t+\tau}|x_t)\, p(x_t|x_{t-\tau})\, dx_t \quad . \tag{5.37}$$

This is called the *Chapman–Kolmogorov* equation. It can be rewritten by multiplying both sides by $p(x_{t-\tau})$ and then integrating over $x_{t-\tau}$:

$$p(x_{t+\tau}|x_{t-\tau})\, p(x_{t-\tau}) = \int_{-\infty}^{\infty} p(x_{t+\tau}|x_t)\, p(x_t|x_{t-\tau})\, p(x_{t-\tau})\, dx_t$$

$$p(x_{t+\tau}, x_{t-\tau}) = \int_{-\infty}^{\infty} p(x_{t+\tau}|x_t)\, p(x_t, x_{t-\tau})\, dx_t$$

$$p(x_{t+\tau}) = \int_{-\infty}^{\infty} p(x_{t+\tau}|x_t)\, p(x_t)\, dx_t$$

$$= \int_{-\infty}^{\infty} p(x_{t+\tau}, x_t)\, dx_t \quad . \tag{5.38}$$

For a Markov chain with $N$ states $x = (1, \ldots, N)$ this becomes

$$p(x_{t+1}) = \sum_{x_t=1}^{N} p(x_{t+1}|x_t)\, p(x_t) \quad . \tag{5.39}$$

If we define an $N$-component vector of the state probabilities $\vec{p}_t = \{p(x_t = 1), \ldots, p(x_t = N)\}$, and a matrix of transition probabilities $\mathbf{P}_{ij} = p(x_{t+1} = i | x_t = j)$, then the update for all of the states can be written as

$$\vec{p}_{t+1} = \mathbf{P} \cdot \vec{p}_t$$
$$\Rightarrow \vec{p}_{t+n} = \mathbf{P}^n \cdot \vec{p}_t \quad . \tag{5.40}$$

The powers of $\mathbf{P}$ hence determine the evolution; in particular, if it's possible to get from every state to every other state then the system is said to be *ergodic* [Reichl, 1984].

It's easy to understand what a Markov model can do. After all, equation (5.40) is a simple linear first-order finite difference equation (Section 2.5). If $\mathbf{P}$ has eigenvectors $\vec{v}_i$ with eigenvalues $\lambda_i$, and the starting distribution is written in terms of the eigenvectors as

$$\vec{p}_t = \sum_i \alpha_i \vec{v}_i \quad , \tag{5.41}$$

then

$$\vec{p}_{t+n} = \mathbf{P}^n \cdot \sum_i \alpha_i \vec{v}_i$$
$$= \sum_i \alpha_i \lambda_i^n \vec{v}_i \quad . \tag{5.42}$$

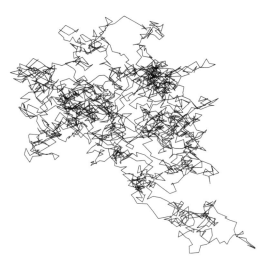

Figure 5.1. 2D Brownian motion.

One of the earliest uses of the Chapman–Kolmogorov equation was in the context of *Brownian motion*. When the Scottish botanist Robert Brown in 1827 used a microscope to look at pollen grains suspended in a solution, he saw them move in a path like the one shown in Figure 5.1. While he originally thought this was a sign of life, it was of course due to the fluctuating impacts of the solvent molecules on the pollen grain. This was a significant piece of evidence for the atomic theory of matter, and in 1905 Einstein developed a quantitative description of Brownian motion that could predict the distribution for how far a particle would travel in a given time.

For 1D Brownian motion (the generalization from 1D to higher dimensions is straightforward), let $f(x,t)\,dx$ be the probability to find the particle between $x$ and $x+dx$ at time $t$. Because it is a probability, $f$ must be normalized, $\int_{-\infty}^{\infty} f(x,t)\,dx = 1$. Brownian motion arises from the many impacts of the fluid molecules on the particle; let $\tau$ be a time that is long compared to these impacts, but short compared to the time for macroscopic motion of the particle. We can then define $p_\tau(x|x')\,dx$, which is the probability for the particle to move from $x'$ to $x$ in a time $\tau$ due to the fluctuating impacts on it. This can be used in the Chapman–Kolmogorov equation to write an update equation for $f$:

$$f(x, t+\tau)\,dx = dx \int_{-\infty}^{\infty} p_\tau(x|x') f(x', t)\,dx' \quad . \tag{5.43}$$

If the fluid is isotropic and homogeneous, then $p_\tau(x|x')$ will depend only on the position difference $x = x' + \delta$, so we can then write $p_\tau(x|x') = p_\tau(\delta)$, where by symmetry $p(\delta) = p(-\delta)$. Then the Chapman-Kolmogorov equation becomes

$$f(x, t+\tau)\,dx = dx \int_{-\infty}^{\infty} p_\tau(\delta) f(x+\delta, t)\,d\delta \quad . \tag{5.44}$$

Since $\tau$ has been chosen to be very small compared to the time scale for macroscopic changes in the particle position, we can expand $f$ in a Taylor series in $x$ and $t$ and keep the lowest-order terms:

$$\left[ f(x,t) + \frac{\partial f}{\partial t} \tau + \mathcal{O}(\tau^2) \right] dx$$

$$= dx \int_{-\infty}^{\infty} p_\tau(\delta) \left[ f(x,t) + \frac{\partial f}{\partial x} \delta + \frac{1}{2} \frac{\partial^2 f}{\partial x^2} \delta^2 + \cdots \right] d\delta$$

$$= dx\, f(x,t) \underbrace{\int_{-\infty}^{\infty} p_\tau(\delta)\, d\delta}_{1} + dx\, \frac{\partial f}{\partial x} \underbrace{\int_{-\infty}^{\infty} p_\tau(\delta)\, \delta\, d\delta}_{0} +$$

$$dx\, \frac{1}{2} \frac{\partial^2 f}{\partial x^2} \underbrace{\int_{-\infty}^{\infty} p_\tau(\delta) \delta^2\, d\delta}_{\langle \delta^2 \rangle} + \mathcal{O}(\delta^3) \quad . \tag{5.45}$$

In the last line, the first integral is 1 because $p_\tau(\delta)$ must be normalized, the second one vanishes because it is the first moment of a symmetrical function $p_\tau(\delta) = p_\tau(-\delta)$, and the last integral is the variance. Cancelling out the $f(x,t)$ on both sides, we're left with

$$\frac{\partial f}{\partial t} = \underbrace{\frac{\langle \delta^2 \rangle}{2\tau}}_{\equiv D} \frac{\partial^2 f}{\partial x^2} \quad . \tag{5.46}$$

We've found that the evolution of the probability distribution for Brownian motion is governed by the familiar diffusion equation. In the context of stochastic processes, this is called a *Wiener process*. The diffusion equation is a particular case of the *Fokker–Plank* PDE which governs the evolution of a probability distribution for an underlying Markov process.

We obtained the diffusion equation for a system in which time and space are continuous, but for many problems they are discrete. Consider a 1D random walker that at each time step $t_i$ can hop from a site $x_n$ one unit to the left or right, $x_{n\pm1}$. The change in probability $p(x_n, t_i)$ to find it at a point is then equal to the sum of the probabilities for it to hop into the point, minus the sum of probabilities to hop off the point

$$p(x_n, t_i) - p(x_n, t_{i-1}) = \frac{1}{2}[p(x_{n+1}, t_{i-1}) + p(x_{n-1}, t_{i-1})] - p(x_n, t_{i-1})$$

$$p(x_n, t_i) = \frac{1}{2}[p(x_{n+1}, t_{i-1}) + p(x_{n-1}, t_{i-1})] \quad . \tag{5.47}$$

This can be rewritten suggestively by subtracting $p(x_n, t_{i-1})$ from both sides and rearranging:

$$\frac{p(x_n, t_i) - p(x_n, t_{i-1})}{\delta_t} = \underbrace{\frac{\delta_x^2}{2\delta_t}}_{\equiv D} \left[ \frac{p(x_{n+1}, t_{i-1}) - 2p(x_n, t_{i-1}) + p(x_{n-1}, t_{i-1})}{\delta_x^2} \right] \tag{5.48}$$

These are discrete approximations of partial derivatives (Chapter 7). If we take the limit $\delta_t, \delta_x \to 0$ (keeping $D$ finite), we find once again that

$$\frac{\partial p}{\partial t} = D \frac{\partial^2 p}{\partial x^2} \quad . \tag{5.49}$$

More generally, the time derivative of the probability to be at a state $x_n$ is equal to the sum over states $x_m$ of the probability to be at $x_m$ times the rate $W_{x_m \to x_n}$ at which transitions are made from there to $x_n$, minus the probability to be at $x_n$ times the rate at which transitions are made back to $x_m$:

$$\frac{\partial p(x_n, t)}{\partial t} = \sum_m W_{x_m \to x_n} p(x_m, t) - W_{x_n \to x_m} p(x_n, t) \quad . \tag{5.50}$$

This is called the *Master equation*. For a stationary solution the transition rate between two sites is equal in both directions, a condition called *detailed balance*.

### 5.2.2 Stochastic Differential Equations

An alternative analysis of Brownian motion was first done by Langevin in terms of a stochastic differential equation. A particle moving in a fluid feels a drag force, and as long as the velocity is not too great the force is given by the Stokes drag formula

$$\vec{F} = -6\pi\mu a \vec{v} \quad , \tag{5.51}$$

where $\mu$ is the viscosity of the fluid, $a$ is the diameter of the particle, and $\vec{v}$ is the velocity of the particle [Batchelor, 1967]. In addition to this force, we can model Brownian motion by including a fluctuating force $\eta$ that is due to the molecular impacts on the particle. In terms of these forces, $\vec{F} = m\vec{a}$ for the particle becomes (in 1D):

$$m\frac{d^2x}{dt^2} = -6\pi\mu a \frac{dx}{dt} + \eta \quad . \tag{5.52}$$

This is an example of what is now called a *Langevin equation*. Because $\eta$ is a random variable, $x$ becomes one, much like the promotion of operator types in a computer program. Therefore we cannot solve for $x$ directly; we must instead use this differential equation to solve for observable quantities that depend on it. To do this, first recognize that

$$\frac{d(x^2)}{dt} = 2x\frac{dx}{dt} \tag{5.53}$$

and

$$\frac{d^2(x^2)}{dt^2} = 2\left(\frac{dx}{dt}\right)^2 + 2x\frac{d^2x}{dt^2} = 2v^2 + 2x\frac{d^2x}{dt^2} \quad . \tag{5.54}$$

Using this, if we multiply both sides of equation (5.52) by $x$ we can rewrite it as

$$\frac{m}{2}\frac{d^2(x^2)}{dt^2} - mv^2 = -3\pi\mu a \frac{d(x^2)}{dt} + \eta x \quad . \tag{5.55}$$

Next, let's take the time expectation value

$$\frac{m}{2}\frac{d^2\langle x^2\rangle}{dt^2} + 3\pi\mu a \frac{d\langle x^2\rangle}{dt} = \underbrace{m\langle v^2\rangle}_{kT} + \underbrace{\langle \eta x\rangle}_{0} \quad . \tag{5.56}$$

In the first term on the right hand side, we've used the fact that the particle is in thermodynamic equilibrium with the fluid to apply the *Equipartition Theorem* [Gershenfeld,

1999a], which tells us that

$$\frac{1}{2}m\langle v^2\rangle = \frac{D}{2}kT \quad , \tag{5.57}$$

where $D$ is the dimension (1 in this case), $k$ is Boltzmann's constant $1.38 \times 10^{-23}$ (J/K), and $T$ is the temperature (in Kelvin). The second term vanishes because the rapidly fluctuating noise term is uncorrelated with the slowly moving particle. Therefore,

$$\frac{d^2\langle x^2\rangle}{dt^2} + \frac{3\pi\mu a}{m}\frac{d\langle x^2\rangle}{dt} = \frac{2kT}{m} \quad . \tag{5.58}$$

This is now an ordinary differential equation for the variance, which can easily be solved to find

$$\langle x^2\rangle = Ae^{-6\pi\mu at/m} + \frac{kT}{3\pi\mu a}t \quad . \tag{5.59}$$

The first term is a rapidly decaying exponential transient, leaving us with

$$\langle x^2\rangle = \frac{kT}{3\pi\mu a}t \quad . \tag{5.60}$$

This result agrees with the form of Einstein's calculation (Problem 5.3), even though we got here by a very different path. Solving more general stochastic differential equations, and justifying assumptions such as throwing out the $\langle \eta x\rangle$ term, requires extending ordinary calculus to integrals of stochastic functions. This is done by the Ito and the Stratonovich calculus [Gardiner, 1990].

## 5.3 RANDOM NUMBER GENERATORS

There is a frequent and apparently paradoxical need to use a computer to generate random numbers. In modeling a stochastic system it is necessary to include a source of noise, but computers are (hopefully) not noisy. One solution is to attach a computer peripheral that performs a quantum measurement (which as far as we know can be completely random), or perhaps measures the molecular fluctuations in your coffee cup (which are extremely close to random), but for most people this is not a convenient option. Instead, a more reasonable alternative is to use an algorithm that produces pseudo–random numbers that appear to be more random than can be detected by your application. There is a large difference in what is required to fool a player of a video game and a cryptographic analyst of a one–time pad. There is a corresponding broad range of choices for random number generators, based on how sensitive your problem is to the hidden order that must be present in any deterministic algorithm. While these are numerical rather than analytical methods, and so rightfully belong in the next part of this book, they are so closely connected with the rest of this chapter that it is more natural to include them here.

Why discuss random number generators when most every programming language has one built in? There are two reasons: portability and reliability. By explicitly including an algorithm for generating needed random numbers, a program will be sure to give the same answer whatever system it is run on. And built-in generators range from being much too complex for simple tasks to much too simple for complex needs.

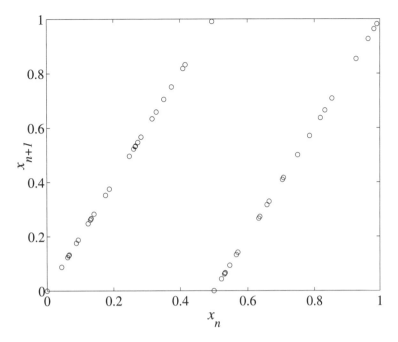

Figure 5.2. 1000 points from the map $x_{n+1} = 2x_n$ (mod 1).

### 5.3.1 Linear Congruential

*Linear congruential* random number generators, and more sophisticated variants, are the most common technique used for producing random numbers. The simplest example is the map

$$x_{n+1} = 2x_n \text{ (mod 1)} \quad . \tag{5.61}$$

Starting with an initial value for $x_0$ (chosen to be between 0 and 1), this generates a series of new values. The procedure is to multiply the old value by 2, take the part that is left after dividing by 1 (i.e., the fractional part), and use this as the new value. This string of numbers is our first candidate as a random number generator. But how random are the successive values of $x$? Not very; Figure 5.2 plots $x_{n+1}$ versus $x_n$ for a series of 1000 points. Two things are immediately apparent: the points lie on two lines rather than being uniformly distributed (as they should be for two independent random numbers), and it doesn't look like there are 1000 points in the figure. The first problem can easily be explained by looking back at the definition of the map, which shows that successive pairs of points lie on a line of slope 2, which gets wrapped around because of the *mod* operator. To understand the second problem, consider $x$ written in a fractional binary expansion (where each digit to the right of the binary point stands for $2^{-1}, 2^{-2}, \ldots$). Each iteration of this map shifts all the digits one place to the left, and throws out the digit that crosses the binary point. This means that it brings up all the bits of the starting position, and finally settles down to a fixed point at $x = 0$ when all the bits are used up.

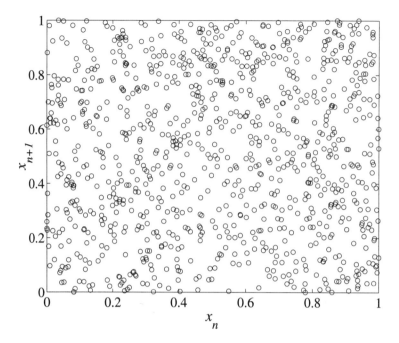

Figure 5.3. 1000 points from the map $x_{n+1} = 8121x_n + 28411 \pmod{134456}$.

This bad example can be generalized to the class of maps

$$x_{n+1} = ax_n + b \pmod{c} \quad . \tag{5.62}$$

The value of $a$ determines the slope of the lines that the points are on and how many lines there will be ($a = 2$ gave us 2 lines). We want this to be as large as possible, so that the lines fill the space as densely as possible. Then $b$ and $c$ must be chosen relative to $a$ so that the period of the map is as long as possible (it doesn't repeat after a few iterations), there are no fixed points that it can get stuck at, and the digits are otherwise as random as they can be. Choosing optimal values for $a$, $b$, and $c$ is a surprisingly subtle problem, but good choices have been worked out as a function of the machine word size used to represent $x$ [Knuth, 1981]. For the common case of a 32-bit integer, with the leading bit used as a sign bit, an optimal choice is

$$x_{n+1} = 8121x_n + 28411 \pmod{134456} \quad . \tag{5.63}$$

Iterating this map produces a string of integers between 1 and 134456 (or a fraction between 0 and 1 if the integer is divided by 134456), shown in Figure 5.3. This now appears to be much more random, and is adequate when there is a simple need for some numbers that "look" random.

This is still not a great generator, because there are only 134456 distinct possible values, and so in a string that long it is possible to detect the predictability. It's also easy to see here why the bits of $x$ are not equally random: if $x_n$ is even then $x_{n+1}$ will be odd, and *vice versa*, so the lowest order bit of $x$ simply oscillates at each step. Not very random. To further improve such a simple linear congruential generator, it is possible

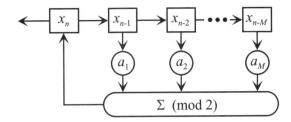

Figure 5.4. A linear feedback shift register.

to add degrees of freedom by techniques such as running multiple generators in parallel and using them to shuffle entries in a large array [Press et al., 1992].

### 5.3.2 Linear Feedback

Linear congruential generators have the problem that all of the bits in each number are usually not equally random; *linear feedback shift registers* (*LFSRs*) provide a powerful alternative that can be used to generate truly psuedo–random bits. A binary linear feedback shift register is specified by a *recursion relation*

$$x_n = \sum_{i=1}^{M} a_i x_{n-i} \ (\text{mod } 2) \quad . \tag{5.64}$$

This can be viewed as a series of registers through which the bits are shifted, with taps specified by the $a_i$'s that select the values to be added mod 2 (Figure 5.4).

If the taps are properly chosen, the bits that come out of the end of the shift register are as random as possible. This means that the power spectrum is flat (up to the repeat time, which is $2^M - 1$ for a register with $M$ steps), all possible substrings of bits occur equally often, and so forth. Such a *maximal* LFSR is designed by taking the $z$-transform of the recursion relation, and finding the taps that make this polynomial have no smaller polynomial factors [Simon et al., 1994]. Table 5.1 gives a (non-unique) choice for maximal taps for a range of register lengths. For example, for order 12 the tap values are 1, 4, 6, 12, so the recursion relation is

$$x_n = x_{n-1} + x_{n-4} + x_{n-6} + x_{n-12} \quad . \tag{5.65}$$

Because the recurrence time is exponential in the length of the register, a surprisingly modest LFSR can have an extremely long period and be hard to distinguish from random (Problem 5.2).

The most sophisticated techniques for making numbers appear random are associated with cryptography, since detectable deviations from randomness can be used to help break a code. In fact, the best way of all to improve a random number generator is to run its output through a cryptographic encoding scheme [Simmons, 1992], which pass much more sophisticated tests of randomness than the algorithms we have covered (which are child's play for a cryptanalyst). This of course does come at the expense of much more computational effort.

Table 5.1. *For an LFSR* $x_n = \sum_{i=1}^{M} a_i x_{n-i}$ (mod 2), *lag i values for which* $a_i = 1$ *for the given order M (all of the other $a_i$'s are 0).*

| M | i | M | i | M | i |
|---|---|---|---|---|---|
| 2 | 1, 2 | 13 | 1, 3, 4, 13 | 24 | 1, 2, 7, 24 |
| 3 | 1, 3 | 14 | 1, 6, 10, 14 | 25 | 3, 25 |
| 4 | 1, 4 | 15 | 1, 15 | 26 | 1, 2, 6, 26 |
| 5 | 2, 5 | 16 | 1, 3, 12, 16 | 27 | 1, 2, 5, 27 |
| 6 | 1, 6 | 17 | 3, 17 | 28 | 3, 28 |
| 7 | 3, 7 | 18 | 7, 18 | 29 | 2, 29 |
| 8 | 2, 3, 4, 8 | 19 | 1, 2, 5, 19 | 30 | 1, 2, 23, 30 |
| 9 | 4, 9 | 20 | 3, 20 | 31 | 3, 31 |
| 10 | 3, 10 | 21 | 2, 21 | 32 | 1, 2, 22, 32 |
| 11 | 2, 11 | 22 | 1, 22 | 33 | 13, 33 |
| 12 | 1, 4, 6, 12 | 23 | 5, 23 | 34 | 1, 2, 27, 34 |

## 5.4 SELECTED REFERENCES

[Feller, 1968] Feller, William (1968). *An Introduction to Probability Theory and its Applications*. 3rd edn. New York, NY: Wiley.

A classis reference for probability theory.

[Gardiner, 1990] Gardiner, C.W. (1990). *Handbook of Stochastic Methods*. 2nd edn. New York, NY: Springer-Verlag.

This is a beautiful survey of techniques for working with stochastic systems.

[Knuth, 1981] Knuth, Donald E. (1981). *Semi-Numerical Algorithms*. 2nd edn. *The Art of Computer Programming*, vol. 2. Reading, MA: Addison-Wesley.

The standard starting point for questions about generating and testing random numbers.

[Press *et al.*, 1992] Press, William H., Teukolsky, Saul A., Vetterling, William T., & Flannery, Brian P. (1992). *Numerical Recipes in C: The Art of Scientific Computing*. 2nd edn. New York, NY: Cambridge University Press.

*Numerical Recipes* has a good collection of practical routines for generating random numbers.

## 5.5 PROBLEMS

(5.1) (a) Work out the first three cumulants $C_1, C_2$, and $C_3$.

(b) Evaluate the first three cumulants for a Gaussian distribution

$$p(x) = \frac{1}{\sqrt{2\pi\sigma^2}} e^{-(x-\bar{x})^2/2\sigma^2} \quad . \tag{5.66}$$

(5.2) (a) For an order 4 maximal LFSR write down the bit sequence.

(b) If an LFSR has a clock rate of 1 GHZ, how long must the register be for the time between repeats to be the age of the universe ($\sim 10^{10}$ years)?

(5.3) (a) Use a Fourier transform to solve the diffusion equation (5.46) (assume that the initial condition is a normalized delta function at the origin).
(b) What is the variance as a function of time?
(c) How is the diffusion coefficient for Brownian motion related to the viscosity of a fluid?
(d) Write a program (including the random number generator) to plot the position as a function of time of a random walker in 1D that at each time step has an equal probability of making a step of $\pm 1$. Plot an ensemble of 10 trajectories, each 1000 points long, and overlay error bars of width $3\sigma(t)$ on the plot.
(e) What fraction of the trajectories should be contained in the error bars?

*Part Two*
Numerical Models

By now it should be evident that the fraction of differential equations that can be solved exactly with a sane amount of effort is quite small, and that once we stray too far from linearity special techniques are needed for there to be any hope of writing down a closed-form solution. In this second part of the book we will turn to the appealing alternative: numerical solutions. Although the widespread access to fast computers has perhaps led to an over-reliance on numerical answers when there are other possibilities, and a corresponding false sense of security about the possibility of serious numerical problems or errors, it is now possible without too much trouble to find solutions to most equations that are routinely encountered.

An important issue is when the result of a numerical calculation can be trusted. It's (almost) always possible to produce some kind of number, but it's much harder to produce a meaningful number. One good rule of thumb is that the result should not depend on the algorithm parameters (for example, decreasing a step size should give the same answer), otherwise the result provides information about the algorithm rather than the underlying problem. Another crucial sanity test is to check the algorithm on a problem with a known exact solution to make sure that it agrees. Even so, there's still a chance that a subtle difference between the algorithm and the underlying problem can lead to fundamentally different results. It is common to assume that numerical errors appear as a small random noise term, but it is possible that the errors are very correlated rather than random [Sauer *et al.*, 1997]. There have been a number of attempts at developing techniques to bound the exact solution with upper and lower estimates, or to provide some measure of the existence and proximity of the exact solution; these are promising, but far from universally applicable.

A principle that is so important for numerical methods that it doesn't even have a name (and is frequently ignored) is that if you know your system has some conservation laws then you should choose variables that automatically conserve them. Otherwise, your conservation laws will be ignored by the solution, or will need to be explicitly enforced. Not only does this make your solution more accurate and reduce the computational effort, it can help tame numerical instabilities that might otherwise occur.

More generally, remember that numerical and analytical solutions should not be viewed as exclusive alternatives, but rather as complementary approaches. Many important ideas have come at this interface, such as *solitons*. These nonlinear nondispersive waves first appeared as unusual stable structures in numerical simulations of plasmas; this observation motivated and then guided the successful search for corresponding analytical solutions [Miles, 1981, Zabusky, 1981].

There is a continuum between numerical and analytical solutions, and a trade-off between the need for computer power and mathematical insight. The more that an algorithm takes advantage of knowledge about an equation, the less work that the computer need do. However, given a fast computer and a hard problem a less clever algorithm may be preferable. In the early days of numerical mathematics, algorithms were implemented by people called *calculators* (sometimes large rooms full of them) using slide rules, tables, or arithmetic machines. This put a large premium on reducing the number of steps needed, and hence on developing algorithms that maximize step sizes and minimize the number of needed function evaluations. As desktop workstations now begin to exceed the speed of recent supercomputers, less efficient algorithms that have other desirable properties (such as simplicity or reliability) can be used. We will cover the most important algorithms that are straightforward to understand and implement; for these as well as everything else related to numerical methods [Press *et al.*, 1992] is a great starting point for more information. Another interesting reference is [Acton, 1990], a revised edition of a beautiful book (with a sneaky cover) that has had a large impact on many people working in the field of numerical analysis.

Some caution is needed in delving more deeply into the literature on numerical analysis. Often the algorithms that are most useful in practice are the ones that are least amenable to proving rigorous results about their error and convergence properties, and hence are less well represented. It's necessary to condition what is known against a usually unstated prior of what kinds of problems are likely to be encountered.

To a computational complexity theorist, the most important distinction in deciding if a problem is tractable is the difference between those that can be solved in a number of steps that is polynomial in the size of the problem, such as sorting, and those that cannot, such as factoring (which is believed to be exponential) [Lewis & Papadimitriou, 1981]. For example, finding prime factors requires $\mathcal{O}(e^{N/3})$ steps for an $N$-digit number, while on a quantum computer (if you're fortunate enough to know how to build one), it requires $\mathcal{O}(N^3)$ [Ekert & Jozsa, 1996]. If we assume a computer speed of 100 MHz, for a 100-digit number this difference corresponds to a time of 35 days versus 0.01 seconds. But to a working numerical analyst, the relevant distinction is really between $\mathcal{O}(N^2)$ algorithms and faster ones, which in practice is the distinction between what is feasible and what is not for nontrivial problems. Naively, a Fourier transform requires $\mathcal{O}(N^2)$ steps because it needs a matrix multiplication, but by taking advantage of the structure in the calculation the *Fast Fourier Transform* algorithm reduces it to $\mathcal{O}(N \log N)$ (Section 11.2). While this difference might not seem as remarkable as the difference between exponential and polynomial time, in practice it is profound. For example, for $N = 10$, an $N^2$ algorithm requires 100 steps, and an $N \log N$ algorithm requires 33 steps, not much difference. But for $N = 10^9$, an $N^2$ algorithm requires $10^{18}$ steps ($10^5$ days at 100 MHz), while $N \log N$ requires $3 \times 10^{10}$ steps (300 seconds), quite a difference indeed! This issue of the scaling of an algorithm with problem size will recur throughout the coming chapters, and is one of the most important lessons in all of numerical analysis. There are endless of examples of promising new algorithms that do not survive the scaling up to nontrivial problems.

# 6 Finite Differences: Ordinary Differential Equations

## 6.1 NUMERICAL APPROXIMATIONS

This chapter will consider the problem of finding the numerical solution to the first-order (usually nonlinear) differential equation

$$\frac{dy}{dx} = f(x, y) \quad . \tag{6.1}$$

Because of the presence of $y$ on the right hand side we can't simply integrate $f(x)\,dx$; we'll need some kind of iterative procedure to calculate a new value of $y$ and use it to evaluate $f$. Fortunately there are techniques for solving ODEs that are relatively straightforward to implement and use, and that are broadly applicable. If $\dot{y} = dy/dx$ also appears on the right hand side, $f(x, y, \dot{y}) = 0$, this becomes a *differential-algebraic equation* (*DAE*), a still harder problem that usually requires more complex algorithms matched to the problem [Brenan et al., 1996].

The restriction to first-order equations actually isn't much of a restriction at all. The algorithms for solving a single first-order equation will immediately generalize to systems of equations

$$\frac{dy_1}{dx} = f_1(x, y_1, \ldots, y_N)$$

$$\frac{dy_2}{dx} = f_2(x, y_1, \ldots, y_N)$$

$$\vdots$$

$$\frac{dy_N}{dx} = f_N(x, y_1, \ldots, y_N) \quad . \tag{6.2}$$

And, since a higher-order differential equation of the form

$$\frac{d^N y}{dx^N} = f\left(x, y, \frac{dy}{dx}, \ldots, \frac{d^{N-1} y}{dx^{N-1}}\right) \tag{6.3}$$

can be written as a system of first-order equations (equation (2.15))

$$y^{(1)} \equiv \frac{dy}{dx} \quad \cdots \quad y^{(N-1)} \equiv \frac{d^{N-1} y}{dx^{N-1}}$$

$$\frac{dy^{(1)}}{dx} = y^{(2)}$$

$$\vdots$$

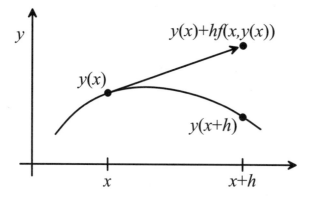

Figure 6.1. An Euler method.

$$\frac{dy^{(N-1)}}{dx} = y^{(N)}$$
$$\frac{dy^{(N)}}{dx} = f(x, y^{(1)}, \ldots, y^{(N-1)}) \quad , \tag{6.4}$$

we will also be able to solve higher-order equations.

We would like to find an approximate formula to relate $y(x+h)$ to $y(x)$ for some small step $h$. An obvious way to do this is through the Taylor expansion of $y$

$$\begin{aligned}
y(x+h) &= y(x) + h\frac{dy}{dx}\bigg|_x + \frac{h^2}{2}\frac{d^2y}{dx^2}\bigg|_x + \mathcal{O}(h^3) \\
&= y(x) + hf(x, y(x)) + \frac{h^2}{2}\frac{d}{dx}f(x, y(x)) + \mathcal{O}(h^3) \\
&= y(x) + hf(x, y(x)) + \frac{h^2}{2}\left[\frac{\partial f}{\partial x} + \frac{\partial f}{\partial y}\frac{dy}{dx}\right] + \mathcal{O}(h^3) \\
&= y(x) + hf(x, y(x)) + \frac{h^2}{2}\left[\frac{\partial f}{\partial x} + f\frac{\partial f}{\partial y}\right] + \mathcal{O}(h^3) \ .
\end{aligned} \tag{6.5}$$

An approximation scheme must match these terms to agree with the Taylor expansion up to the desired order.

The first two terms of the expansion

$$y(x+h) = y(x) + hf(x, y(x)) \tag{6.6}$$

can be used to find $y(x+h)$ given $y(x)$, and this step can then be repeated to find $y(x+2h) = y(x+h) + hf(x+h, y(x+h))$, and so forth. This is the simplest algorithm for solving differential equations, called *Euler's method* (shown in Figure 6.1). It is simple to understand and simple to program; its clarity is matched only by its dreadful performance. At each step the error is $\mathcal{O}(h^2)$ (the lowest-order term where the Euler method differs from the Taylor expansion of $y(x)$ is the $h^2$ term), and so a very small step size is needed for a reasonably accurate solution. Even worse, the errors can accumulate so rapidly that the numerical solution actually becomes unstable and blows up. Consider

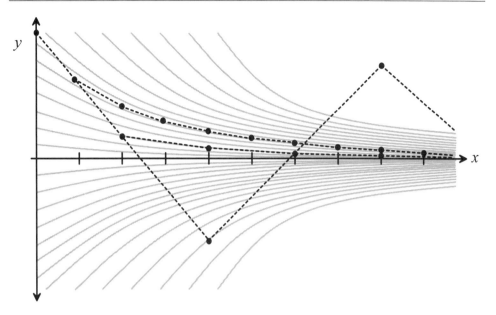

Figure 6.2. Origin of oscillation in the Euler method. The gray lines show the family of solutions of $dy/dx = Ay$, and the dotted lines show the numerical solution for various step sizes.

the simple differential equation

$$\frac{dy}{dx} = Ay \quad . \tag{6.7}$$

The exact solution is $y(x) = e^{Ax}$; if $A < 0$ then $\lim_{x \to \infty} y(x) = 0$. The Euler approximation for this equation is

$$y(x + h) = y(x) + hAy(x)$$
$$= (1 + hA)y(x) \quad . \tag{6.8}$$

This is a first-order difference equation, with a solution equal to $y(x) = (1+hA)^{x/h} y(0)$. If $A > 0$ this solution diverges, as it should. If $0 > hA > -1$, then the solution properly decays to zero. But look what happens as $hA$ becomes even more negative. If $-1 > hA > -2$, the magnitude of the solution still decays, but it has picked up an oscillation that is not in the original equation. Even worse, if $-2 > hA$, then the magnitude of the solution will diverge! We've solved an equation, but it now has nothing to do with our original differential equation.

Figure 6.2 shows why the Euler method is such a poor approximation. Since the derivative is evaluated only at the beginning of the interval, if an overly ambitious step size is chosen the extrapolation can overshoot so far that the solution changes sign. A natural improvement is to use an Euler step in order to estimate the slope in the middle of the interval, and then use this slope to update $y$:

$$y(x+h) = y(x) + hf\left[x + \frac{h}{2},\ y(x) + \frac{h}{2}f(x, y(x))\right] \quad . \tag{6.9}$$

The error made by this approximation can be found by doing a Taylor expansion on $f$

as a function of $h$:

$$f\left[x + \frac{h}{2}, \; y(x) + \frac{h}{2}f(x,y(x))\right]$$

$$= f(x,y(x)) + h\frac{d}{dh}f\left[x + \frac{h}{2}, \; y(x) + \frac{h}{2}f(x,y(x))\right]_{h=0} + \mathcal{O}(h^2)$$

$$= f(x,y(x)) + h\left[\frac{1}{2}\frac{\partial f}{\partial x} + \frac{1}{2}f(x,y(x))\frac{\partial f}{\partial y}\right] + \mathcal{O}(h^2) \qquad (6.10)$$

and so equation (6.9) becomes

$$y(x+h) = y(x) + hf(x,y(x)) + \frac{h^2}{2}\left[\frac{\partial f}{\partial x} + f\frac{\partial f}{\partial y}\right] + \mathcal{O}(h^3) \quad . \qquad (6.11)$$

Comparing this with equation (6.5), we see that this is exactly the expansion of the solution of the differential equation up to second order. This is called the *second-order Runge–Kutta* or the *midpoint* method. We have found a way to evaluate the function that gives us an answer that is correct to second order, but that does not require explicitly working out the Taylor expansion. Problem 6.2 shows the benefit of improving the approximation order.

## 6.2 RUNGE–KUTTA METHODS

This procedure can be carried out to use more function evaluations to match higher-order terms in the Taylor expansion. The derivation rapidly becomes very tedious, and there is no unique solution for a given order, but by far the most common approximation is the *fourth-order Runge–Kutta* approximation

$$k_1 = hf(x, y(x))$$

$$k_2 = hf\left(x + \frac{h}{2}, \; y(x) + \frac{k_1}{2}\right)$$

$$k_3 = hf\left(x + \frac{h}{2}, \; y(x) + \frac{k_2}{2}\right)$$

$$k_4 = hf(x + h, \; y(x) + k_3)$$

$$y(x+h) = y(x) + \frac{k_1}{6} + \frac{k_2}{3} + \frac{k_3}{3} + \frac{k_4}{6} + \mathcal{O}(h^5) \quad . \qquad (6.12)$$

In the midpoint method we improved the accuracy by evaluating the function in the middle of the interval. The fourth-order Runge–Kutta formula improves on that by using two evaluations in the middle of the interval, and one at the end of the interval, in order to make the solution correct out to the fourth-order term in the Taylor series. For a system of equations, this becomes

$$k_{1,i} = hf_i(x, y_1, \ldots, y_N)$$

$$k_{2,i} = hf_i\left(x + \frac{h}{2}, \; y_1 + \frac{k_{1,1}}{2}, \ldots, y_N + \frac{k_{1,N}}{2}\right)$$

$$k_{3,i} = hf_i\left(x + \frac{h}{2}, \; y_1 + \frac{k_{2,1}}{2}, \ldots, y_N + \frac{k_{2,N}}{2}\right)$$

$$k_{4,i} = hf_i(x+h,\ y_1+k_{3,1},\ldots,y_N+k_{3,N})$$
$$y_i(x+h) = y_i(x) + \frac{k_{1,i}}{6} + \frac{k_{2,i}}{3} + \frac{k_{3,i}}{3} + \frac{k_{4,i}}{6} + \mathcal{O}(h^5) \quad . \tag{6.13}$$

The compactness of this fourth-order formula provides a nice compromise between implementation and execution effort. Although still higher-order approximations are possible, the next section will look at smarter ways to improve the approximation error.

For a given problem, how is the step size $h$ chosen? Obviously, it must reflect the desired accuracy of the final answer. But, how can we estimate the accuracy if we don't know the answer? The simplest solution is to keep reducing the step size until the solution does not change within the desired tolerance. We can be more clever than that, but this is always a good sanity check: the result of a numerical calculation should not depend on the algorithm parameters.

A more intelligent approach is to consider how the approximation error depends on the step size. In a fourth-order method, the deviation between the approximate value found in a single full step $y_{\text{full}}(x+h)$ and the correct value $y_{\text{true}}(x+h)$,

$$y_{\text{full}}(x+h) - y_{\text{true}}(x+h) = h^5 \varphi(x) + \mathcal{O}(h^6) \quad , \tag{6.14}$$

consists of $h^5$ times a quantity $\varphi$ which is approximately constant over the interval (to order $h^5$; this is the definition of the order of the error). If, instead of a single step $h$, two smaller steps of $h/2$ are made, the error after the first half-step is

$$y_{\text{half}}(x+h/2) - y_{\text{true}}(x+h/2) = \left(\frac{h}{2}\right)^5 \varphi(x) + \mathcal{O}(h^6) \tag{6.15}$$

and then after the second half-step it is approximately

$$y_{\text{half}}(x+h/2+h/2) - y_{\text{true}}(x+h) = 2\left(\frac{h}{2}\right)^5 \varphi(x) + \mathcal{O}(h^6) \tag{6.16}$$

(it is conventional to assume that the errors at each step add, although in fact this is the worst case and the combined errors for some problems might conspire to be better than that). Therefore, the difference between $y(x+h)$ calculated in a full step of $h$ and in two half-steps of $h/2$ is

$$y_{\text{full}}(x+h) - y_{\text{half}}(x+h/2+h/2) = h^5\varphi(x) - 2\left(\frac{h}{2}\right)^5 \varphi(x) + \mathcal{O}(h^6)$$
$$\approx h^5 \varphi \quad . \tag{6.17}$$

The difference between a single step and two half-steps provides an estimate of the local error in the step. Such an error estimate can be used to guide an automatic stepper routine: after making a full step and comparing the result to that from two half-steps, if the error is larger than an upper threshold then the step size is decreased, and if the error is smaller than a lower threshold then it is increased. This allows the program to move to large steps if the function is smooth and featureless, and then to drop down to small steps in regions where the function is complicated.

Note that the local error estimates cannot trivially be added up to get a global error estimate at the end of the calculation. Although it is common to assume that the local errors can be combined as uncorrelated random variables, there are many cases where they are very correlated and lead the solution away from the correct answer (for example, in the context of solving chaotic equations see [Dawson et al., 1994]). And remember that there's a hardware limit to the error that can be achieved; single-precision floating-point numbers typically have about six significant digits, and double-precision numbers have twelve digits.

Having done the extra work needed to make the local error estimate, we can also improve our approximation by combining the results. Equations (6.14) and (6.16) are two equations in the two unknowns $y_{\text{true}}$ and $\varphi$ (ignoring terms $\mathcal{O}(h^6)$), which can easily be solved to find

$$y_{\text{true}}(x+h) = y_2 + \frac{y_2 - y_1}{15} + \mathcal{O}(h^6) \quad . \tag{6.18}$$

After checking the error on the fourth-order method by making a full step and two half-steps, this lets the error be reduced by making use of the two estimates for $y(x+h)$ (although to this order we cannot estimate the error in the improved approximation).

A refinement on this step-doubling method uses six function evaluations to give a fifth-order Runge–Kutta approximation, and a different combination of the same six values for a fourth-order approximation (this is due to Fehlberg [Press et al., 1992]). Although the functional form is more complex than the standard fourth-order method, this permits an error estimate to be made without needing any extra function evaluations, a desirable trade-off if the function evaluations are computationally costly.

The details of implementing an adaptive interval updating scheme depend on the nature of the equation being integrated. If there are rough regions expected it is crucial to throw away a step with a large error and try again with a smaller step; if the answer is expected to be smooth and execution time is a problem, the point can be saved and the reduced step applied to the following interval. Similarly, small changes in the step size help the routine fine-tune its step, but large changes are needed if the solution varies enormously. If the factors used to increase and decrease the step are incommensurate then it is possible to reach any step size, otherwise the step size will be limited to a rational subset (such as powers of 2).

## 6.3 BEYOND RUNGE–KUTTA

The combination of a fourth-order Runge–Kutta solver with an adaptive interval stepper is easy to program, easy to use, and can handle most any reasonably well-behaved problem. For this reason it is a workhorse for solving differential equations. This section introduces two important alternatives. At best, they can find more accurate solutions with larger steps and fewer function evaluations, but they are also fussier and can fail catastrophically. Runge–Kutta is always a good starting point; these fancier algorithms should be considered if the execution time or accuracy need to be improved.

The first, *predictor-corrector* methods, start by recognizing that a step in solving the first-order differential equation

$$\frac{dy}{dx} = f(x, y(x)) \tag{6.19}$$

can formally be written as an intergral over an interval $h$

$$y(x+h) = y(x) + \int_x^{x+h} f(x, y(x))\, dx \quad . \tag{6.20}$$

The problem with this integral is that to evaluate it we need to know $y(x)$, but that is what we're trying to solve for in the first place. All is not lost, however: we do know the history of $f(x, y(x))$ before the interval that we are trying to step over. We ignored this history in the Runge–Kutta methods and just used values in the interval, but if the function is not varying too wildly we can do better and extrapolate over the interval. A common way to do the extrapolation is to assume a polynomial form for $f$ (Chapter 12 will look in detail at other ways to approximate functions). For example, for a third-order method, we assume that locally

$$f(x, y(x)) = a + bx + cx^2 \quad . \tag{6.21}$$

This can easily be integrated:

$$\int_x^{x+h} f(x, y(x))\, dx = ah + bxh + \frac{1}{2}bh^2 + cx^2 h + cxh^2 + \frac{1}{3}ch^3 \quad . \tag{6.22}$$

Although it's possible to fit a polynomial at each step to determine the coefficients ($a, b$, and $c$ here), we can get the same answer by judicious function evaluations. If we cleverly guess that we can write the integral as a sum of past values of the function, weighted by unknown coefficients $(\alpha, \beta, \gamma)$,

$$\int_x^{x+h} f(x, y(x))\, dx \stackrel{?}{=} h\{\alpha f[x, y(x)] + \beta f[x-h, y(x-h)]$$
$$+ \gamma f[x-2h, y(x-2h)]\} \quad , \tag{6.23}$$

then plugging in equation (6.21) on the right hand side shows that

$$\int_x^{x+h} f(x, y(x))\, dx = ah(\alpha + \beta + \gamma) + bxh(\alpha + \beta + \gamma) +$$
$$bh^2(-\beta - 2\gamma) + cx^2 h(\alpha + \beta + \gamma) + cxh^2(-2\beta - 4\gamma) + ch^3(\beta + 4\gamma) \quad . \tag{6.24}$$

Equation (6.24) will agree with equation (6.22) if

$$\alpha + \beta + \gamma = 1$$
$$-\beta - 2\gamma = \frac{1}{2}$$
$$\beta + 4\gamma = \frac{1}{3} \quad . \tag{6.25}$$

These equations are easily solved to find $\alpha = 23/12$, $\beta = -4/3$, and $\gamma = 5/12$, or

$$y_p(x+h) = y(x) +$$
$$\frac{h}{12}\{23f[x, y(x)] - 16f[x-h, y(x-h)] + 5f[x-2h, y(x-2h)]\} \quad . \tag{6.26}$$

This gives an estimate of $y(x+h)$ based on extrapolating the history of $f$; for this reason it is called a *predictor* step. Doing an integral with function evaluations like this is an example of *numerical quadrature*.

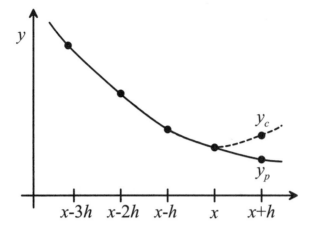

Figure 6.3. A predictor-corrector method.

Although we could use the predicted $y(x+h)$ as the input to a new predictor step, that would try to solve the differential equation by repeated polynomial extrapolations (a bad idea). But we can reapply the differential equation to find an improved estimate based on the prediction; this is called a *corrector* step (Figure 6.3). Since we have an estimate of $y$ at the end of the interval, for the corrector we can look for an implicit numerical quadrature formula that uses it. For the third-order example, we want

$$\int_x^{x+h} f(x, y(x))\, dx \stackrel{?}{=} h\{\alpha' f[x+h, y(x+h)] + \beta' f[x, y(x)]$$
$$+ \gamma' f[x-h, y(x-h)]\} \quad . \tag{6.27}$$

Repeating the preceeding calculation gives $\alpha' = 5/12$, $\beta' = 2/3$, and $\gamma' = -1/12$, or

$$y_c(x+h) = y(x) +$$
$$\frac{h}{12}\{5f[x+h, y(x+h)] + 8f[x, y(x)] - f[x-h, y(x-h)]\} \quad . \tag{6.28}$$

The result from the corrector step can be used in a new predictor step, which is then corrected, and so forth. Getting this iteration going will require a set of starting values, which can be provided by a self-starting method such as Runge–Kutta. And the difference between the predictor and corrector steps provides a local error estimate. This is an example of an *Adams–Bashforth–Moulton* method; it's common to use the fourth-order form,

$$y_p(x+h) = y(x) + \frac{h}{24}\{55f[x, y(x)] - 59f[x-h, y(x-h)]$$
$$+ 37f[x-2h, y(x-2h)] - 9f[x-3h, y(x-3h)]\} \tag{6.29}$$

and

$$y_c(x+h) = y(x) + \frac{h}{24}\{9f[x+h, y_p(x+h)] + 19f[x, y(x)]$$
$$- 5f[x-h, y(x-h)] + f[x-2h, y(x-2h)]\} \quad . \tag{6.30}$$

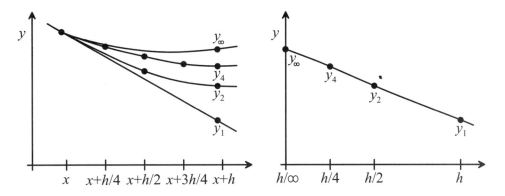

Figure 6.4. Richardson extrapolation.

When polynomial extrapolation is justified, a predictor-corrector routine can make significantly longer steps than a Runge–Kutta routine because it doesn't ignore points that have already been calculated (but it will fail around singularities or discontinuities that are poorly fit by a polynomial).

The idea of extrapolation, plus the step-doubling routine that we used for Runge–Kutta error control, hints at the final numerical method: *Richardson extrapolation*. We saw that two steps of $h/2$ give a smaller final error than one step of $h$. Four steps of $h/4$ give a smaller error still, and $\infty$ steps of $h/\infty$ would be even better (exact, in fact). Although it's not very practical to plan on taking infinitely many steps, the sequence leading up to it can give us insight into the infinite limit. The idea is to calculate the value at the end of the interval many times with successively finer steps, and then fit a function to extrapolate to the magical limit of an infinitely small step size (Figure 6.4). The *Bulirsch–Stoer* method uses polynomials or ratios of polynomials to do the extrapolation. Setting this up requires a more complex algorithm with many more internal parameters than Runge–Kutta uses, but in return a *much* larger step size can be used if the solution is not too complex. Predictor-corrector methods are older and better studied than Richardson extrapolation methods, but it is reasonable to believe that it is easier to predict the convergence of a sequence rather than the extrapolation of a complicated function and so extrapolation methods are becoming more common.

We have so far assumed that the differential equation being solved is reasonably well behaved. A particularly nasty source of problems is *stiff* differential equations, which typically arise when a problem has vastly different time or length scales. Consider the following example:

$$\frac{d^2y}{dx^2} - 10^6 y = 0 \quad . \tag{6.31}$$

This is easily solved to find the general solution

$$y = Ae^{1000t} + Be^{-1000t} \quad . \tag{6.32}$$

Consider what will happen if you give this equation to an unsuspecting differential equation solver with the initial conditions on $y$ and $\dot{y}$ chosen so that $A = 0$. It will start stepping along, making its usually harmless small errors at each step. However, a small error in $y$ (and therefore $\dot{y}$) means that a tiny bit of the other solution will creep in, and as soon as $A \neq 0$ then $\exp(1000t)$ will annihilate $\exp(-1000t)$ and the solution will blow up. The first thing to check when you run into a stiff differential equation is whether the variables can be rescaled so that their orders are comparable. Beyond that there is a range of special techniques for stiff differential equations that contain the different solutions; see [Gear, 1971].

This chapter has exclusively considered initial value problems. Sometimes *boundary-value* problems arise, in which values are known in the middle or the end of the interval. A classic example that was an important application for early computers was gunnery problems that seek initial conditions to launch a shell to land on a target. This is a harder task, and there are no simple solutions. One class of techniques, fittingly called *shooting methods*, sends multiple solutions across the interval and then tries to iteratively update its guess for the initial conditions that satisfy the boundary conditions. The other common approach is to use *finite elements* to discretize the entire interval to be solved and calculate it in parallel (Chapter 8).

## 6.4 SELECTED REFERENCES

[Press et al., 1992] Press, William H., Teukolsky, Saul A., Vetterling, William T., & Flannery, Brian P. (1992). *Numerical Recipes in C: The Art of Scientific Computing*. 2nd edn. New York, NY: Cambridge University Press.

As in so many other areas, the best first place to turn for numerical methods.

[Gear, 1971] Gear, C. William (1971). *Numerical Initial Value Problems in Ordinary Differential Equations*. Englewood Cliffs, NJ: Prentice-Hall.

[Stoer & Bulirsch, 1993] Stoer, J., & Bulirsch, R. (1993). *Introduction to Numerical Analysis*. 2nd edn. New York, NY: Springer-Verlag. Translated by R. Bartels, W. Gautschi, and C. Witzgall.

These two are classic texts for differential equations.

[Young & Gregory, 1988] Young, David M., & Gregory, Robert Todd (1988). *A Survey of Numerical Mathematics*. New York, NY: Dover Publications. 2 volumes.

This is almost as broad in scope as [Press et al., 1992], but has more mathematical analysis of the algorithms in return for less practical guidance.

## 6.5 PROBLEMS

(6.1) What is the second-order approximation error of the *Heun* method, which averages the slope at the beginning and the end of the interval?

$$y(x+h) = y(x) + \frac{h}{2}\{f(x,y(x)) + f[x+h, y(x) + hf(x,y(x))]\} \qquad (6.33)$$

(6.2) For a simple harmonic oscillator $\ddot{y}+y = 0$, with initial conditions $y(0) = 1$, $\dot{y}(0) = 0$, find $y(t)$ from $t = 0$ to $100\pi$. Use an Euler method and a fixed-step fourth-order Runge–Kutta method. For each method, what step size is needed for the average error over the interval to be less than 0.001? Check also the error in the value and slope at the last point.

(6.3) If the step size for a fourth-order Runge–Kutta algorithm is decreased by a factor of 10, by approximately how much will the local error decrease? Verify this functional relationship by writing an adaptive variable stepper to find the average step size necessary for a given local error over the interval in the preceding problem. Check also the dependence of the average actual error, and the final error, on the step size.

# 7 Finite Differences: Partial Differential Equations

The world is defined by structure in space and time, and it is forever changing in complex ways that can't be solved exactly. Therefore the numerical solution of partial differential equations leads to some of the most important, and computationally intensive, tasks in all of numerical analysis (such as forecasting the weather). This chapter introduces *finite difference* techniques; the next two will look at other ways to discretize partial differential equations (finite elements and cellular automata). Just as we used a Taylor expansion to derive a numerical approximation for ordinary differential equations, the same procedure can be applied to partial differential equations. Because the discretization must be done in space as well as time, there are many more possible strategies for finding good (and bad) approximations.

We will start with two degrees of freedom, say one spatial variable $x$ and a time $t$. Given a function $u(x,t)$, its spatial derivatives are found from the Taylor expansion

$$u(x + \Delta x, t) = u(x, t) + \Delta x \left.\frac{\partial u}{\partial x}\right|_{x,t} + \frac{(\Delta x)^2}{2!} \left.\frac{\partial^2 u}{\partial x^2}\right|_{x,t} + \mathcal{O}[(\Delta x)^3] \quad . \tag{7.1}$$

The first partial derivative can be approximated by the *forward difference*

$$\frac{u(x + \Delta x, t) - u(x, t)}{\Delta x} = \left.\frac{\partial u}{\partial x}\right|_{x,t} + \mathcal{O}[\Delta x] \quad . \tag{7.2}$$

If we replace $\Delta x$ with $-\Delta x$, this becomes the equally reasonable *backwards difference* approximation

$$\frac{u(x, t) - u(x - \Delta x, t)}{\Delta x} = \left.\frac{\partial u}{\partial x}\right|_{x,t} + \mathcal{O}[\Delta x] \quad . \tag{7.3}$$

Both have first-order errors. The order of the approximation can be raised to second order by taking the difference between two time steps, which subtracts out the quadratic term:

$$\frac{u(x + \Delta x, t) - u(x - \Delta x, t)}{2\Delta x} = \left.\frac{\partial u}{\partial x}\right|_{x,t} + \mathcal{O}[(\Delta x)^2] \quad . \tag{7.4}$$

Although this might appear always to be preferable, we will see that it can have surprising undesirable stability properties.

The straightforward finite difference approximation to the second partial derivative

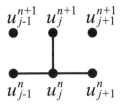

Figure 7.1. A computational cluster.

also has a second-order error:

$$\frac{1}{\Delta x}\left[\frac{u(x+\Delta x,t)-u(x,t)}{\Delta x}-\frac{u(x,t)-u(x-\Delta x,t)}{\Delta x}\right]=$$

$$\frac{u(x+\Delta x,t)-2u(x,t)+u(x-\Delta x,t)}{(\Delta x)^2}=\left.\frac{\partial^2 u}{\partial x^2}\right|_{x,t}+\mathcal{O}[(\Delta x)^2] \quad . \quad (7.5)$$

Numerical methods for partial differential equations are usually classified by the characteristics for the equation that they apply to (Chapter 3), which measure how information from the boundary conditions influences the solution. Characteristics can even be used as the basis for numerical solvers [Ames, 1992], but here we will simply use them as convenient labels for the most common cases: a wave equation (hyperbolic characteristics), diffusive processes (parabolic), and boundary value problems (elliptic). More complex systems can have some or all of these elements.

## 7.1 HYPERBOLIC EQUATIONS: WAVES

To see how the stability of the solution depends on the finite difference scheme, let's start with a simple first-order hyperbolic PDE for a conserved quantity in one dimension

$$\frac{\partial u}{\partial t}=-v\frac{\partial u}{\partial x} \quad . \quad (7.6)$$

Substitution readily shows that this is solved by any function of the form

$$u=f(x-vt) \quad . \quad (7.7)$$

Writing $u(j\Delta x, n\Delta t) = u_j^n$ to make the notation clearer, a simple discretization is first-order in time and second-order in space:

$$\frac{u_j^{n+1}-u_j^n}{\Delta t}=-v\left(\frac{u_{j+1}^n-u_{j-1}^n}{2\Delta x}\right)$$

$$u_j^{n+1}=u_j^n-\frac{v\Delta t}{2\Delta x}(u_{j+1}^n-u_{j-1}^n) \quad (7.8)$$

(using a first-order spatial approximation would make it asymmetrical). It can be convenient to represent such approximations by drawing the cluster of values used in the update rule (Figure 7.1). Given an initial distribution $u_j^n$, it is straightforward to iterate this rule forward in time.

To analyze the stability of a finite difference scheme, the *von Neumann stability analysis* locally linearizes the equations (if they are not linear) and then looks at the growth of the linear modes

$$u_j^n = A(k)^n e^{ikj} \tag{7.9}$$

which have an oscillatory dependence on space and an exponential dependence on time. Plugging in this ansatz gives a solution to the finite difference equation for $A(k)$. If $|A(k)| > 1$ for some $k$, then these modes will diverge and the scheme will be unstable (remember that the exact solution (7.7) does not diverge). For equation (7.1) this gives

$$A^{n+1} e^{ikj} = A^n e^{ikj} - \frac{v\Delta t}{2\Delta x}\left(A^n e^{ik(j+1)} - A^n e^{ik(j-1)}\right)$$

$$A = 1 - \frac{v\Delta t}{2\Delta x}\left(e^{ik} - e^{-ik}\right)$$

$$= 1 - i\frac{v\Delta t}{\Delta x}\sin k \quad . \tag{7.10}$$

The absolute magnitude of this is always greater than 1, and so this scheme is always unstable. Any initial condition will diverge!

This disturbing behavior in such a sensible approximation is easily corrected with the *Lax method*, which averages the neighbors for the time derivative:

$$u_j^{n+1} = \frac{1}{2}(u_{j+1}^n + u_{j-1}^n) - \frac{v\Delta t}{2\Delta x}(u_{j+1}^n - u_{j-1}^n) \quad . \tag{7.11}$$

Repeating the stability analysis shows that the amplitude of a solution is

$$A = \cos k - i\frac{v\Delta t}{\Delta x}\sin k \quad . \tag{7.12}$$

Requiring that the magnitude be less than 1,

$$|A|^2 = \cos^2 k + \left(\frac{v\Delta t}{\Delta x}\right)^2 \sin^2 k \leq 1$$

$$\Rightarrow \frac{|v|\Delta t}{\Delta x} \leq 1 \quad . \tag{7.13}$$

This is the *Courant–Friedrichs–Levy* stability criterion, and it will recur for a number of other schemes. It says that the velocity at which information propogates within the numerical algorithm ($\Delta x/\Delta t$) must be faster than the velocity of the solution $v$. For space and time steps that satisfy this condition, the Lax method will be stable. Otherwise, there is a "numerical boom" as the real solution tries to out-run the rate at which the numerical solution can advance. The lateral averaging for the time derivative in the Lax method helps the numerical information propagate, compared to the unstable approximation that we started with (equation (7.1)). The origin of this stability becomes clearer if the Lax method is rewritten by subtracting $u_j^n$ from both sides:

$$\frac{u_j^{n+1} - u_j^n}{\Delta t} = -v\left(\frac{u_{j+1}^n - u_{j-1}^n}{2\Delta x}\right) + \frac{1}{2\Delta t}(u_{j+1}^n - 2u_j^n + u_{j-1}^n) \quad . \tag{7.14}$$

This is just our original equation (7.1), with an extra fictitious diffusion term added that depends on the discretization:

$$\frac{\partial u}{\partial t} = -v\frac{\partial u}{\partial x} + \frac{(\Delta x)^2}{2\Delta t}\frac{\partial^2 u}{\partial x^2} \quad . \tag{7.15}$$

This is an example of an artificial *numerical dissipation*, which can occur (and even be added intentionally) in stable schemes. In this case it is good, because it serves to damp out the spurious high-frequency modes ($k \sim 1$) while preserving the desired long wavelength solutions. In other cases it might be a problem if the goal is to look at the long-term behavior of a nondissipative system.

The Lax method cures the stability problem and is accurate to second order in space, but it is only first-order in time. This means that $v\Delta t$ will need to be much smaller than $\Delta x$ to have the same accuracy in time and space (even though a much larger time step will be stable). A natural improvement is to go to second order in time:

$$u_j^{n+1} = u_j^{n-1} - \frac{v\Delta t}{\Delta x}\left(u_{j+1}^n - u_{j-1}^n\right) \quad . \tag{7.16}$$

The stability analysis for this equation now leads to a quadratic polynomial for the amplitude, giving two solutions

$$A = -i\frac{v\Delta t}{\Delta x}\sin(k) \pm \sqrt{1 - \left[\frac{v\Delta t}{\Delta x}\sin(k)\right]^2} \quad . \tag{7.17}$$

If $|v|\Delta t/\Delta x \leq 1$ then the radical will be real, and $|A|^2 = 1$ independent of $k$. The Courant condition applies again, but now there is no dependence of the amplitude on the spatial wavelength $k$ and so there is no artificial damping (unlike the Lax method). This is called the *leapfrog method* because it separates the space into two interpenetrating lattices that do not influence each other ($u_j^{n+1}$ does not depend on $u_j^n$). Numerical round-off errors can lead to a divergence of the sublattices over long times, requiring the addition of an artificial coupling term.

Problem 7.1 considers the finite difference approximaton to the wave equation.

## 7.2 PARABOLIC EQUATIONS: DIFFUSION

We will next look for finite difference approximations for the 1D diffusion equation

$$\frac{\partial u}{\partial t} = \frac{\partial}{\partial x}\left(D\frac{\partial u}{\partial x}\right) \quad , \tag{7.18}$$

and will assume that the diffusion coefficient is constant

$$\frac{\partial u}{\partial t} = D\frac{\partial^2 u}{\partial x^2} \quad . \tag{7.19}$$

The methods to be described will have natural generalizations when $D$ is not constant.

The straightforward discretization is

$$\frac{u_j^{n+1} - u_j^n}{\Delta t} = D \left[ \frac{u_{j+1}^n - 2u_j^n + u_{j-1}^n}{(\Delta x)^2} \right]$$

$$u_j^{n+1} = u_j^n + \frac{D \Delta t}{(\Delta x)^2} [u_{j+1}^n - 2u_j^n + u_{j-1}^n] \quad . \tag{7.20}$$

Solving the stability analysis,

$$A = 1 + \frac{D \Delta t}{(\Delta x)^2} \underbrace{[e^{ik} - 2 + e^{-ik}]}_{\substack{2 \cos k - 2 \\ 2\left(2\cos^2 \frac{k}{2} - 1\right) - 2}}$$

$$= 1 - \frac{4 D \Delta t}{(\Delta x)^2} \sin^2 \frac{k}{2}$$

$$|A| \leq 1 \Rightarrow \frac{4 D \Delta t}{(\Delta x)^2} \leq 2 \Rightarrow \frac{2 D \Delta t}{(\Delta x)^2} \leq 1 \quad . \tag{7.21}$$

The method is stable for small step sizes, but since for a diffusive process the time $t$ to expand a distance $L$ is roughly $t \sim L^2/D$, the number of time steps required to model this will be $\sim L^2/(\Delta x)^2$ (i.e., a *very* large number).

The stability can be improved by evaluating the space derivative forwards in time:

$$\frac{u_j^{n+1} - u_j^n}{\Delta t} = D \left( \frac{u_{j+1}^{n+1} - 2u_j^{n+1} + u_{j-1}^{n+1}}{(\Delta x)^2} \right)$$

$$u_j^{n+1} - \frac{D \Delta t}{(\Delta x)^2} [u_{j+1}^{n+1} - 2u_j^{n+1} + u_{j-1}^{n+1}] = u_j^n \quad . \tag{7.22}$$

The stability analysis for this is

$$A - \frac{D \Delta t}{(\Delta x)^2} [A e^{ik} - 2A + A e^{-ik}] = 1$$

$$A \left[1 + \frac{4 D \Delta t}{(\Delta x)^2} \sin^2 \frac{k}{2} \right] = 1$$

$$A = \frac{1}{1 + \frac{4 D \Delta t}{(\Delta x)^2} \sin^2 \frac{k}{2}} \leq 1 \quad . \tag{7.23}$$

This scheme is stable for all step sizes, but might appear to be useless: how can we implement it since we don't know the forward values used in the space derivative? These future values are implicitly determined by the past values, and the trick is to recognize that the full set of equations can be inverted. The stability follows because peeking into the future in this way helps move information through the solution more quickly.

The boundary conditions are typically given as either *fixed* ($u_1$ and $u_N$ are specified) or *periodic* ($u_1 = u_{N+1}$, so that the system does not have edges). If we assume fixed boundary conditions and define $\alpha = D \Delta t/(\Delta x)^2$, then equation (7.22) can be written as a matrix problem

$$\begin{pmatrix} 1 & 0 & 0 & \cdots & & & 0 \\ -\alpha & 1+2\alpha & -\alpha & 0 & \cdots & & 0 \\ \ddots & \ddots & \ddots & & & & \\ \vdots & 0 & -\alpha & 1+2\alpha & -\alpha & 0 & \vdots \\ & & \ddots & \ddots & \ddots & & \\ 0 & \cdots & 0 & -\alpha & 1+2\alpha & -\alpha \\ 0 & \cdots & & 0 & 0 & 1 \end{pmatrix} \begin{pmatrix} u_1^{n+1} \\ u_2^{n+1} \\ \vdots \\ u_i^{n+1} \\ \vdots \\ u_{N-1}^{n+1} \\ u_N^{n+1} \end{pmatrix} = \begin{pmatrix} u_1^n \\ u_2^n \\ \vdots \\ u_i^n \\ \vdots \\ u_{N-1}^n \\ u_N^n \end{pmatrix}$$

This is a *tridiagonal* matrix (all the elements are zero, except for the diagonal and the adjacent elements), and it can easily be inverted to find $u^{n+1}$ in terms of $u^n$ without doing all of the work needed to invert an arbitrary matrix.

The system of equations corresponding to an arbitrary tridiagonal matrix is

$$b_1 u_1 + c_1 u_2 = d_1$$
$$a_i u_{i-1} + b_i u_i + c_i u_{i+1} = d_i \quad (2 \le i \le N-1) \quad .$$
$$a_N u_{N-1} + b_N u_N = d_N \quad (7.24)$$

For us, $a = c = -\alpha$, $b = 1+2\alpha$, and the $d$'s are the starting $u$'s. These can be solved in two passes. In a system of equations, multiplying one equation by a constant and adding it to another one does not change the solution. If we multiply the first row by $-a_2/b_1$ and add it to the second row this will eliminate the $a_2$ term. If we then divide the second row by the new $b_2$ term, and repeat these steps (*Gauss elimination*) down the matrix, we will get a new matrix with zeros below the diagonal and ones on the diagonal (this is called an *upper-diagonal* matrix). Then, a reverse pass back up the matrix that multiplies the new $N$th row by the new values for $-c_{N-1}/b_N$ and adds it to the previous row, and so forth, converts the matrix to a diagonal one and the solution can be read off. Using primes for the values after the forward pass, a bit of algebra shows that

$$c_1' = \frac{c_1}{b_1} \quad d_1' = \frac{d_1}{b_1}$$

$$c_{i+1}' = \frac{c_{i+1}}{b_{i+1} - a_{i+1} c_i'} \quad d_{i+1}' = \frac{d_{i+1} - a_{i+1} d_i'}{b_{i+1} - a_{i+1} c_i'} \quad . \quad (7.25)$$

Then, the reverse pass gives

$$u_N = d_N'$$

$$u_i = d_i' - c_i' u_{i+1} \quad . \quad (7.26)$$

This is an $\mathcal{O}(N)$ steps algorithm, and so there is little performance penalty for using an implicit discretization instead of an explicit one.

The accuracy can be improved to second order in time by averaging the spatial derivative at the beginning and the end of the interval:

$$\frac{u_j^{n+1} - u_j^n}{\Delta t} = \frac{D}{2(\Delta x)^2} [(u_{j+1}^{n+1} - 2u_j^{n+1} + u_{j-1}^{n+1}) + (u_{j+1}^n - 2u_j^n + u_{j-1}^n)] \quad . \quad (7.27)$$

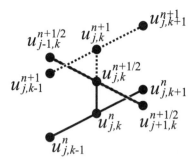

Figure 7.2. Computational clusters for alternating implicit steps for the $x$ (solid) and $y$ (dashed) coordinates.

This is called the *Crank–Nicholson* method, and the stability analysis shows that it is stable for any time step:

$$A = \frac{1 - (2D\Delta t/(\Delta x)^2)\sin^2(kx/2)}{1 + (2D\Delta t/(\Delta x)^2)\sin^2(kx/2)} \quad . \tag{7.28}$$

To solve a diffusion problem in higher dimensions we could make it implicit in all of the dimensions. This works, but results in a banded matrix that is no longer tridiagonal. Although it can be inverted (with much more effort), a simple trick recovers tridiagonal matrices. This is the *Alternating-Direction Implicit* method (*ADI*), which is an example of the general principle of *Operator Splitting*. In 2D, instead of advancing both coordinates in one step, the $x$ coordinates are advanced in a first implicit (tridiagonal) step of $\Delta t/2$, and then in a second implicit step of $\Delta t/2$ the new $y$ coordinates are found. For the implicit method this is

$$\begin{aligned}
u_{j,k}^{n+1/2} &= u_{j,k}^n + \frac{D\Delta t}{2(\Delta x)^2}\,(u_{j+1,k}^{n+1/2} - 2u_{j,k}^{n+1/2} + u_{j-1,k}^{n+1/2} + \\
&\quad u_{j,k+1}^n - 2u_{j,k}^n + u_{j,k-1}^n) \\
u_{j,k}^{n+1} &= u_{j,k}^{n+1/2} + \frac{D\Delta t}{2(\Delta x)^2}\,(u_{j+1,k}^{n+1/2} - 2u_{j,k}^{n+1/2} + u_{j-1,k}^{n+1/2} + \\
&\quad u_{j,k-1}^{n+1} - 2u_{j,k}^{n+1} + u_{j,k+1}^{n+1})
\end{aligned} \tag{7.29}$$

(shown in Figure 7.2).

## 7.3 ELLIPTIC EQUATIONS: BOUNDARY VALUES

The remaining class of partial differential equations to be discussed are of the form of the elliptic boundary value problem

$$\nabla^2 u = \rho \quad . \tag{7.30}$$

This is *Poisson's equation*; if the source term $\rho = 0$ then it becomes *Laplace's equation*. The boundary condition can be specified by giving the value of $u$ on the boundary (*Dirichlet conditions*), the value of the normal derivative of $u$ on the boundary (*Neumann conditions*), or a mixture of these. Poisson's equation and Laplace's equation are among the most commonly solved numerical equations because they apply to so many different areas. It's instructive to look at some important examples to see how they arise:

- *Heat Flow*

  The heat flux $\vec{F}$ in a material is proportional to the gradient of the temperature $T$ by the thermal conductivity $K$:

  $$\vec{F} = -K\nabla T \quad . \tag{7.31}$$

  The change of heat $Q$ in a volume is related to the temperature change by the specific heat $C$ and the density $\rho$ by

  $$\frac{dQ}{dt} = \int_V C\rho \frac{\partial T}{\partial t}\, dV \quad, \tag{7.32}$$

  and it is also equal to the surface integral of the heat flux

  $$\frac{dQ}{dt} = -\int_S \vec{F} \cdot d\vec{A} \quad . \tag{7.33}$$

  Equating these and using Gauss' theorem,

  $$\int C\rho \frac{\partial T}{\partial t}\, dV = -\int \vec{F} \cdot d\vec{A}$$
  $$= -\int \nabla \cdot \vec{F}\, dV$$
  $$= \int K\nabla^2 T\, dV$$
  $$\Rightarrow \nabla^2 T = \frac{1}{\kappa} \frac{\partial T}{\partial t} \quad, \tag{7.34}$$

  where $\kappa = K/C\rho$ is the thermal diffusivity. This is a diffusion equation, and for a steady-state problem the time derivative of $T$ will vanish, leaving Laplace's equation for the temperature distribution.

- *Fluid Flow*

  The continuity equation for a fluid of density $\rho$ and velocity $\vec{v}$ is

  $$\frac{\partial \rho}{\partial t} + \nabla \cdot \rho\vec{v} = 0 \quad . \tag{7.35}$$

  If the density is constant this reduces to

  $$\nabla \cdot \vec{v} = 0 \quad . \tag{7.36}$$

  A second condition on the velocity field is that if it starts out irrotational ($\nabla \times \vec{v} = 0$) it will remain irrotational (for example, this will be the case if at $-\infty$ the flow is uniform; see any fluids text such as [Batchelor, 1967] for a derivation). If the curl of a vector field vanishes, it can be written as the gradient of a potential:

  $$\nabla \times \vec{v} = 0 \Rightarrow \vec{v} = -\nabla \varphi \quad . \tag{7.37}$$

Combining this with the continuity equation gives Laplace's equation again

$$\nabla \cdot (\nabla \varphi) = \nabla^2 \varphi = 0 \quad . \tag{7.38}$$

- *Electric Fields*
  In MKS units, the electric field $\vec{E}$ is determined in terms of the charge density $\rho$, the magnetic field $\vec{B}$, and the polarizability $\epsilon$ by

$$\nabla \cdot \epsilon \vec{E} = \rho \qquad \nabla \times \vec{E} = -\frac{\partial \vec{B}}{\partial t} \quad . \tag{7.39}$$

For a steady-state problem, the time derivative of $\vec{B}$ is zero, and so the electric field is the gradient of a potential $\varphi$

$$\nabla \times \vec{E} = 0 \Rightarrow \vec{E} = -\nabla \varphi \quad . \tag{7.40}$$

Combining this with the charge equation and assuming that $\epsilon$ does not depend on position gives Poisson's equation

$$\nabla^2 \varphi = -\frac{\rho}{\epsilon} \quad . \tag{7.41}$$

A problem closely related to Laplace's equation is *Helmholtz's Equation*, which we found by separating out the time dependence in a wave equation:

$$\nabla^2 u = \frac{1}{c^2} \frac{\partial^2 u}{\partial t^2}$$

$$u = A e^{i\omega t} \Rightarrow e^{i\omega t} \nabla^2 A = -\frac{\omega^2}{c^2} A e^{i\omega t}$$

$$\nabla^2 A + \underbrace{\frac{\omega^2}{c^2}}_{k^2} A = 0 \quad . \tag{7.42}$$

The obvious finite difference approximation for Poisson's equation is (in 1D)

$$\frac{u_{j+1} - 2u_j + u_{j-1}}{(\Delta x)^2} = \rho_j \quad . \tag{7.43}$$

Unlike the initial value problems we have been studying, this has no time dependence. It can be written as a matrix problem which can be solved exactly

$$\mathbf{A} \cdot \vec{u} = (\Delta x)^2 \vec{\rho} \quad \Rightarrow \quad \vec{u} = (\Delta x)^2 \mathbf{A}^{-1} \cdot \vec{\rho} \quad , \tag{7.44}$$

where

$$\mathbf{A} = \begin{pmatrix} -2 & 1 & 0 & \cdots & & 0 & 1 \\ 1 & -2 & 1 & 0 & \cdots & & 0 \\ 0 & \ddots & \ddots & \ddots & & & \\ \vdots & 0 & 1 & -2 & 1 & 0 & \vdots \\ & & \ddots & \ddots & \ddots & & 0 \\ 0 & & \cdots & 0 & 1 & -2 & 1 \\ 1 & 0 & & \cdots & 0 & 1 & -2 \end{pmatrix} \quad \text{and} \quad \vec{u} = \begin{pmatrix} u_1 \\ u_2 \\ \vdots \\ u_i \\ \vdots \\ u_{N-1} \\ u_N \end{pmatrix} \tag{7.45}$$

(for periodic boundary conditions; with fixed boundary conditions the upper-right and lower-left corner elements would be zero). For Helmholtz's equation, the matrix problem requires finding the eigenvalues and eigenvectors of **A** to determine the modes.

In 2D, the finite difference approximation is

$$\frac{u_{j+1,k} + u_{j-1,k} + u_{j,k+1} + u_{j,k-1} - 4u_{j,k}}{(\Delta x)^2} = \rho_{j,k} \quad . \tag{7.46}$$

This can be also be solved by a matrix inversion:

$$\mathbf{A} \cdot \begin{pmatrix} u_{1,1} \\ u_{2,1} \\ \vdots \\ u_{N,1} \\ u_{1,2} \\ \vdots \\ u_{N,2} \\ \vdots \\ u_{N,N} \end{pmatrix} = \Delta x^2 \begin{pmatrix} \rho_{1,1} \\ \rho_{2,1} \\ \vdots \\ \rho_{N,1} \\ \rho_{1,2} \\ \vdots \\ \rho_{N,2} \\ \vdots \\ \rho_{N,N} \end{pmatrix} \quad , \tag{7.47}$$

where

$$\mathbf{A} = \begin{pmatrix} \ddots & & & & \ddots & \ddots & \ddots & & & & \ddots & & \\ & 0 & 1 & 0 & \cdots & 0 & 1 & -4 & 1 & 0 & \cdots & 0 & 1 & 0 & \cdots \\ \cdots & 0 & 1 & 0 & \cdots & 0 & 1 & -4 & 1 & 0 & \cdots & 0 & 1 & 0 & \cdots \\ & \cdots & 0 & 1 & 0 & \cdots & 0 & 1 & -4 & 1 & 0 & \cdots & 0 & 1 & 0 \\ & & & & \ddots & & \ddots & \ddots & \ddots & & & & \ddots & & \end{pmatrix}$$
(7.48)

The generalization to higher dimensions is straightforward, making a column vector from a multidimensional array by sequentially reading down the axes.

For fixed boundary conditions, the matrix (7.45) is tridiagonal, which we saw in Section 7.2 is easy to invert. For periodic boundary conditions it is almost tridiagonal, but with the extra corner elements added. Fortunately, if a matrix is related to another one by adding the outer product of two vectors

$$\mathbf{A} \rightarrow (\mathbf{A} + \vec{u} \otimes \vec{v}) \qquad (\vec{u} \otimes \vec{v})_{i,j} \equiv u_i v_j \tag{7.49}$$

then there is a simple relationship between their inverses. In our case, $\vec{u} = \vec{v} = (1, 0, \ldots, 0, 1)$, and $-1$ is subtracted from the upper-left and lower-right elements of **A** before the inversion. The inverses are related by the *Sherman–Morrison* formula, derived by doing a power series expansion of the inverse and then using the associativity of the inner and outer products

$$\begin{aligned}
(\mathbf{A} + \vec{u} \otimes \vec{v})^{-1} &= (\mathbf{A} \cdot (1 + \mathbf{A}^{-1} \cdot \vec{u} \otimes \vec{v}))^{-1} \\
&= (1 + \mathbf{A}^{-1} \cdot \vec{u} \otimes \vec{v})^{-1} \cdot \mathbf{A}^{-1} \\
&= [1 - (\mathbf{A}^{-1} \cdot \vec{u} \otimes \vec{v}) + \\
&\quad (\mathbf{A}^{-1} \cdot \vec{u} \otimes \vec{v}) \cdot (\mathbf{A}^{-1} \cdot \vec{u} \otimes \vec{v}) - \cdots] \cdot \mathbf{A}^{-1} \\
&= [1 - (\mathbf{A}^{-1} \cdot \vec{u} \otimes \vec{v}) + \\
&\quad \mathbf{A}^{-1} \cdot \vec{u} \otimes \underbrace{(\vec{v} \cdot \mathbf{A}^{-1} \cdot \vec{u})}_{\beta} \otimes \vec{v} - \cdots] \cdot \mathbf{A}^{-1} \\
&= [1 - (\mathbf{A}^{-1} \cdot \vec{u} \otimes \vec{v})(1 - \beta + \beta^2 - \cdots)] \cdot \mathbf{A}^{-1} \\
&= \mathbf{A}^{-1} - \frac{(\mathbf{A}^{-1} \cdot \vec{u}) \otimes (\vec{v} \cdot \mathbf{A}^{-1})}{1 + \beta} \quad .
\end{aligned} \qquad (7.50)$$

Matrix (7.48) is banded-diagonal, having diagonal rows of 1s $N$ elements on either side of the diagonal, and in higher dimensions there will be more bands. Although these are sparse matrices and it can be possible to solve them directly (see [Press *et al.*, 1992] for techniques), the effort required in two or more dimensions can quickly become prohibitive.

One alternative that is applicable for constant-coefficient linear problems is *Fourier Transform* methods. In 2D, the discrete Fourier transform of the field is

$$\hat{u}_{m,n} = \sum_{j=0}^{M-1} \sum_{k=0}^{N-1} u_{j,k} e^{2\pi i m j/M} e^{2\pi i n k/N} \quad , \qquad (7.51)$$

and the inverse transform is

$$u_{j,k} = \frac{1}{N^2} \sum_{m=0}^{M-1} \sum_{n=0}^{N-1} \hat{u}_{m,n} e^{-2\pi i j m/M} e^{-2\pi i k n/N} \quad . \qquad (7.52)$$

Plugging the transforms of $u$ and $\rho$ into equation (7.46), and recognizing that the transform of a function can vanish everywhere only if the function itself is equal to zero, gives

$$\hat{u}_{m,n} \left( e^{2\pi i m/M} + e^{-2\pi i m/N} + e^{2\pi i n/M} + e^{-2\pi i n/N} - 4 \right)$$

$$= \hat{\rho}_{m,n} (\Delta x)^2 \quad . \qquad (7.53)$$

Rearranging terms and simplifying the complex exponentials,

$$\hat{u}_{m,n} = \frac{\hat{\rho}_{m,n}(\Delta x)^2}{2 \cos \frac{2\pi m}{M} + 2 \cos \frac{2\pi n}{N} - 4} \quad . \qquad (7.54)$$

Therefore, the forward transform of the source term can be calculated, this can be used to find the $\hat{u}$, and then the inverse transform can be taken to find $u$. This solution imposes periodic boundary conditions; some other boundary conditions can be imposed by choosing the form of the expansion (for example, using only sines if the solution vanishes on the boundary).

The Fourier transform method is so simple only for linear constant-coefficient problems with boundary conditions along the coordinate axes; some kind of iterative approximation process is needed for more general problems. An important class of techniques is found by remembering that Poisson's equation is the steady-state solution of a diffusion problem (whether or not it originally arises from diffusion)

$$\frac{\partial u}{\partial t} = \nabla^2 u - \rho \quad . \tag{7.55}$$

This means that all of the techniques for solving diffusion problems can be applied here, with the asymptotic answer giving the solution to Poisson's equation.

The simplest is *Jacobi's method*, which just takes forward time differences

$$u_{j,k}^{n+1} = u_{j,k}^n + \frac{\Delta t}{(\Delta x)^2}(u_{j+1,k}^n + u_{j-1,k}^n + u_{j,k+1}^n + u_{j,k-1}^n - 4u_{j,k}^n) - \Delta t \rho_{j,k} \quad . \tag{7.56}$$

Here the time step does not have any physical significance; we just want the largest possible step that converges to the solution. In 2D the Courant condition is $\Delta t/(\Delta x)^2 \leq 1/4$, leading to

$$u_{j,k}^{n+1} = \frac{1}{4}(u_{j+1,k}^n + u_{j-1,k}^n + u_{j,k+1}^n + u_{j,k-1}^n) - \frac{(\Delta x)^2}{4}\rho_{j,k} \quad . \tag{7.57}$$

This has a very natural interpretation: starting from a random guess, at each time step each lattice site is set to the average of its neighbors (and a source term is added). This process is repeated until the solution stops changing, a technique called *relaxation*. A related algorithm, the *Gauss–Seidel* method, uses updated values as soon as they become available

$$u_{j,k}^{n+1} = \frac{1}{4}(u_{j+1,k}^n + u_{j-1,k}^{n+1} + u_{j,k+1}^n + u_{j,k-1}^{n+1}) - \frac{(\Delta x)^2}{4}\rho_{j,k} \tag{7.58}$$

(assuming that the updating proceeds down rows).

These both work, but the convergence is too slow for them to be useful. This can be seen by rewriting them in terms of the matrix problem

$$\mathbf{A} \cdot \vec{u} = \vec{\rho} \tag{7.59}$$

and then separating $\mathbf{A}$ into lower-triangular, diagonal, and upper-triangular parts

$$(\mathbf{L} + \mathbf{D} + \mathbf{U}) \cdot \vec{u} = \vec{\rho} \quad . \tag{7.60}$$

For the Jacobi method, the lower- and upper-triangular parts are moved to the right hand side:

$$\mathbf{D} \cdot \vec{u}_{n+1} = -(\mathbf{L} + \mathbf{U}) \cdot \vec{u}_n + \vec{\rho} \quad . \tag{7.61}$$

The convergence rate will be determined by the eigenvalues of the iteration matrix $-\mathbf{D}^{-1} \cdot (\mathbf{L} + \mathbf{U})$. The magnitude of all of the eigenvalues must be less than 1 for stability, and the largest eigenvalue determines the overall convergence rate (the largest eigenvalue is called the *spectral radius* $\rho_s$). For a large $N \times N$ square lattice problem, the spectral radius is asymptotically equal to [Ames, 1992]

$$\rho_{\text{Jacobi}} \simeq 1 - \frac{\pi^2}{2N^2} \quad . \tag{7.62}$$

Therefore, reducing the error by a factor of 10 requires $-\ln 10/\ln \rho_s \simeq N^2$ steps. In the Gauss–Seidel method, the lower-triangular part is moved over to the left side:

$$(\mathbf{L} + \mathbf{D}) \cdot \vec{u}_{n+1} = -\mathbf{U} \cdot \vec{u}_n + \vec{\rho} \quad . \tag{7.63}$$

For the square 2D lattice, this has a spectral radius of

$$\rho_{\text{Gauss-Seidel}} \simeq 1 - \frac{\pi^2}{N^2} \quad , \tag{7.64}$$

and so the number of steps needed to reduce the error by a factor of 10 is half that required by the Jacobi method. For a $100 \times 100$ lattice, both of these methods require $\sim 10^4$ steps for an improvement of a factor of 10 in the answer, which is usually prohibitive. The Gauss–Seidel method is preferable to the Jacobi method because it converges faster and does not require auxiliary storage, but something better than both is needed.

The Gauss–Seidel method can be rewritten in a suggestive form as follows:

$$\begin{aligned}
(\mathbf{L} + \mathbf{D}) \cdot \vec{u}_{n+1} &= -\mathbf{U} \cdot \vec{u}_n + \vec{\rho} \\
\vec{u}_{n+1} &= (\mathbf{L} + \mathbf{D})^{-1}[-\mathbf{U} \cdot \vec{u}_n + \vec{\rho}] \\
&= \vec{u}_n - (\mathbf{L} + \mathbf{D})^{-1} \cdot [\mathbf{U} \cdot \vec{u}_n - \vec{\rho}] - \vec{u}_n \\
&= \vec{u}_n - (\mathbf{L} + \mathbf{D})^{-1} \cdot [(\mathbf{L} + \mathbf{D} + \mathbf{U}) \cdot \vec{u}_n - \vec{\rho}] \\
&= \vec{u}_n - (\mathbf{L} + \mathbf{D})^{-1} \cdot [\mathbf{A} \cdot \vec{u}_n - \vec{\rho}] \\
&= \vec{u}_n - (\mathbf{L} + \mathbf{D})^{-1} \cdot \vec{E}_n \quad ,
\end{aligned} \tag{7.65}$$

where $\vec{E}_n$ is the error at the $n$th time step. In each update, the error gets multiplied by $(\mathbf{L} + \mathbf{D})^{-1}$ and subtracted from the state. The idea of *Successive Over-Relaxation* (*SOR*) is to extrapolate this correction and subtract a larger change

$$\vec{u}_{n+1} = \vec{u}_n - \alpha(\mathbf{L} + \mathbf{D})^{-1} \cdot \vec{E}_n \quad . \tag{7.66}$$

It can be shown that this converges for $0 < \alpha < 2$ [Ames, 1992]. When $\alpha = 1$ this is just the Gauss–Seidel method, $\alpha < 1$ is underrelaxation (which slows the convergence), and $1 < \alpha < 2$ is overrelaxation. The convergence rate depends on the value of $\alpha$; choosing a value that is too large is as bad as choosing one that is too small because the solution will overshoot the final value. The optimal relaxation rate is

$$\alpha = \frac{2}{1 + \sqrt{1 - \rho_{\text{Jacobi}}^2}} \quad , \tag{7.67}$$

which leads to an asymptotic spectral radius of

$$\rho_{\text{SOR}} \simeq 1 - \frac{2\pi}{N} \quad . \tag{7.68}$$

This reduces the number of steps needed to reduce the error by a factor of 10 to $\mathcal{O}(N)$, which is now proportional to the grid size rather than the square of the grid size. Written out in components, for the 2D problem SOR is

$$u_{j,k}^{n+1} = (1 - \alpha)u_{j,k}^n + \frac{\alpha}{4}(u_{j+1,k}^n + u_{j-1,k}^{n+1} + u_{j,k+1}^n + u_{j,k-1}^{n+1}) - \frac{\alpha(\Delta x)^2}{4}\rho_{j,k} \tag{7.69}$$

SOR is very easy to program, but does require determining the relaxation parameter $\alpha$ (although this can be estimated empirically, since if $\alpha$ is too large the solution will oscillate). An alternative is to use ADI (which permits larger time steps), but that converges at roughly the same rate. For large problems that require repeated fast solution both techniques have been superseded by *multigrid* methods [Press et al., 1992], which find the final solution on $N$ grid points in $\mathcal{O}(N)$ steps. These methods are based on iteratively coarse-graining the problem to produce a simpler one that can be solved quickly, and then interpolating to find the approximate solution at higher resolution. This is analogous to the Richardson extrapolation methods for ODEs, but because of the extra dimensions they are more complicated to implement.

## 7.4 SELECTED REFERENCES

[Ames, 1992] Ames, William F. (1992). *Numerical Methods for Partial Differential Equations.* 3rd edn. Boston, MA: Academic Press.

Ames is a classic reference for numerical methods for PDEs, with a bit more emphasis on mathematical rigor than on practical advice.

## 7.5 PROBLEMS

(7.1) Consider the 1D wave equation

$$\frac{\partial^2 u}{\partial t^2} = v^2 \frac{\partial^2 u}{\partial x^2} \quad . \tag{7.70}$$

(a) Write down the straightforward finite-difference approximation.
(b) What order approximation is this in time and in space?
(c) Use the von Neumann stability criterion to find the mode amplitudes.
(d) Use this to find a condition on the velocity, time step, and space step for stability (hint: consider the product of the two amplitude solutions).
(e) Do different modes decay at different rates for the stable case?
(f) Numerically solve the wave equation for the evolution from an initial condition with $u = 0$ except for one nonzero node, and verify the stability criterion.
(g) If the equation is replaced by

$$\frac{\partial^2 u}{\partial t^2} = v^2 \frac{\partial^2 u}{\partial x^2} + \gamma \frac{\partial}{\partial t} \frac{\partial^2 u}{\partial x^2} \quad , \tag{7.71}$$

assume that

$$u(x, t) = A e^{i(kx - \omega t)} \tag{7.72}$$

and find a relationship between $k$ and $\omega$, and simplify it for small $\gamma$. Comment on the relationship to the preceeding question.

(h) Repeat the numerical solution of the wave equation with the same initial conditions, but include the damping term.

(7.2) Write a program to solve a 1D diffusion problem on a lattice of 500 sites, with an initial condition of zero at all the sites, except the central site which starts at the value 1.0. Take $D = \Delta x = 1$, and use fixed boundary conditions set equal to zero.

   (a) Use the explicit finite difference scheme, and look at the behavior for $\Delta t = 1$, 0.5, and 0.1. What step size is required by the Courant condition?
   (b) Now repeat this using implicit finite differences and compare the stability.

(7.3) Use ADI to solve a 2D diffusion problem on a lattice, starting with randomly seeded values.

(7.4) Use SOR to solve Laplace's equation in 2D, with boundary conditions $u_{j,1} = u_{1,k} = 0$, $u_{N,k} = -1$, $u_{j,N} = 1$, and explore how the convergence rate depends on $\alpha$, and how the best choice for $\alpha$ depends on the lattice size.

# 8 Finite Elements

We have seen how to use finite differences to approximate partial differential equations on a lattice, and how to analyze and improve the stability and accuracy of these approximations. As powerful as these ideas are, there are two important cases where they do not directly apply: problems that are most naturally described in terms of a spatially inhomogeneous grid, and problems that are posed in terms of a variational principle. For example, in studying the deformations of an auto body, it can be most natural to describe it in terms of finding the minimum energy configuration instead of a partial differential equation, and for computational efficiency it is certainly important to match the location of the solution nodes to the shape of the body.

These limitations with finite differences can be solved by the use of *finite element* methods. They start with general analytical expansion techniques for finding approximate solutions to partial differential equations (the method of *weighted residuals* for problems that are posed by differential equations, and the *Rayleigh–Ritz* method for variational problems), and then find a numerical solution by using local basis functions with the spatial components of the field as the expansion weights. Instead of discretizing the space in which the problem is posed, this discretizes the form of the function used to represent the solution.

Because these problems are so important in engineering practice they consume an enormous number of CPU cycles. Finite element solvers are not easy to write; most people use dedicated packages. In addition to the core routines for solving large sparse matrix problems and systems of ordinary differential equations, it is necessary to specify the input geometry and then visualize the output results. There are many good general commercial packages available such as ANSYS (http://www.ansys.com/) and MSC (http://www.macsch.com/), as well as specialized ones such as Maxwell for electromagnetic fields (http://www.ansoft.com/). These can easily cost $10,000–$100,000; there are also many available research packages.

## 8.1 WEIGHTED RESIDUALS

The method of weighted residuals converts a partial differential equation into a system of ordinary differential equations (or an algebraic equation if there is no time dependence). Let $u(\vec{x}, t)$ be a field variable (such as temperature) that depends on space and possibly on time, to be found as the solution of a given partial differential equation specified by a differential operator $D$, possibly with a source term $f$:

$$D[u(\vec{x}, t)] = f(\vec{x}, t) \quad . \tag{8.1}$$

An example is Poisson's equation $\nabla^2 u = \rho$. We will assume that $u$ is a scalar, but the following discussion is easily extended to vector $\vec{u}$.

If $\tilde{u}(\vec{x}, t)$ is an approximate solution, its *residual* $R$ is defined to be the deviation from the correct value

$$R(\vec{x}, t) = D[\tilde{u}(\vec{x}, t)] - f(\vec{x}, t) \quad . \tag{8.2}$$

Clearly a good approximate solution for $u(\vec{x}, t)$ will make the residual as small as possible, but there are many ways to define "small."

The first step in developing finite differences was to consider $\vec{x}$ and $t$ to be discrete variables. Here, we will keep them continuous and instead write $u(\vec{x}, t)$ as a discrete sum of a set of expansion weights $a_i$ times (for now arbitrary) basis functions $\varphi_i$

$$u(\vec{x}, t) \approx \sum_i a_i(t) \varphi_i(\vec{x}) \tag{8.3}$$

(Chapter 12 will examine the general problem of expanding a function in terms of basis functions). We require that the $\varphi_i$ be a *complete* set that can represent any function (up to the order of the approximation), but they need not be *orthogonal*. We will assume that the basis functions have been chosen to satisfy the boundary conditions; if this is not the case then there will be extra boundary equations in the following derivation.

The residual will be identically equal to zero only for the correct solution. For our approximate solution we will attempt the easier task of making the residual small in some average sense. There are many strategies for weighting the residual in the average to determine the best choice for the expansion coefficients $a_i$, the three most important ones for finite elements being:

- **Collocation** In *collocation*, the residual is set equal to zero at $N$ sites

$$R(\vec{x}_i) = 0 \quad (i = 1, \ldots, N) \quad . \tag{8.4}$$

This gives a system of $N$ equations for the $N$ unknown $a_i$'s. This is straightforward, but says nothing about the value of the residual away from the points where it is evaluated.

- **Least Squares** In the *least squares* method of weighted residuals, the square of the residual is integrated over the domain of the problem, and the expansion coefficients are sought that minimize this integral:

$$\frac{\partial}{\partial a_i} \int R(\vec{x})^2 \, d\vec{x} = 0 \quad . \tag{8.5}$$

- **Galerkin** General *Galerkin* methods choose a family of weighting functions $w_i(\vec{x})$ to use in evaluating the residual

$$\int R(\vec{x}) w_i(\vec{x}) \, d\vec{x} = 0 \quad . \tag{8.6}$$

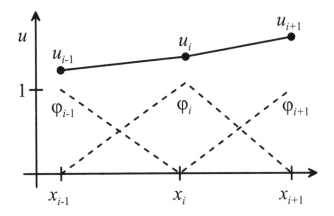

Figure 8.1. The 1D finite element "hat" function.

While any one such equation does not imply that the absolute value of the residual is small, if the set of weighting functions are nontrivially related then this can provide a tight constraint on the magnitude of the residual. The most common choice for the weighting functions is just the basis functions themselves

$$w_i(\vec{x}) = \frac{\partial u}{\partial a_i} = \varphi_i(\vec{x}) \quad . \tag{8.7}$$

This is called the *Bubnov–Galerkin* method, or sometimes just the Galerkin method. In the *Fourier–Galerkin* method a Fourier expansion is used for the basis functions (the famous chaotic *Lorenz set* of differential equations were found as a Fourier–Galerkin approximation to atmospheric convection [Lorenz, 1963], Section 16.3).

Of all these methods, the Galerkin techniques are the most common because of the convenient formulation they lead to.

These techniques for weighting residuals do not yet have anything to say about finite elements; they apply equally well to any family of basis functions $\varphi_i$, and can be used to find analytical as well as numerical approximations. To be successful, however, the basis functions need to be chosen with care. In coming chapters we will see that global functions generally do a terrible job of fitting local behavior. The trick in finite elements is to recognize that the basis functions can be chosen so that they are nonzero only in small regions (the elements), and further can be defined so that the unknown expansion coefficients $a_i$ are just the values of the field variable $u$ (and as many derivatives as are needed for the problem) evaluated at desired locations, with the basis functions interpolating between these values. For example, in 1D the simplest such expansion is piecewise linear (the *hat* functions, shown in Figure 8.1):

$$\varphi_i = \begin{cases} \dfrac{x - x_{i-1}}{x_i - x_{i-1}} & x_{i-1} \leq x < x_i \\ \dfrac{x_{i+1} - x}{x_{i+1} - x_i} & x_i \leq x < x_{i+1} \\ 0 & x < x_{i-1} \text{ or } x \geq x_{i+1} \end{cases} \quad . \tag{8.8}$$

Since $\varphi_i(x_i) = 1$ and $\varphi_j(x_i) = 0$ for all $j \neq i$,

$$u(x_i) = \sum_i a_i \varphi_i(x_i) \qquad (8.9)$$
$$= a_i \quad ,$$

therefore the expansion coefficient $a_i$ at the element point $x_i$ is just the field value $u_i = u(x_i)$. In one element $x_i \leq x < x_{i+1}$ the field is piecewise linearly interpolated as

$$u(x) = u_i \frac{x_{i+1} - x}{x_{i+1} - x_i} + u_{i+1}\frac{x - x_i}{x_{i+1} - x_i} \quad . \qquad (8.10)$$

If a finite element expansion is used in one of the residual weighting strategies the result is a set of algebraic equations for the unknown coefficients $a_i$, or a set of ordinary differential equations if the problem is time dependent. Unlike finite differences, we are now free to put the approximation nodes $x_i$ wherever is appropriate for the problem.

Let's return to the simple first-order flux PDE

$$\frac{\partial u}{\partial t} = -v\frac{\partial u}{\partial x} \qquad (8.11)$$

to see how the Galerkin method is applied. For each basis function $\varphi_j$ there is a weighted residual equation integrated over the problem domain

$$\int \left(\frac{\partial u}{\partial t} + v\frac{\partial u}{\partial x}\right) \varphi_j \, dx = 0 \qquad (8.12)$$

(in this case the source term $f = 0$). Plugging in

$$u(x,t) = \sum_i a_i(t)\varphi_i(x) \qquad (8.13)$$

gives

$$\sum_i \int \left(\frac{da_i}{dt}\varphi_i\varphi_j + va_i\varphi_j\frac{d\varphi_i}{dx}\right) dx = 0 \quad . \qquad (8.14)$$

This can be written in matrix form as

$$\mathbf{A} \cdot \frac{d\vec{a}}{dt} + \mathbf{B} \cdot \vec{a} = \vec{0} \qquad (8.15)$$

where

$$A_{ij} = \int \varphi_i \varphi_j \, dx \qquad (8.16)$$

and

$$B_{ij} = v\int \varphi_j \frac{d\varphi_i}{dx} \, dx \qquad (8.17)$$

are matrices that depend only on the basis functions, and the vector $\vec{a}$ is the set of expansion coefficients. This is now a system of ordinary differential equations that can be solved with the methods that we studied in Chapter 6. Since each basis function overlaps only with its immediate neighbors, the **A** and **B** matrices are very sparse and so they can be solved efficiently (this is the main job of a finite element package, and much of numerical linear algebra).

The linear interpolation of the hat functions breaks the solution space up into many elements, each of which has a single degree of freedom $u_i$. On the other hand, if the basis functions in equation (8.3) extend over the entire domain, then this can be viewed as a single large element that has many degrees of freedom (the $a_i$'s). There is a range of options between these extremes, and part of the art of applying finite elements is balancing the use of more-accurate large complicated elements with the use of less-accurate simple small elements. As with solving ODEs, the simplest linear elements are a poor approximation and it is usually advantageous to use more complex elements. It's also necessary to make sure that the elements have enough degrees of freedom to be able to satisfy the residual weighting equation, which for polynomial expansions will depend on the order of the PDE. For example, the bending of stiff plates has a term that depends on the fourth spatial derivative of the displacement. Since the fourth derivative of a linear element is zero, a linear element cannot satisfy this (other than the trivial solution $u = 0$).

It is common to use polynomials for more complicated elements. Within a 1D element $x_i \leq x < x_{i+1}$, the piecewise linear approximation is made up of a superposition of two *shape functions*

$$\psi_1 = \frac{x - x_i}{x_{i+1} - x_i} \quad \psi_2 = \frac{x_{i+1} - x}{x_{i+1} - x_i} \quad , \tag{8.18}$$

the parts of the basis functions that are nonzero in the element. These functions have the desirable property of being equal to 1 at one of the element boundaries and vanishing at the other. In addition, they sum to 1 everywhere in the interval, so that if $u_i = u_{i+1}$ then the approximation is a constant over the interval. *Lagrange interpolation* generalizes this to a set of $N$ $N$th-order normalized polynomials defined to vanish at all but one of $N$ sites:

$$\begin{aligned}\psi_1 &= \frac{(x_2 - x)(x_3 - x)\cdots(x_N - x)}{(x_2 - x_1)(x_3 - x_1)\cdots(x_N - x_1)} \\ \psi_2 &= \frac{(x_1 - x)(x_3 - x)\cdots(x_N - x)}{(x_1 - x_2)(x_3 - x_2)\cdots(x_N - x_2)} \\ &\vdots \\ \psi_N &= \frac{(x_1 - x)(x_2 - x)\cdots(x_{N-1} - x)}{(x_1 - x_N)(x_2 - x_N)\cdots(x_{N-1} - x_N)} \end{aligned} \tag{8.19}$$

These can be used in higher-order elements.

A more intuitive way to define polynomial basis functions is in terms of the value of the field and its derivatives at the boundary of the element. For example, in 1D for a cubic polynomial

$$u = a_0 + a_1 x + a_2 x^2 + a_3 x^3 \tag{8.20}$$

defined over the interval $0 \leq x < h$, a trivial calculation shows that the $a$'s are related to the boundary values $u_0, u_h$ and the boundary slopes $\dot{u}_0, \dot{u}_h$ by

$$\begin{pmatrix} u_0 \\ \dot{u}_0 \\ u_h \\ \dot{u}_h \end{pmatrix} = \begin{bmatrix} 1 & 0 & 0 & 0 \\ 0 & 1 & 0 & 0 \\ 1 & h & h^2 & h^3 \\ 0 & 1 & 2h & 3h^2 \end{bmatrix} \begin{pmatrix} a_0 \\ a_1 \\ a_2 \\ a_3 \end{pmatrix} \quad . \tag{8.21}$$

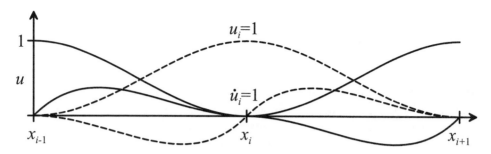

Figure 8.2. 1D finite element polynomial basis functions defined in terms of the value and derivative of the function at the element boundaries.

Given desired values for the $u$'s and $\dot{u}$'s, this can be inverted to find the $a$'s. In particular, indexed in terms of $(u_0, \dot{u}_0, u_h, \dot{u}_h)$, the four shape functions in this representation are $(1, 0, 0, 0)$, $(0, 1, 0, 0)$, $(0, 0, 1, 0)$, $(0, 0, 0, 1)$. These give the element four degrees of freedom, but if we impose continuity on the function and slope across the element boundaries this is reduced to two degrees of freedom. Figure 8.2 shows how the element shape functions can be assembled into two basis functions, one having the value of $u_i$ for the expansion coefficient, and the other $\dot{u}_i$. These handy functions, used for *Hermite interpolation*, will return when we look at *splines* in Section 12.1.2.

The generalization of these elements to higher dimensions is straightforward. For example, in 2D a triangular element with corners at $(x_0, y_0)$, $(x_1, y_1)$, and $(x_2, y_2)$ can be interpolated by three bilinear basis functions

$$\varphi_0 = \frac{(x_1 - x)(y_2 - y)}{(x_1 - x_0)(y_2 - y_0)}$$
$$\varphi_1 = \frac{(x_2 - x)(y_0 - y)}{(x_2 - x_1)(y_0 - y_1)}$$
$$\varphi_2 = \frac{(x_0 - x)(y_1 - y)}{(x_0 - x_2)(y_1 - y_2)} \quad . \tag{8.22}$$

To use these elements it is necessary to cover space with a triangulation (a mesh of triangles); such *mesh generation* is a major task of a finite element environment. In 3D a tetrahedron is the primitive element, and so forth.

It is useful to integrate by parts in the Galerkin method to reduce the order of the highest spatial derivative and therefore the required element order. For example, if we start with a 1D wave equation

$$\frac{\partial^2 u}{\partial t^2} = v^2 \frac{\partial^2 u}{\partial x^2} \tag{8.23}$$

defined in the interval (0,1), the Galerkin expansion is

$$\sum_i \int_0^1 \left( \frac{d^2 a_i}{dt^2} \varphi_i \varphi_j - v^2 a_i \varphi_j \frac{d^2 \varphi_i}{dx^2} \right) dx = 0 \quad . \tag{8.24}$$

The second term can be integrated by parts:

$$\sum_i \frac{d^2 a_i}{dt^2} \underbrace{\int_0^1 \varphi_i \varphi_j \, dx}_{\mathbf{A}_{ij}} + \sum_i a_i \underbrace{\left[ \int_0^1 v^2 \frac{d\varphi_i}{dx} \frac{d\varphi_j}{dx} \, dx - v^2 \varphi_j \frac{d\varphi_i}{dx} \bigg|_0^1 \right]}_{\mathbf{B}_{ij}} = 0$$

$$\mathbf{A} \cdot \frac{d^2 \vec{a}}{dt^2} + \mathbf{B} \cdot \vec{a} = \vec{0} \quad . \tag{8.25}$$

This is now a second-order matrix differential equation for the vector of expansion coefficients $\vec{a}$, and because of the integration by parts the maximum spatial derivative is single rather than double (and so first-order elements can be used). Since this is a linear problem we can go further and assume a periodic time dependence

$$\vec{a}(t) = \vec{a}_0 e^{i\omega t} \tag{8.26}$$

to find an eigenvalue problem for the modes

$$\omega^2 \mathbf{A} \cdot \vec{a}_0 = \mathbf{B} \cdot \vec{a}_0 \quad . \tag{8.27}$$

In higher dimensions, Green's theorem can be used to reduce the order of the equation by relating area or volume integrals to integrals over boundaries [Wyld, 1976].

## 8.2 RAYLEIGH–RITZ VARIATIONAL METHODS

In Chapter 4 we saw how many physical problems are most naturally posed in terms of a variational integral, and then saw how to use Euler's equation to find a differential equation associated with a variational principle. In this section we will look at finite element methods that start directly from a variational integral $\mathcal{I}$ and find the field distribution that makes an integral extremal

$$\delta \mathcal{I} = \delta \int F[u(\vec{x}, t)] \, d\vec{x} = 0 \quad . \tag{8.28}$$

$F$ might be the energy, or action, or time, and there can be other integral constraints added with Lagrange multipliers, such as the path length.

One of the most important applications of finite elements is to structural mechanics problems, for which the integral to be minimized is the potential energy in a structure. For example, ignoring shear, the potential energy $V$ of a beam bent from its equilibrium configuration $u(x) = 0$ by a lateral force $f(x)$ is given in terms of the elasticity modulus $E$ and the cross-sectional moment of intertia $I$ by

$$V = \int_0^L \left( \frac{1}{2} EI \left( \frac{d^2 u}{dx^2} \right)^2 - u(x) f(x) \right) dx \quad . \tag{8.29}$$

The first term is the elastic energy stored in the beam by bending, and the second term is the work done against the applied force. The equilibrium of the beam is given by $\delta V = 0$.

As with the method of weighted residuals, we will approximate $u$ by an expansion in basis functions

$$u(\vec{x}, t) = \sum_i a_i(t)\varphi_i(\vec{x}) \quad , \tag{8.30}$$

and try to find the best choice for the $a_i$'s. If the integral is extremal, then its partial derivative with respect to all of the $a_i$'s must vanish:

$$\frac{\partial \mathcal{I}}{\partial a_i} = 0 \quad (i = 1, \ldots, N) \tag{8.31}$$

(with the equilibrium stability determined by whether this is a maximum or a minimum). This *Rayleigh–Ritz* set of equations applies whether the $\varphi_i(\vec{x})$ are defined globally or locally, but this becomes a prescription for a finite element method if each $\varphi_i$ is nonzero only in a local neighborhood and has the field variables and as many derivatives as needed as the expansion coefficients (as we saw with the method of weighted residuals). If a problem already has a finite set of coordinates (such as the generalized coordinates of a Lagrangian), the Rayleigh–Ritz method can be used directly without needing an approximate expansion in basis functions.

For example, in 1D a mass on a spring has a potential energy $kx^2/2$, and if there is an applied force $-F$ the work done in moving against the force is $-Fx$, so the equilibrium position is easily found to be

$$\frac{\partial}{\partial x}\left(\frac{1}{2}kx^2 - Fx\right) = 0 \quad \Rightarrow \quad x = \frac{F}{k} \quad . \tag{8.32}$$

For the more difficult continuous case of equation (8.29), plugging in the expansion (8.30) and asking that the energy be extremal gives

$$0 = \frac{\partial}{\partial a_j} \int_0^L \left[\frac{1}{2}EI\left(\sum_i a_i \frac{d^2\varphi_i}{dx^2}\right)^2 - \sum_i a_i \varphi_i(x)f(x)\right] dx$$

$$= \sum_i a_i \underbrace{\int_0^L EI \frac{d^2\varphi_i}{dx^2} \frac{d^2\varphi_j}{dx^2} dx}_{\mathbf{A}_{ij}} - \underbrace{\int_0^L \varphi_j(x)f(x) dx}_{\vec{b}_j}$$

$$= \mathbf{A} \cdot \vec{a} - \vec{b}$$

$$\Rightarrow \vec{a} = \mathbf{A}^{-1} \cdot \vec{b} \quad . \tag{8.33}$$

Because of the second derivative, quadratic elements are needed.

The Rayleigh–Ritz method has converted this variational problem into an algebraic one. More complex finite element problems result in nonlinear equations to find the coefficients; solving these requires the search techniques covered in Chapter 13.

## 8.3 SELECTED REFERENCES

[Cook et al., 1989] Cook, Robert D., Malkus, David S., & Plesha, Michael E. (1989). *Concepts and Applications of Finite Element Analysis*. 3rd edn. New York, NY: Wiley.

[Bathe, 1996] Bathe, Klaus-Jürgen (1996). *Finite Element Procedures*. Englewood Cliffs, NJ: Prentice-Hall.

There are many finite element texts with similar titles; these two have a good balance between rigor and practice.

[Ames, 1992] Ames, William F. (1992). *Numerical Methods for Partial Differential Equations*. 3rd edn. Boston, MA: Academic Press.

While Ames does not have much on the practical use of finite elements, he has a good introduction to the basic approximations used.

## 8.4 PROBLEMS

(8.1) Use the Galerkin method to find a system of differential equations to approximate the wave equation

$$\frac{\partial^2 u}{\partial t^2} = v^2 \frac{\partial^2 u}{\partial x^2} + \gamma \frac{\partial}{\partial t} \frac{\partial^2 u}{\partial x^2} \quad . \tag{8.34}$$

Take the solution domain to be the interval $[0, 1]$.

(8.2) Evaluate the matrix coefficients from the previous problem for linear hat basis functions, using elements with a fixed size of $h$.

(8.3) Now find the matrix coefficients for Hermite polynomial interpolation basis functions, once again using elements with a fixed size of $h$. A symbolic math environment is useful for this problem.

# 9 Cellular Automata and Lattice Gases

We started our discussion of partial differential equations by considering how they arise as continuum approximations to discrete systems (such as cars on a highway, or masses connected by springs). We then studied how to discretize continuum problems to develop numerical algorithms for their solution. In retrospect this might appear silly, and in some ways it is: why bother with the continuum description at all, if the goal is to solve the problem with a digital computer rather than an analog pencil?

Consider a gas. Under ordinary circumstances, it can be described by the continuous distribution of pressure, temperature, and velocity. However, on short length scales (for example, when studying the head of a disk drive flying a micron above the platter), this approximation breaks down and it becomes necessary to keep track of the individual gas molecules. If the energy is high enough (say, in a nuclear explosion) even that is not enough; it's necessary to add the excited internal degrees of freedom in the molecules. Or if the gas is very cold, quantum interactions between the molecules become significant and it's necessary to include the molecular wave functions. Such *molecular dynamics* calculations are done routinely, but require a great deal of effort: for the full quantum case, it's a challenge to calculate the dynamics for as long as a picosecond.

Any gas can be described by such a microscopic description, but that is usually too much work. Is it possible to stop between the microscopic and macroscopic levels, without having to pass through a continuum description? The answer is a resounding yes. This is the subject of *lattice gases*, or more generally, of *cellular automata* (*CA*), which in the appropriate limits reduce to both the microscopic and the macroscopic descriptions. The idea was first developed by Ulam and von Neumann in the early 1950s, and has since flourished with the development of the necessary theoretical techniques and computational hardware.

A classic example of a CA is Conway's *Game of Life* [Gardner, 1970], in which occupied sites on a grid get born, survive, or die in succeeding generations based on the number of occupied neighboring sites. Any CA has the same elements: a set of connected sites, states that are allowed on the sites, and a rule for how they are updated. We will start by studying a slightly more complicated system that recovers the Navier–Stokes fluid equations, and then will consider the more general question of how cellular automata relate to computation.

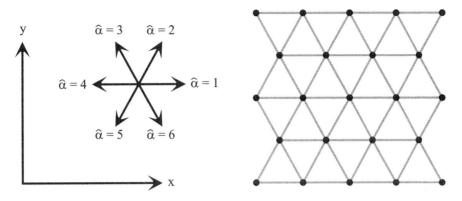

Figure 9.1. Direction indices for a triangular lattice.

## 9.1 LATTICE GASES AND FLUIDS

Hydrodynamics was one of the early successes of the theory of cellular automata, and remains one of the areas that is best developed and most important for practical applications. A cellular automata model of a fluid (traditionally called a lattice gas) is specified by the geometry of a lattice, by the discrete states permitted at each site, and by an update rule for how the states change based on their neighbors. Both partial differential equations and molecular dynamics models use real numbers (for the values of the fields, or for the particle positions and velocities). A lattice gas discretizes everything so that just a few bits describe the state of each site on a lattice, and the dynamics reduce to a simple look-up table based on the values of the neighboring sites. Each site can be considered to be a parcel of fluid. The rules for the sites implement a "cartoon" version of the underlying microscopic dynamics, but should be viewed as operating on a longer length scale than individual particles. We will see that the conservation laws that the rules satisfy determine the form of the equivalent partial differential equations for a large lattice, and that the details of the rules set the parameter values.

A historically important example of a lattice gas is the FHP rule (named after its inventors, Frisch, Hasslacher, and Pomeau [Frisch *et al.*, 1986]). This operates in 2D on a triangular lattice, and the state of each site is specified by six bits (Figure 9.1). Each bit represents a particle on one of the six links around the site, given by the unit vectors $\hat{\alpha}$. On each link a particle can either be present or absent, and all particles have the same unit velocity: in the absence of collisions, they travel one lattice step ahead in one time step. The simple update rule proceeds in two stages, chosen to conserve particle number and momentum (Figure 9.2). First, collisions are handled. At the beginning of the step a particle on a link is considered to be approaching the site, and after the collision step a particle on a link is taken to be leaving the site. If a site has two particles approaching head-on, they scatter. A random choice is made between the two possible outgoing directions that conserve momentum (always choosing one of them would break the symmetry of the lattice). Because of the large number of sites in a lattice gas it is usually adequate to approximate the random decision by simply switching between the two outgoing choices on alternate site updates. If three particles approach symmetrically, they scatter. In all other configurations not shown the particles pass through the collision

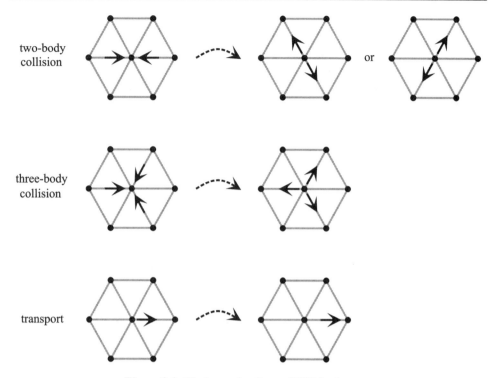

Figure 9.2. Update rules for an FHP lattice gas.

unchanged. After the collision step there is a transport step, in which each particle moves by one unit in the direction that it is pointing and arrives at the site at the far end of the link. While these rules might appear to be somewhat arbitrary, it will turn out that the details will not matter for the form of the governing equations, just the symmetries and conservation laws.

A simpler rule related to FHP is HPP (named after Hardy, de Pazzis, and Pomeau [Hardy et al., 1976]), which operates on a square lattice. Each site is specified by four bits, and direction-changing collisions are allowed only when two particles meet head-on (unlike FHP, here there is only one possible choice for the exit directions after scattering). We will see that HPP and FHP, although apparently quite similar, behave very differently.

Let's label time by an integer $T$, the lattice sites by a vector $\vec{X}$, and the lattice directions by a unit vector $\hat{\alpha}$. If we start an ensemble of equivalent lattices off with the same update rule but different random initial conditions, we can define $f_\alpha(\vec{X}, T)$ to be the fraction of sites $\vec{X}$ at time $T$ with a particle on link $\hat{\alpha}$. In the limit of a large ensemble, this fraction becomes the probability to find a particle on that link. Defining this probability will let us make a connection between the lattice gas and partial differential equations.

At each time step, in the absence of collisions, the fraction of particles at site $\vec{X}$ at time $T$ pointing in direction $\hat{\alpha}$ will move one step in that direction:

$$f_\alpha(\vec{X} + \hat{\alpha}, T + 1) = f_\alpha(\vec{X}, T) \quad . \tag{9.1}$$

Let's introduce new rescaled variables $\vec{x} = \delta_x \vec{X}$ and $t = \delta_t T$ in terms of the (small) space step $\delta_x$ and time step $\delta_t$. Substituting in these variables, collision-free transport becomes

$$f_\alpha(\vec{x} + \delta_x\hat{\alpha}, t + \delta_t) - f_\alpha(\vec{x}, t) = 0 \quad . \tag{9.2}$$

If the probability $f_\alpha$ varies slowly compared to $\delta_x$ and $\delta_t$, we can expand equation (9.2) in $\delta_x$ and $\delta_t$:

$$\frac{\partial f_\alpha(\vec{x}, t)}{\partial t} \delta_t + \hat{\alpha} \cdot \nabla f_\alpha(\vec{x}, t) \, \delta_x + \mathcal{O}(\delta^2) = 0 \quad . \tag{9.3}$$

Choosing to scale the variables so that $\delta_x = \delta_t$, to first order this becomes

$$\frac{\partial f_\alpha(\vec{x}, t)}{\partial t} + \hat{\alpha} \cdot \nabla f_\alpha(\vec{x}, t) = 0 \quad . \tag{9.4}$$

This equation says that the time rate of change of the fraction of particles at a point is equal to the difference in the rate at which they arrive and leave the point by straight transport (i.e., if there is a gradient in $f_\alpha$ more particles will arrive from one side then the other). If collisions are allowed, the time rate of change of $f_\alpha$ will depend on both the spatial gradient and on a collision term $\Omega_\alpha$ scattering particles in or out from other directions

$$\frac{\partial f_\alpha(\vec{x}, t)}{\partial t} + \hat{\alpha} \cdot \nabla f_\alpha(\vec{x}, t) = \Omega_\alpha(\vec{x}, t) \quad . \tag{9.5}$$

$f_\alpha$ is the distribution function to find a particle. The collision term $\Omega_\alpha$ will in general depend on the distribution function for pairs of particles as well as the one-particle distribution function, and these in turn will depend on the three-particle distribution functions, and so forth. This is called the *BBGKY hierarchy* of equations (Bogolyubov, Born, Green, Kirkwood, Yvon [Boer & Uhlenbeck, 1961]). The *Boltzmann equation* approximates this by assuming that $\Omega_\alpha$ depends only on the single-particle distribution functions $f_\alpha$.

We derived equation (9.5) by making use of the fact that particles travel one lattice site in each time step, and then assuming that $f_\alpha$ varies slowly. Now let's add the conservation laws that have been built into the update rules. The total density of particles $\rho$ at a site $\vec{x}$ is just the sum over the probability to find one in each direction

$$\sum_\alpha f_\alpha(\vec{x}) = \rho(\vec{x}) \quad , \tag{9.6}$$

and the momentum density is the sum of the probabilities times their (unit) velocities

$$\sum_\alpha \hat{\alpha} f_\alpha = \rho \vec{v} \quad . \tag{9.7}$$

Since our scattering rules conserve particle number (the particles just get reoriented), the number of particles scattering into and out of a site must balance

$$\sum_\alpha \Omega_\alpha(\vec{x}) = 0 \quad , \tag{9.8}$$

and since the rules conserve momentum then the net momentum change from scattering must vanish

$$\sum_\alpha \hat{\alpha} \Omega_\alpha = 0 \quad . \tag{9.9}$$

Therefore, summing equation (9.5) over directions,

$$\sum_\alpha \left[ \frac{\partial f_\alpha(\vec{x}, t)}{\partial t} + \hat{\alpha} \cdot \nabla f_\alpha(\vec{x}, t) \right] = \sum_\alpha \Omega_\alpha \qquad (9.10)$$

$$\frac{\partial}{\partial t} \sum_\alpha f_\alpha + \sum_\alpha \hat{\alpha} \cdot \nabla f_\alpha = 0$$

$$\frac{\partial \rho}{\partial t} + \nabla \cdot (\rho \vec{v}) = 0 \quad . \qquad (9.11)$$

This is the familiar equation for the continuity of a fluid, and has arisen here because we've chosen scattering rules that conserve mass. A second equation comes from momentum conservation, multiplying equation (9.5) by $\hat{\alpha}$ and summing over directions

$$\frac{\partial}{\partial t} \sum_\alpha \hat{\alpha} f_\alpha + \sum_\alpha \hat{\alpha} (\hat{\alpha} \cdot \nabla f_\alpha) = 0 \quad . \qquad (9.12)$$

The $i$th component of this vector equation is

$$\frac{\partial}{\partial t} \sum_\alpha \hat{\alpha}_i f_\alpha + \sum_\alpha \sum_j \hat{\alpha}_i \hat{\alpha}_j \frac{\partial f_\alpha}{\partial x_j} = 0 \quad . \qquad (9.13)$$

Defining the momentum flux density tensor by

$$\Pi_{ij} \equiv \sum_\alpha \hat{\alpha}_i \hat{\alpha}_j f_\alpha \quad , \qquad (9.14)$$

this becomes

$$\frac{\partial \rho v_i}{\partial t} + \sum_j \frac{\partial \Pi_{ij}}{\partial x_j} = 0 \quad . \qquad (9.15)$$

We now have two equations, (9.11) and (9.15), in three unknowns, $\rho$, $\vec{v}$, and $\Pi$. To eliminate $\Pi$ we can find the continuum form of the momentum flux density tensor by using a *Chapman–Enskog expansion* [Huang, 1987], a standard technique for finding approximate solutions to the Boltzmann equation. We will assume that $f_\alpha$ depends only on $\vec{v}$ and $\rho$ and their spatial derivatives (and not on time explicitly), and so will do an expansion in all possible scalars that can be formed from them. The lowest-order terms of the deviation from the equilibrium uniform configuration are

$$f_\alpha = \frac{\rho}{6} \left( 1 + 2\hat{\alpha} \cdot \vec{v} + A \left[ (\hat{\alpha} \cdot \vec{v})^2 - \frac{1}{2} |\vec{v}|^2 \right] \right.$$
$$\left. + B \left[ (\hat{\alpha} \cdot \nabla)(\hat{\alpha} \cdot \vec{v}) - \frac{1}{2} \nabla \cdot \vec{v} \right] + \cdots \right) \quad . \qquad (9.16)$$

The terms have been grouped this way to guarantee that the solution satisfies the density and momentum equations (9.6) and (9.7) (this can be verified by writing out the components of each term). In this derivation the only features of the FHP rule that we've used are the conservation laws for mass and momentum, and so all rules with these features will have the same form of the momentum flux density tensor (to this order), differering only in the value of the coefficients $A$ and $B$.

The *Navier–Stokes* governing equation for a $d$-dimensional fluid is

$$\frac{\partial \rho \vec{v}}{\partial t} + \mu \rho (\vec{v} \cdot \nabla) \vec{v} = -\nabla p + \eta \nabla^2 \vec{v} + (\zeta + \frac{\eta}{d}) \nabla (\nabla \cdot \vec{v}) \quad , \tag{9.17}$$

where $p$ is the pressure, $\eta$ is the shear viscosity, $\zeta$ is the bulk viscosity, and $\mu = 1$ for translational invariance [Batchelor, 1967]. Using the Chapman-Enskog expansion to evaluate the momentum flux density tensor in equation (9.15) and comparing it with the Navier–Stokes equation shows that they agree if $\zeta = 0$, $\eta = \rho \nu = -\rho B/8$ ($\nu$ is the kinematic viscosity), $\mu = A/4$, and $p = \rho/2$. Further, in the Boltzmann approximation it is possible (with a rather involved calculation) to find the values of $A$ and $B$ for a given CA rule [Wolfram, 1986].

The simple conservation laws built into our lattice gas have led to the full Navier–Stokes equation; the particular rule determines the effective viscosity of the fluid. While the details of this calculation are complicated, there are some simple and important conclusions. The viscosity for the square lattice (HPP model) turns out to depend on direction and hence is not appropriate for most fluids, but the viscosity of the triangular lattice (FHP model) is completely isotropic. The simulations in Problems 9.1 and 9.2 will show just how striking the difference is. This profound implication of the lattice symmetry was not appreciated in the early days of lattice gas models, and helped point the way towards the realization that a simple lattice gas model could in fact be very general. In 3D the situation is more difficult because there is not a 3D lattice that gives an isotropic viscosity. However, it can still be achieved by using a more complex tiling that is not translationally periodic, or a cut through a higher-dimensional lattice such as the 4D Face-Centered Hyper-Cubic (FCHC) rule [Frisch *et al.*, 1987]. It is not possible to reduce the viscosity in a simple model like FHP (an attribute that is needed for modeling a problem such as air flow), but this can be done by adding more than one particle type or by increasing the size of the neighborhood used for the rule [Dubrulle *et al.*, 1991].

Lattice gas models have not displaced the most sophisticated conventional numerical hydrodynamics codes, which are much more mature, and which make it easier to adjust the numerical parameters. But with much less effort they can be surprisingly competitive, and the required hardware, software, and theory are all advancing rapidly [Rothman & Zaleski, 1997].

## 9.2 CELLULAR AUTOMATA AND COMPUTING

Simulating a cellular automata is an unusual type of computation. Rather than the many kinds of memory, instructions, and processors in a conventional computer, it requires just storage for the bits on the lattice, and an implementation of the local update rule. This is an extreme form of a *SIMD* (Single Instruction Multiple Data) parallel computer. The update can be performed by using the bits at each site as an index into a look-up table, so the architecture reduces to memory cycling through a look-up table (where the sequence of the memory retrieval determines the lattice geometry). This means that relatively modest hardware, such as the special purpose CAM computers [Toffoli & Margolus, 1991], can match or exceed the performance of the largest supercomputers for CA problems.

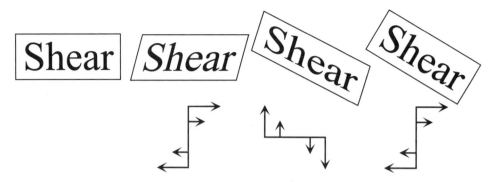

Figure 9.3. Rotation by three shears.

We've seen that a cellular automata computer can simulate fluids. A surprising number of other physical systems also can be modeled by cellular automata rules; an important example is the rendering of 3D graphics. An obvious possibility is to replace the ray-tracing calculations used in computer graphics by the propagation of lattice photons. Less obviously, global rotations in 2D and 3D can be done by shear operations, which can be implemented efficiently on a CA-style computer (Figure 9.3) [Toffoli & Quick, 1997]. Therefore, by using cellular automata 3D graphics can feasibly be done from a first-principles description.

Perhaps even more surprisingly, cellular automata are computationally universal: they can compute anything. Consider the elements in Figure 9.4 (the underlying lattice is not shown). This is similar to a lattice gas: billiard balls move on a lattice with unit velocity, and scatter off of each other and from walls. Two balls colliding generates the AND function, and if one of the streams of balls is continuous it generates the NOT function of the other input. These two elements are sufficent to build up all of logic [Hill & Peterson, 1993]. Memory can be implemented by delays, and wiring by various walls to guide the balls. The balls can be represented by four bits per site (one for each direction), with one extra bit per site needed to represent the walls. This kind of computing, developed by Edward Fredkin, Norman Margolus, Tomaso Toffoli, and collaborators [Fredkin & Toffoli, 1982], has many interesting features. No information is ever destroyed, which means that it is reversible (it can be run backwards to produce inputs from outputs) [Bennett, 1988], and which in turn means that it can be performed (in theory) with arbitrarily little dissipation [Landauer, 1961]. Reversibility is also essential for designing quantum cellular automata, since quantum evolution is reversible. A quantum CA is much like a classical CA, but it permits the sites to be in a superposition of their possible states [Lloyd, 1993]. This is a promising architecture for building quantum computers [Gershenfeld & Chuang, 1997].

For some, cellular automata are much more than just an amusing alternative to traditional models of computation. Most physical theories are based on real numbers. This means that a finite volume of space contains an infinite amount of information, since its state must be specified with real numbers. But if there is an energetic cost to creating information (as there is in most theories), then this implies an infinite amount of energy in a finite space. This is obviously unacceptable; something must bound the information content of space. While such a notion can arise in quantum field theories, CAs start as

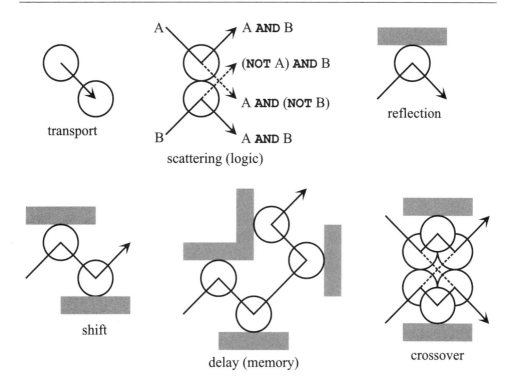

Figure 9.4. Billiard ball logic.

discrete theories and therefore do not have this problem, and so in many ways they are more satisfying than differential equations as a way to specify governing equations. There is nothing less basic about them than differential equations; which is more "fundamental" depends on whether you are solving a problem with a pencil or a computer.

## 9.3 SELECTED REFERENCES

[Hasslacher, 1987] Hasslacher, Brosl (1987). Discrete Fluids. *Los Alamos Science*, 175–217.

A very good introductory paper for lattice gases.

[Doolen et al., 1990] Doolen, Gary D. Frisch, Uriel, Hasslacher, Brosl, Orszag, Steven, & Wolfram, Stephen (eds) (1990). *Lattice Gas Methods for Partial Differential Equations*. Santa Fe Institute Studies in the Sciences of Complexity. Reading, MA: Addison-Wesley.

This collection includes many of the important articles, including [Wolfram, 1986] and [Frisch et al., 1987], which work out the connection between lattice gases and hydrodynamics.

[Rothman & Zaleski, 1994] Rothman, Daniel H., & Zaleski, Stéphane (1994). Lattice-Gas Models of Phase Separation: Interfaces, Phase Transitions, and Multiphase Flow. *Review of Modern Physics*, **66**, 1417–1479.

Reviews the basic theory and extends it to fluids with multiple components.

## 9.4 PROBLEMS

(9.1) Simulate the HPP lattice gas model. Take as a starting condition a randomly seeded lattice, but completely fill a square region in the center and observe how it evolves. Use periodic boundary conditions. Now repeat the calculation, but leave the lattice empty outside of the central filled square.

(9.2) Simulate the FHP model, with the same conditions as the previous model, and describe the difference. Alternate between the two possible two-body rules on alternate site updates.

*Part Three*
Observational Models

The first two parts of this book have covered techniques to explore the behavior of a model but have assumed that the model is known. Frequently, finding the model in the first place is the most difficult, interesting, and important question. Models can come from introspection or observation (or both); here we turn to the problem of inferring a model from measured data. The model may be used to characterize and classify the data, to generalize from the measurements in order to make predictions about new observations, or most ambitiously to learn something about the rules underlying the observed behavior.

In any data-driven modeling effort the two central tasks are always choosing the functional form of the model, and using the data to determine the adjustable parameters of the model. These are closely connected but can lead to different kinds of errors. *Model mismatch* errors are those that arise from a model that is unable to represent the data, and *model estimation* errors come from using incorrect values for the model parameters. Decreasing one kind of error is likely to increases the other kind. This is called a *bias/variance tradeoff* – if you want less bias in the estimate of a model parameter, it usually costs you more variance. A more flexible model that can better represent the data may also be more easily led astray by noise in the data. Each of the coming chapters covers some kind of optimization to minimize these errors. Since one person's data may be another's noise, it will be important throughout to keep an eye on which principle is being used to optimize what quantity with respect to which error measure, and on how the "best" solution is defined. After all, everything is optimal with respect to something, or conversely no one thing is optimal with respect to everything (the *No Free Lunch* theorem [Wolpert & Macready, 1995, Macready & Wolpert, 1996]).

Chapter 10 introduces the canon of function fitting. Chapters 11–14 revisit these standard choices in order to uncover the power of less-familiar alternatives, and Chapters 15 and 16 look at the essential role that time can play in modeling.

# 10 Function Fitting

The goal of function fitting is to choose values for the parameters in a function to best describe a set of data. There are many possible reasons to do this. If a specific meaningful form for the function with a small number of free parameters is known in advance, this is called *parametric fitting*, and finding the parameter values themselves may be the goal (for example, measuring an exponential decay constant to determine a reaction rate). If the form of the function is not known, and so a very flexible function with many free parameters is used, this becomes *nonparametric fitting* (although this distinction is often vague). One reason to do nonparametric fitting is to try to find and describe underlying trends in noisy data, in which case the fit must build in some prior beliefs and posterior observations about what defines the difference between the signal and the noise. In *function approximation* there is no noise; the goal is to take known (and perhaps laboriously calculated) values of a function and find a function that is easier to evaluate and that can interpolate between, or extrapolate beyond, the known values.

It's useful to view function fitting in a context such as the *Minimum Description Length* principle (*MDL*) [Rissanen, 1986], or the related *Algorithmic Information Theory* [Chaitin, 1990]. One extreme is to report the observed data itself as your model. This is not hard to do, but the "model" is very large and has no ability to generalize. Another extreme is report the smallest amount of information possible needed to describe the data, but this may require a great deal of supporting documentation about how to use the model (your tidy computer program is of no use to a Martian unless it comes with a complete description of a computer that can run it). The best model typically lies between these extremes: there is some amount of information about the data, and some about the model architecture. According to MDL, the sum of these two kinds of information taken together should be as small as possible. While this idea cannot be applied directly (like so many other attractive ideas, it includes a solution to the *halting problem* of deciding if an arbitrary program will terminate, which is known to be impossible [Turing, 1936, Chaitin, 1994]), it is a useful guiding principle that can be made explicit given specific assumptions about a problem.

In this chapter we will look at the basic features of function fitting: the general principles by which data can be used to constrain a model, the (often overlooked) connection with the choice of an error measure, how to fit a model with linear and nonlinear parameters, and the limits on what we can expect to learn from fitting. We will not be particularly concerned with the functional form used; the next three chapters will look in much more detail at the representation of data, functions, and optimization strategies.

## 10.1 MODEL ESTIMATION

The general fitting problem has three ingredients: a model architecture (which we'll call $m$) that has a set of adjustable parameters $\varphi$, and measured data $d$. The goal is to find values of the parameters that lead to the best agreement (in some sense) between the predictions of the model and the data. An example of $m$ might be a polynomial of a given order, where the $\varphi$ are the coefficients.

A reasonable way to go about finding the best coefficients is to ask for the $\varphi$ that are most likely given the choice of the model and the measured data. This means that we want to find the $\varphi$ that maximizes $p(\varphi|d, m)$. Using Bayes' rule (equation (5.11)), our job is then to find

$$\max_{\varphi} p(\varphi|d, m) = \max_{\varphi} \frac{p(d|\varphi, m)\, p(\varphi|m)}{p(d|m)}$$
$$= \max_{\varphi} \frac{p(d|\varphi, m)\, p(\varphi|m)}{\int_{\varphi} p(d|\varphi, m)\, p(\varphi|m)\, d\varphi}$$
$$= \max_{\varphi} \frac{\text{likelihood} \times \text{prior}}{\text{evidence}} \quad . \qquad (10.1)$$

The probability has been factored into three terms. The *likelihood* measures the match between the data and the predictions of the model with the coefficents, based on an *error model*. The *prior* introduces advance beliefs about which values of the coefficients are reasonable and which are not. And the *evidence* measures how well the model can describe the data.

If you solve equation (10.1) then you are an official card-carrying *Bayesian* [Bernardo & Smith, 1994]. The reason that there are not too many of them around is that solving equation (10.1) represents a lot of work. First of all, it's necessary to explicitly put priors on every parameter that is used. Then, the integration for the evidence is over all values of all the parameters, which can be an enormous computational task for a large model. Although efficient techniques have been developed for these kinds of integrals using *Monte-Carlo* sampling techniques that replace exact integration with a probabilistic approximation [Besag et al., 1995], they are still computationally intensive. Finally, the maximization over parameters is just an inner loop; the best description is given by a maximization over model architectures as well. Full Bayesian model estimation is not a step to be taken lightly; it should be done after it's clear that a more tractable approach cannot do the job.

Much of the work goes into the integration for the evidence term. But this does not affect a single maximization over $\varphi$; it comes in making comparisons among competing models. If we decide in advance that we are going to stick with one model architecture then equation (10.1) can be simplified by dropping the conditioning on the model:

$$\max_{\varphi} p(\varphi|d) = \max_{\varphi} \frac{p(d|\varphi)\, p(\varphi)}{p(d)} \quad . \qquad (10.2)$$

Now the evidence term has become a simple prior on the likelihood of the data set. Even this can usually be dropped; it's relevant only in combining multiple data sets of varying pedigrees. Finding parameters with (10.2) is called *Maximum A Posteriori* estimation (*MAP*).

MAP still requires putting a prior on the parameters. This is a very powerful idea, to be explored in the next chapter, but if we make the simplest choice of a uniform prior $p(\varphi) = p(d) = 1$ then we're left with

$$\max_{\varphi} p(\varphi|d) = \max_{\varphi} p(d|\varphi) \quad . \tag{10.3}$$

This is the easiest kind of model estimation of all, called *Maximum Likelihood* (*ML*). That is what we will now apply.

## 10.2 LEAST SQUARES

Let's assume that we are given a set of $N$ noisy measurements of a quantity $y_n$ as a function of a variable $x_n$, and we seek to find values for coefficients $\varphi$ in a function $y_n = y(x_n, \varphi)$ that describes their relationship (the generalization to vector variables will be straightforward). In Section 5.1.2 we learned that in the absence of any other information the *Central Limit Theorem* tells us that the most reasonable choice for the distribution of a random variable is Gaussian, and so we will make that choice for the distribution of errors in $y_n$. Problem 10.1 will use an entropy argument to reach the same conclusion. In practice, many systems choose to ignore this insight and have non-Gaussian distributions. The real reason why this assumption is so commonly made (frequently implictly) is that it leads to a particularly simple and useful error model: *least squares*.

If the errors have a Gaussian distribution around the true value $y(x_n, \varphi)$ then the probability to observe a value between $y$ and $y + dy$ is given by a Gaussian centered on the true value

$$p(y) \, dy = \frac{1}{\sqrt{2\pi\sigma_n^2}} \, e^{-[y-y(x_n,\varphi)]^2/(2\sigma_n^2)} \, dy \quad . \tag{10.4}$$

The variance $\sigma_n^2$ might depend on quantities such as the noise in a photodetector or the number of samples that are measured.

We will further assume that the errors between samples are independent as well as identically distributed (*iid*). This means that the probability to see the entire data set is given by the product of the probabilities to see each point,

$$p(\text{data}|\text{model}) = \prod_{n=1}^{N} \frac{1}{\sqrt{2\pi\sigma_n^2}} \, e^{-[y_n - y(x_n,\varphi)]^2/(2\sigma_n^2)} \, dy \quad . \tag{10.5}$$

We seek the $\varphi$ that maximizes this probability. If $p$ is maximal then so is its logarithm (the *log-likelihood*), and since the log of a product is equal to the sum of the logs, this becomes

$$-\log p(\text{data}|\text{model}) = \sum_{n=1}^{N} \frac{[y_n - y(x_n, \varphi)]^2}{2\sigma_n^2} + \frac{1}{2} \log(2\pi\sigma_n^2) \quad . \tag{10.6}$$

Because we've moved the minus sign to the left hand side we now want to find the $\varphi$ that minimizes the right hand side. The first term measures the distance between the data and the model, and the second one catches us if we try to cheat and make a model with a huge variance that explains everything equally well (or poorly). We can drop the second term since it does not depend on the parameters $\varphi$ that we are adjusting, and so we want to find the values that satisfy

$$\min_{\varphi} \sum_{n=1}^{N} \frac{[y_n - y(x_n, \varphi)]^2}{2\sigma_n^2} \quad . \tag{10.7}$$

If the variances $\sigma_n^2$ are constant (in particular, if we set $\sigma_n^2 = 1$ when we have no idea at all what it should be) this reduces to

$$\min_{\varphi} \sum_{n=1}^{N} [y_n - y(x_n, \varphi)]^2 \quad . \tag{10.8}$$

This is the familiar *least squares* error measure. It is the maximum likelihood estimator for data with normally distributed errors, but it is used much more broadly because it is simple, convenient, and frequently not too far off from an optimal choice for a particular problem. An example of where least squares might be a bad choice for an error measure is a bi-modal data set that has two peaks. The least squares error is minimized by a point between the peaks, but such a point has very little probability of actually occurring in the data set.

Instead of the square of the deviation between the model and the data, other powers can be used as an error measure. The first power (the magnitude of the difference) is the maximum likelihood estimate if the errors are distributed exponentially, and higher powers place more emphasis on outliers.

## 10.3 LINEAR LEAST SQUARES

Once we've chosen our error measure we need to find the parameters for the distribution that minimizes it. Perhaps the most important example of such a technique is *linear least squares*, because it is straightforward to implement and broadly applicable.

To do a least squares fit we will start by expanding our unknown function as a linear sum of $M$ known basis functions $f_m$

$$y(x) = \sum_{m=1}^{M} a_m f_m(x) \quad . \tag{10.9}$$

We want to find the coefficients $a_m$ that minimize the sum of the squared errors between this model and a set of $N$ given observations $y_n(x_n)$. The basis functions $f_m$ need not be orthogonal, but they must not be linear (otherwise the sum would be trivial); it is the coefficients $a_m$ that enter linearly. For example, the $f_m$ could be polynomial terms, with the $a_m$ as the coefficients of the polynomial.

A least squares fit can be written as a matrix problem

$$\begin{pmatrix} f_1(x_1) & f_2(x_1) & \cdots & f_M(x_1) \\ f_1(x_2) & f_2(x_2) & \cdots & f_M(x_2) \\ f_1(x_3) & f_2(x_3) & \cdots & f_M(x_3) \\ \vdots & \vdots & \vdots & \vdots \\ f_1(x_{N-1}) & f_2(x_{N-1}) & \cdots & f_M(x_{N-1}) \\ f_1(x_N) & f_2(x_N) & \cdots & f_M(x_N) \end{pmatrix} \begin{pmatrix} a_1 \\ a_2 \\ \vdots \\ a_M \end{pmatrix} = \begin{pmatrix} y_1 \\ y_2 \\ y_3 \\ \vdots \\ y_{N-1} \\ y_N \end{pmatrix} . \quad (10.10)$$

If we have the same number of free parameters as data points then the matrix will be square, and so the coefficients can be found by inverting the matrix and multiplying it by the observations. As long as the matrix is not singular (which would happen if our basis functions were linearly dependent), this inversion could be done exactly and our fit would pass through all of the data points. If our data are noisy this is a bad idea; we'd like to have many more observations than we have model parameters. We can do this if we use the *pseudo-inverse* of the matrix to minimize the least squared error. The *Singular Value Decomposition (SVD)* is a powerful technique that solves this (as well as many other problems).

### 10.3.1 Singular Value Decomposition

For a general linear equation $\mathbf{A} \cdot \vec{v} = \vec{b}$, the space of all possible vectors $\vec{b}$ for which the equation is solvable is the *range* of $\mathbf{A}$. The dimension of the range (i.e., the number of vectors needed to form a basis of the range) is called the *rank* of $\mathbf{A}$. There is an associated homogeneous problem $\mathbf{A} \cdot \vec{v} = 0$; the vectors $\vec{v}$ that satisfy the homogeneous equation lie in the *nullspace* of $\mathbf{A}$. If there is no nullspace then the matrix is of *full rank* (which is equal to the number of columns of $\mathbf{A}$).

If $\mathbf{A}$ is an arbitrary $N \times M$ matrix, an important result from linear algebra is that it can always be written in the form [Golub & Van Loan, 1996]

$$\begin{pmatrix} \\ \mathbf{A} \\ \\ \end{pmatrix} = \begin{pmatrix} \\ \mathbf{U} \\ \\ \end{pmatrix} \underbrace{\begin{pmatrix} w_1 & & & 0 \\ & w_2 & & \\ & & \ddots & \\ 0 & & & w_M \end{pmatrix}}_{\mathbf{W}} \begin{pmatrix} \mathbf{V}^T \end{pmatrix} \quad (10.11)$$

where $\mathbf{U}$ and $\mathbf{V}$ are *orthogonal* matrices whose inverse is equal to their transpose $\mathbf{U}^T \cdot \mathbf{U} = \mathbf{V} \cdot \mathbf{V}^T = \mathbf{I}$ (where $\mathbf{I}$ is the $M \times M$ identity matrix). This is called the *Singular Value Decomposition (SVD)* of the matrix, and the elements of the $M \times M$ diagonal matrix $\mathbf{W}$ are the *singular values* $w_i$.

The reason that the SVD is so important is that the columns of $\mathbf{U}$ associated with nonzero singular values ($w_i \neq 0$) form an orthonormal basis for the range of $\mathbf{A}$, and the columns of $\mathbf{V}$ associated with $w_i = 0$ form an orthonormal basis for the nullspace of $\mathbf{A}$. The singular values $w_i$ give the lengths of the principal axes of the hyper-ellipsoid defined by $\mathbf{A} \cdot \vec{x}$, where $\vec{x}$ lies on a hyper-sphere $|\vec{x}|^2 = 1$.

In terms of the SVD, the solution to $\mathbf{A} \cdot \vec{v} = \vec{b}$ is

$$\vec{v} = \mathbf{V} \cdot \mathbf{W}^{-1} \cdot \mathbf{U}^T \cdot \vec{b} \quad . \tag{10.12}$$

Since $\mathbf{W}$ is diagonal, its inverse $\mathbf{W}^{-1}$ is also diagonal and is found by replacing each diagonal element by its inverse. This sounds like a recipe for disaster since we just saw that $w_i = 0$ for elements of the nullspace. The odd solution to this problem is simply to declare that $1/0 = 0$, and set to zero the diagonal elements of $\mathbf{W}^{-1}$ corresponding to $w_i = 0$. To see that this apparent nonsense works, let's look for another solution to $\mathbf{A} \cdot \vec{v} = \vec{b}$. Assume that $\vec{v}$ is found according to the prescription for zeroing singular values in equation (10.12). We can add an arbitrary vector $\vec{v}'$ from the nullspace ($\mathbf{A} \cdot \vec{v}' = 0$) and still have a solution. The magnitude of the sum of these vectors is

$$\begin{aligned} |\vec{v} + \vec{v}'| &= |\mathbf{V} \cdot \mathbf{W}^{-1} \cdot \mathbf{U}^T \cdot \vec{b} + \vec{v}'| \\ &= |\mathbf{V} \cdot (\mathbf{W}^{-1} \cdot \mathbf{U}^T \cdot \vec{b} + \mathbf{V}^T \cdot \vec{v}')| \\ &= |\mathbf{W}^{-1} \cdot \mathbf{U}^T \cdot \vec{b} + \mathbf{V}^T \cdot \vec{v}'| \quad . \end{aligned} \tag{10.13}$$

Multiplication by $\mathbf{V}$ was eliminated in the last line because $\mathbf{V}$ is an orthonormal matrix and hence does not change the magnitude of the answer. The magnitude of $|\vec{v} + \vec{v}'|$ is made up of the sum of two vectors. The first one will have its $i$th element equal to 0 for every vanishing singular value $w_i = 0$ because of the rule for zeroing these elements of $\mathbf{W}^{-1}$. On the other hand, since $\vec{v}'$ is in the nullspace,

$$\begin{aligned} \mathbf{A} \cdot \vec{v}' &= \vec{0} \\ \mathbf{U} \cdot \mathbf{W} \cdot \mathbf{V}^T \cdot \vec{v}' &= \vec{0} \\ \mathbf{W} \cdot \mathbf{V}^T \cdot \vec{v}' &= \mathbf{U}^T \cdot \vec{0} \\ \mathbf{W} \cdot \mathbf{V}^T \cdot \vec{v}' &= \vec{0} \quad . \end{aligned} \tag{10.14}$$

This means that the $i$th component of the vector $\mathbf{V}^T \cdot \vec{v}'$ must equal 0 for every $w_i \neq 0$. Returning to the last line of equation (10.13), the left hand term can be nonzero only if $w_i \neq 0$, and the right hand term can be nonzero only if $w_i = 0$: these two vectors are orthogonal. Therefore, the magnitude of their sum is a minimum if $\vec{v}' = \vec{0}$. Adding any component from the nullspace to $\vec{x}$ increases its magnitude. This means that for an underdetermined problem (i.e., one in which there is a nullspace), the SVD along with the rule for zeroing elements in $\mathbf{W}^{-1}$ chooses the answer with the smallest magnitude (i.e., no component in the nullspace).

Now let's look at what the SVD does for an overdetermined problem. This is the case that we care about for fitting data, where we have more measurements than free parameters. We can no longer hope for an exact solution, but we can look for one that minimizes the *residual*

$$|\mathbf{A} \cdot \vec{v} - \vec{b}| = \sqrt{(\mathbf{A} \cdot \vec{v} - \vec{b})^2} \quad . \tag{10.15}$$

Let's once again choose $\vec{v}$ by zeroing singular values in equation (10.12), and see what happens to the residual if we add an arbitrary vector $\vec{v}'$ to it. This adds an error of $\vec{b}' = \mathbf{A} \cdot \vec{v}'$:

$$\begin{aligned} |\mathbf{A} \cdot (\vec{v} + \vec{v}') - \vec{b}| &= |\mathbf{A} \cdot \vec{v} + \vec{b}' - \vec{b}| \\ &= |(\mathbf{U} \cdot \mathbf{W} \cdot \mathbf{V}^T) \cdot (\mathbf{V} \cdot \mathbf{W}^{-1} \cdot \mathbf{U}^T \cdot \vec{b}) + \vec{b}' - \vec{b}| \end{aligned}$$

$$= |(\mathbf{U} \cdot \mathbf{W} \cdot \mathbf{W}^{-1} \cdot \mathbf{U}^T - 1) \cdot \vec{b} + \vec{b}'|$$
$$= |\mathbf{U} \cdot [(\mathbf{W} \cdot \mathbf{W}^{-1} - 1) \cdot \mathbf{U}^T \cdot \vec{b} + \mathbf{U}^T \cdot \vec{b}']|$$
$$= |(\mathbf{W} \cdot \mathbf{W}^{-1} - 1) \cdot \mathbf{U}^T \cdot \vec{b} + \mathbf{U}^T \cdot \vec{b}'| \quad . \tag{10.16}$$

Here again the magnitude is the sum of two vectors. $(\mathbf{W} \cdot \mathbf{W}^{-1} - 1)$ is a diagonal matrix, with nonzero entries for $w_i = 0$, and so the elements of the left term can be nonzero only where $w_i = 0$. The right hand term can be rewritten as follows:

$$\mathbf{A} \cdot \vec{v}' = \vec{b}'$$
$$\mathbf{U} \cdot \mathbf{W} \cdot \mathbf{V}^T \cdot \vec{v}' = \vec{b}'$$
$$\mathbf{W} \cdot \mathbf{V}^T \cdot \vec{v}' = \mathbf{U}^T \cdot \vec{b}' \quad . \tag{10.17}$$

The $i$th component can be nonzero only where $w_i \neq 0$. Once again, we have the sum of two vectors, one of which is nonzero only where $w_i = 0$, and the other where $w_i \neq 0$, and so these vectors are orthogonal. Therefore, the magnitude of the residual is a minimum if $\vec{v}' = 0$, which is the choice that SVD makes. Thus, the SVD finds the vector that minimizes the least squares residual for an overdetermined problem.

The computational cost of finding the SVD of an $N \times M$ matrix is $\mathcal{O}(NM^2 + M^3)$. This is comparable to the ordinary inversion of an $M \times M$ matrix, which is $\mathcal{O}(M^3)$, but the prefactor is larger. Because of its great practical significance, good SVD implementations are available in most mathematical packages. We can now see that it is ideal for solving equation (10.10). If the basis functions are chosen to be polynomials, the matrix to be inverted is called a *Vandermonde* matrix. For example, let's say that we want to fit a 2D bilinear model $z = a_0 + a_1 x + a_2 y + a_3 xy$. Then we must invert

$$\begin{pmatrix} 1 & x_1 & y_1 & x_1 y_1 \\ 1 & x_2 & y_2 & x_2 y_2 \\ 1 & x_3 & y_3 & x_3 y_3 \\ \vdots & \vdots & \cdots & \vdots \\ 1 & x_{N-1} & y_{N-1} & x_{N-1} y_{N-1} \\ 1 & x_N & y_N & x_N y_N \end{pmatrix} \begin{pmatrix} a_0 \\ a_1 \\ a_2 \\ a_3 \end{pmatrix} = \begin{pmatrix} z_1 \\ z_2 \\ z_3 \\ \vdots \\ z_{N-1} \\ z_N \end{pmatrix} \quad . \tag{10.18}$$

If the matrix is square ($M = N$), the solution can go through all of the data points. These are interpolating polynomials, like those we used for finite elements. A rectangular matrix ($M < N$) is the relevant case for fitting data. If there are any singular values near zero (given noise and the finite numerical precision they won't be exactly zero) it means that some of our basis functions are nearly linearly dependent and should be removed from the fit. If they are left in, SVD will find the set of coefficients with the smallest overall magnitude, but it is best for numerical accuracy (and convenience in using the fit) to remove the terms with small singular values. The SVD inverse will then provide the coefficients that give the best least squares fit. Choosing where to cut off the spectrum of singular values depends on the context; a reasonable choice is the largest singular value weighted by the computer's numerical precision, or by the fraction of noise in the data [Golub & Van Loan, 1996].

In addition to removing small singular values to eliminate terms which are weakly determined by the data, another good idea is to scale the expansion terms so that the magnitudes of the coefficients are comparable, or even better to rescale the data to have unit variance and zero mean (almost always a good idea in fitting). If 100 is a typical value for $x$ and $y$, then $10^4$ will be a typical value for their product. This means that the coefficient of the $xy$ term must by $\sim 100$ times smaller than the coefficient of the $x$ or $y$ terms. For higher powers this problem will be even worse, ranging from an inconvenience in examining the output from the fit, to a serious loss of numerical precision as a result of multiplying very large numbers by very small numbers.

Beyond finding the fit coefficients we usually want to know the uncertainty in the fit. Let's look at how an error in the measurements $\vec{\eta_b}$ results in an error in the fit $\vec{\eta_v}$:

$$\vec{v} + \vec{\eta_v} = \mathbf{V} \cdot \mathbf{W}^{-1} \cdot \mathbf{U}^T \cdot (\vec{b} + \vec{\eta_b}) \quad . \tag{10.19}$$

Since this is a linear equation, these random variables are related by

$$\vec{\eta_v} = \mathbf{V} \cdot \mathbf{W}^{-1} \cdot \mathbf{U}^T \cdot \vec{\eta_b} \quad . \tag{10.20}$$

The variance of one component of $\vec{\eta_v}$ is then

$$\begin{aligned}
\sigma_{v,i}^2 &= \langle \eta_{v,i} \eta_{v,i} \rangle \\
&= \left\langle \sum_j V_{ij} \frac{1}{w_j} \sum_k U_{kj} \eta_{b,k} \sum_l V_{il} \frac{1}{w_l} \sum_m U_{ml} \eta_{b,m} \right\rangle \\
&= \sum_j \sum_l V_{ij} V_{il} \frac{1}{w_j} \frac{1}{w_l} \sum_k \sum_m U_{kj} U_{ml} \underbrace{\langle \eta_{b,k} \eta_{b,m} \rangle}_{\sigma_b^2 \delta_{km}} \\
&= \sum_j \sum_l V_{ij} V_{il} \frac{1}{w_j} \frac{1}{w_l} \underbrace{\sum_k U_{kj} U_{kl}}_{\delta_{jl}} \sigma_b^2 \\
&= \sigma_b^2 \sum_j \frac{V_{ij}^2}{w_j^2} \quad .
\end{aligned} \tag{10.21}$$

In the second line, we've assumed that the measurement errors are an uncorrelated random variable with a fixed variance $\sigma_b^2$ (if that is not the case, their covariance must be carried through the calculation). In the third line, we've used the orthonormality of $\mathbf{U}$. Therefore, the relative variance in the fit is

$$\frac{\sigma_{v,i}^2}{\sigma_b^2} = \sum_j \frac{V_{ij}^2}{w_j^2} \quad . \tag{10.22}$$

## 10.4 NONLINEAR LEAST SQUARES

Using linear least squares we were able to find the best set of coefficients $\vec{a}$ in a single step (the SVD inversion). The price for this convenience is that the coefficients cannot appear inside the basis functions. For example, we could use Gaussians as our bases, but we would be able to vary only their amplitude and not their location or variance. It would be much more general if we could write

$$y(x) = \sum_{m=1}^{M} f_m(x, \vec{a}_m) \quad , \quad (10.23)$$

where the coefficients are now inside the nonlinear basis functions. We can still seek to minimize the error

$$\chi^2(\vec{a}) = \sum_{n=1}^{N} \left( \frac{y_n - y(x_n, \vec{a})}{\sigma_n} \right)^2 \quad , \quad (10.24)$$

but we will now need to do an iterative search to find the best solution, and we are no longer guaranteed to find it (see Section 12.5).

The basic techniques for nonlinear fitting that we'll cover in this chapter are based on the insight that we may not be able to invert a matrix to find the best solution, but we can evaluate the error locally and then move in a direction that improves it. The gradient of the error is

$$(\nabla \chi^2)_k = \frac{\partial \chi^2}{\partial a_k} = -2 \sum_{n=1}^{N} \frac{y_n - y(x_n, \vec{a})}{\sigma_i^2} \frac{\partial y(x_n, \vec{a})}{\partial a_k} \quad , \quad (10.25)$$

and its second derivative (the *Hessian*) is

$$\mathbf{H}_{kl} = \frac{\partial^2 \chi^2}{\partial a_k \partial a_l} \quad (10.26)$$

$$= 2 \sum_{n=1}^{N} \frac{1}{\sigma_i^2} \left[ \frac{\partial y(x_n, \vec{a})}{\partial a_k} \frac{\partial y(x_n, \vec{a})}{\partial a_l} - [y_n - y(x_n, \vec{a})] \frac{\partial^2 y(x_n, \vec{a})}{\partial a_l \partial a_k} \right] \quad .$$

Since the second term in the Hessian depends on the sum of terms proportional to the residual between the model and the data, which should be small and can change sign, it is customary to drop this term in nonlinear fitting.

From a starting guess for $\vec{a}$, we can update the estimate by the method of *steepest descent* or *gradient descent*, taking a step in the direction in which the error is decreasing most rapidly

$$\vec{a}_{new} = \vec{a}_{old} - \alpha \nabla \chi^2(\vec{a}_{old}) \quad , \quad (10.27)$$

where $\alpha$ determines how big a step we make. On the other hand, $\chi^2$ can be expanded around a point $\vec{a}_0$ to second order as

$$\chi^2(\vec{a}) = \chi^2(\vec{a}_0) + [\nabla \chi^2(\vec{a}_0)] \cdot (\vec{a} - \vec{a}_0) + \frac{1}{2} (\vec{a} - \vec{a}_0) \cdot \mathbf{H} \cdot (\vec{a} - \vec{a}_0) \quad , \quad (10.28)$$

which has a gradient

$$\nabla \chi^2(\vec{a}) = \nabla \chi^2(\vec{a}_0) + \mathbf{H} \cdot (\vec{a} - \vec{a}_0) \quad . \quad (10.29)$$

The minimum ($\nabla \chi^2(\vec{a}) = 0$) can therefore be found by iterating

$$\vec{a}_{new} = \vec{a}_{old} - \mathbf{H}^{-1} \cdot \nabla \chi^2(\vec{a}_{old}) \quad (10.30)$$

(this is *Newton's method*). Either of these techniques let us start with an initial guess for $\vec{a}$ and then successively refine it.

### 10.4.1 Levenberg–Marquardt Method

Far from a minimum Newton's method is completely unreliable: the local slope may shoot the new point further from the minimum than the old one was. On the other hand, near a minimum Newton's method converges very quickly and gradient descent slows to a crawl since the gradient being descended is disappearing. A natural strategy is to use gradient descent far away, and then switch to Newton's method close to a minimum. But how do we decide when to switch between them, and how large should the gradient descent steps be? The *Levenberg–Marquardt method* [Marquardt, 1963] is a clever solution to these questions, and is the most common method used for nonlinear least squares fitting.

We can use the Hessian to measure the curvature of the error surface, taking small gradient descent steps if the surface is curving quickly. Using the diagonal elements alone in the Hessian to measure the curvature is suggested by the observation that this gives the correct units for the scale factor $\alpha$ in equation (10.27):

$$\delta a_i = -\frac{1}{\lambda \mathbf{H}_{ii}} \frac{\partial \chi^2}{\partial a_i} \quad , \tag{10.31}$$

where $\lambda$ is a new dimensionless scale factor. If we use this weighting for gradient descent, we can then combine it with Newton's method by defining a new matrix

$$M_{ii} = \frac{1}{2} \frac{\partial^2 \chi^2}{\partial a_i^2}(1 + \lambda)$$

$$M_{ij} = \frac{1}{2} \frac{\partial^2 \chi^2}{\partial a_i \partial a_j} \quad (i \neq j) \quad . \tag{10.32}$$

If we use this to take steps given by

$$\mathbf{M} \cdot \delta \vec{a} = -\nabla \chi^2 \tag{10.33}$$

or

$$\delta \vec{a} = -\mathbf{M}^{-1} \cdot \nabla \chi^2 \quad , \tag{10.34}$$

when $\lambda = 0$ this just reduces to Newton's method. On the other hand, if $\lambda$ is very large then the diagonal terms will dominate, which is just gradient descent (equation (10.27)). $\lambda$ controls an interpolation between steepest descent and Newton's method. To use the Levenberg–Marquardt method, $\lambda$ starts off moderately large. If the step improves the error, $\lambda$ is decreased (Newton's method is best near a minimum), and if the step increases the error then $\lambda$ is increased (gradient descent is better).

$\chi^2$ can easily have many minima, but the Levenberg–Marquardt method will find only the local minimum closest to the starting condition. For this reason, a crucial sanity check is to plot the fitting function with the starting parameters and compare it with the data. If it isn't even close, it is unlikely that Levenberg–Marquardt will converge to a useful answer. It is possible to improve its performance in these cases by adding some kind of randomness that lets it climb out of small minima. If a function is hard to fit because it has very many local minima, or the parameter space is so large that it is hard to find sane starting values, then a technique that is better at global searching is called for. These extensions will be covered in Chapter 13.

## 10.5 ESTIMATION, FISHER INFORMATION, AND THE CRAMÉR–RAO INEQUALITY

We've seen increasingly powerful techniques to fit functions to data sets. Is there no limit to this cleverness? Unfortunately, and not surprisingly, there is indeed a limit on how much information about unknown parameters can be extracted from a set of measurements. This chapter closes with a view of the information in a probability distribution that sets a limit on the accuracy of measurements.

Let $p_\alpha(x)$ be a probability distribution that depends on a parameter $\alpha$ (such as the variance of a Gaussian); the goal is to estimate the value of $\alpha$ from a series of measurements of $x$. Let $f(x_1, x_2, ..., x_N)$ be the estimator of $\alpha$. It is *biased* if $\langle f(x_1, x_2, \ldots, x_N)\rangle \neq \alpha$, and it is *consistent* if $\lim_{N \to \infty} f(x_1, x_2, \ldots, x_N) = \alpha$. An estimator $f_1$ *dominates* $f_2$ if $\langle (f_1(x_1, x_2, \ldots, x_N) - \alpha)^2 \rangle \leq \langle (f_2(x_1, x_2, \ldots, x_N) - \alpha)^2 \rangle$. This raises the question of what is the minimum variance possible for an unbiased estimator of $\alpha$? The answer is given by the *Cramér–Rao bound*.

Start by defining the *score*:

$$V = \frac{\partial}{\partial \alpha} \log p_\alpha(x) = \frac{\partial_\alpha p_\alpha(x)}{p_\alpha(x)} \quad . \tag{10.35}$$

The expected value of the score is

$$\begin{aligned}
\langle V \rangle &= \int_{-\infty}^{\infty} p_\alpha(x) \frac{\partial_\alpha p_\alpha(x)}{p_\alpha(x)} \, dx \\
&= \int_{-\infty}^{\infty} \partial_\alpha p_\alpha(x) \, dx \\
&= \partial_\alpha \int_{-\infty}^{\infty} p_\alpha(x) \, dx \\
&= \partial_\alpha 1 \\
&= 0 \quad . 
\end{aligned} \tag{10.36}$$

This means that $\sigma^2(V) = \langle V^2 \rangle$. The variance of the score is called the *Fisher information*:

$$J(\alpha) = \langle [\partial_\alpha \log p_\alpha(x)]^2 \rangle \quad . \tag{10.37}$$

The score for a set of independent, identically distributed variables is the sum of the individual scores

$$\begin{aligned}
V(x_1, x_2, \ldots, x_N) &= \partial_\alpha \log p_\alpha(x_1, x_2, \ldots, x_N) \\
&= \sum_{n=1}^{N} \partial_\alpha \log p_\alpha(x_n) \\
&= \sum_{n=1}^{N} V(x_n) \quad ,
\end{aligned} \tag{10.38}$$

and so the Fisher information for the set is

$$\begin{aligned}
J_N(\alpha) &= \langle [\partial_\alpha \log p_\alpha(x_1, x_2, \ldots, x_N)]^2 \rangle \\
&= \langle V^2(x_1, x_2, \ldots, x_N) \rangle \\
&= \left\langle \left( \sum_{n=1}^{N} V(x_n) \right)^2 \right\rangle \\
&= \sum_{n=1}^{N} \langle V^2(x_n) \rangle \\
&= N\, J(\alpha)
\end{aligned} \qquad (10.39)$$

(remember that the individual scores are uncorrelated).

The Cramér–Rao inequality states that the mean square error of an unbiased estimator $f$ of $\alpha$ is lower bounded by the reciprocal of the Fisher information:

$$\sigma^2(f) \geq \frac{1}{J(\alpha)} \quad . \qquad (10.40)$$

To prove this, start with the *Cauchy–Schwarz inequality*

$$\begin{aligned}
\langle (V - \langle V \rangle)(f - \langle f \rangle) \rangle^2 &\leq \langle (V - \langle V \rangle)^2 \rangle \langle (f - \langle f \rangle)^2 \rangle \\
\langle Vf - \langle V \rangle f - V \langle f \rangle + \langle V \rangle \langle f \rangle \rangle^2 &\leq \langle V^2 - 2V\langle V \rangle + \langle V \rangle^2 \rangle \langle (f - \langle f \rangle)^2 \rangle \\
\langle Vf \rangle^2 &\leq \langle V^2 \rangle \langle (f - \langle f \rangle)^2 \rangle \\
\langle Vf \rangle^2 &\leq J(\alpha)\, \sigma^2(f)
\end{aligned} \qquad (10.41)$$

(remember $\langle V \rangle = 0$). The lefthand side equals one:

$$\begin{aligned}
\langle Vf \rangle &= \int_{-\infty}^{\infty} \frac{\partial_\alpha p_\alpha(x)}{p_\alpha(x)} f(x) p_\alpha(x)\, dx \\
&= \int_{-\infty}^{\infty} \partial_\alpha p_\alpha(x) f(x)\, dx \\
&= \partial_\alpha \int_{-\infty}^{\infty} p_\alpha(x) f(x)\, dx \\
&= \partial_\alpha \langle f(x) \rangle \\
&= \partial_\alpha \alpha \\
&= 1 \quad ,
\end{aligned} \qquad (10.42)$$

thus proving the Cramér–Rao inequality.

Just like the information theoretic channel capacity, this sets a lower limit on what is possible but does not provide any guidance in finding the minimum variance unbiased estimator. The inequality measures how much information the distribution provides about a parameter. Not surprisingly, the Fisher information can be related to the entropy of the distribution; this is done by *de Bruijn's identity* [Cover & Thomas, 1991]. Roughly, the entropy measures the volume and the Fisher information measures the surface of the distribution.

One final caution about the Cramér-Rao bound. Not only may it not be reachable in practice, but it may be misleading because it is a bound on unbiased estimators. Unbiased does not necessarily mean better: it is possible for a biased estimator to dominate an unbiased one (as well as have other desirable characteristics). However, just like channel capacity, although it should not be taken too literally it does provide a good rough estimate of what is plausible and what is not.

## 10.6 SELECTED REFERENCES

[Press et al., 1992] Press, William H., Teukolsky, Saul A., Vetterling, William T., & Flannery, Brian P. (1992). *Numerical Recipes in C: The Art of Scientific Computing*. 2nd edn. New York, NY: Cambridge University Press.

*Numerical Recipes* is particularly strong for function fitting.

[Cover & Thomas, 1991] Cover, Thomas M., & Thomas, Joy A. (1991). *Elements of Information Theory*. New York, NY: Wiley.

A good modern introduction to the many connections between information theory and inference.

## 10.7 PROBLEMS

(10.1) Another way to choose among models is to select the one that makes the weakest assumptions about the data; this is the purpose of *maximum entropy* methods. Assume that what is measured is a set of expectation values for functions $f_i$ of a random variable $x$,

$$\langle f_i(x) \rangle = \int_{-\infty}^{\infty} p(x) f_i(x) \, dx \quad . \tag{10.43}$$

(a) Given these measurements, find the compatible normalized probability distribution $p(x)$ that maximizes the entropy

$$S = -\int_{-\infty}^{\infty} p(x) \log p(x) \, dx \quad . \tag{10.44}$$

(b) What is the maximum entropy distribution if we know only the second moment

$$\sigma^2 = \int_{-\infty}^{\infty} p(x) \, x^2 \, dx \quad ? \tag{10.45}$$

(10.2) Now consider the reverse situation. Let's say that we know that a data set $\{x_n\}_{n=1}^{N}$ was drawn from a Gaussian distribution with variance $\sigma^2$ and unknown mean $\mu$. Try to find an optimal estimator of the mean (one that is unbiased and has the smallest possible error in the estimate).

# 11 Transforms

The coming chapters will revisit the ground we just covered, finding significant problems and remarkable capabilities lurking behind apparently innocuous assumptions made in last chapter's introduction to function fitting. Here we ask the easily overlooked question of whether data is best analyzed in the form that it is given (*hint:* the answer is frequently no). This is a question about *representation* – what's the best way to view the data to highlight the features of interest? The goal will be to boil a set of measurements down to a smaller set that is more independent, freeing subsequent analysis from having to rediscover the redundancy. A good representation can go a long way towards solving a difficult problem, and conversely a bad one can doom an otherwise well-intentioned effort.

## 11.1 ORTHOGONAL TRANSFORMS

We will generally be concerned with *orthogonal transformations*. These are ones that are particularly simple to undo, an important feature since we don't want our transformation to throw away information in the data unless we tell it to.

A matrix is *orthogonal* if its inverse is equal to its transpose,

$$\mathbf{M}^T \cdot \mathbf{M} = \mathbf{I} \quad , \tag{11.1}$$

where as usual the *transpose* is denoted by

$$\mathbf{M}^T_{ij} \equiv \mathbf{M}_{ji} \tag{11.2}$$

and **I** is the identity matrix with 1s on the diagonal and 0s elsewhere. Multiplication of a vector by an orthogonal matrix defines an orthogonal transformation on the vector. For a complex matrix the *adjoint* is the complex conjugate of the transpose

$$\mathbf{M}^\dagger_{ij} \equiv \mathbf{M}^*_{ji} \quad . \tag{11.3}$$

If the adjoint is the inverse,

$$\mathbf{M}^\dagger \cdot \mathbf{M} = \mathbf{I} \quad , \tag{11.4}$$

then **M** is *unitary*. The column or row vectors $\vec{v}_i$ of an orthogonal matrix are not only *orthogonal*, $\vec{v}_i \cdot \vec{v}_j = 0$ ($i \neq j$), they are *orthonormal*, $\vec{v}_i \cdot \vec{v}_j = \delta_{ij}$, but by convention the matrix itself is still usually just called orthogonal.

An important property of the adjoint is that it interchanges the order of a product of matrices:

$$(\mathbf{A} \cdot \mathbf{B})^\dagger = \mathbf{B}^\dagger \cdot \mathbf{A}^\dagger \quad . \tag{11.5}$$

Remember also that matrix multiplication is *distributive* $(\mathbf{A} \cdot (\mathbf{B} + \mathbf{C}) = \mathbf{A} \cdot \mathbf{B} + \mathbf{A} \cdot \mathbf{C})$ and *associative* $(\mathbf{A} \cdot (\mathbf{B} \cdot \mathbf{C}) = (\mathbf{A} \cdot \mathbf{B}) \cdot \mathbf{C})$, but need not be *commutative* $(\mathbf{A} \cdot \mathbf{B} \neq \mathbf{B} \cdot \mathbf{A})$.

Now consider a linear transformation on a column vector $\vec{x}$ to a new one $\vec{y} = \mathbf{M} \cdot \vec{x}$. The *Euclidean norm* of $\vec{x}$ is its length as measured by the sum of the square of the elements,

$$|\vec{x}|^2 = \vec{x}^\dagger \cdot \vec{x} = \sum_i x_i^* x_i = \sum_i |x_i|^2 \quad . \tag{11.6}$$

If $\mathbf{M}$ is unitary, then the norm of $\vec{y}$ is

$$\begin{aligned} |\vec{y}|^2 &= |(\mathbf{M} \cdot \vec{x})^\dagger \cdot (\mathbf{M} \cdot \vec{x})| \\ &= |(\vec{x}^\dagger \cdot \mathbf{M}^\dagger) \cdot (\mathbf{M} \cdot \vec{x})| \\ &= |\vec{x}^\dagger \cdot (\mathbf{M}^\dagger \cdot \mathbf{M}) \cdot \vec{x}| \\ &= |\vec{x}^\dagger \cdot \vec{x}| \\ &= |\vec{x}|^2 \quad . \end{aligned} \tag{11.7}$$

A unitary (or orthogonal) transformation preserves the norm of a vector. It rotates a data point to a new location, but can't change its distance from the origin. This means that it can rearrange the points but not do something as nasty as make some of them disappear.

## 11.2 FOURIER TRANSFORMS

The *Discrete Fourier Transformation* (*DFT*) is a familiar example of a unitary transformation (Problem 11.1). Given a data vector $\{x_0, x_1, \ldots, x_{N-1}\}$, the DFT is defined by

$$\begin{aligned} X_f &= \frac{1}{\sqrt{N}} \sum_{n=0}^{N-1} e^{2\pi i f n / N} x_n \\ &\equiv \sum_{n=0}^{N-1} M_{fn} x_n \\ &= \mathbf{M} \cdot \vec{x} \quad , \end{aligned} \tag{11.8}$$

and the corresponding inverse transform by

$$x_n = \frac{1}{\sqrt{N}} \sum_{f=0}^{N-1} e^{-2\pi i f n / N} X_f \quad . \tag{11.9}$$

The $X_f$ are the coefficients for an expansion of the vector in terms of sinusoids.

Computing the DFT requires multiplying the data vector by the transform matrix. Finding one element needs $N$ multiplies and adds, and there are $N$ elements, so this appears to be an $\mathcal{O}(N^2)$ algorithm. Remarkably, and significantly, this is not the case. Notice the the DFT can be split into two sums as follows:

$$\begin{aligned} X_f &= \frac{1}{\sqrt{N}} \sum_{n=0}^{N-1} e^{2\pi i f n/N} x_n \\ &= \frac{1}{\sqrt{N}} \sum_{n=0}^{N/2-1} e^{2\pi i f(2n)/N} x_{2n} + \frac{1}{\sqrt{N}} \sum_{n=0}^{N/2-1} e^{2\pi i f(2n+1)/N} x_{2n+1} \\ &= \frac{1}{\sqrt{N}} \sum_{n=0}^{N/2-1} e^{2\pi i f(2n)/N} x_{2n} + \frac{e^{2\pi i f/N}}{\sqrt{N}} \sum_{n=0}^{N/2-1} e^{2\pi i f(2n)/N} x_{2n+1} \\ &= \frac{1}{\sqrt{N}} \sum_{n=0}^{N/2-1} e^{2\pi i f n/(N/2)} x_{2n} + \frac{e^{2\pi i f/N}}{\sqrt{N}} \sum_{n=0}^{N/2-1} e^{2\pi i f n/(N/2)} x_{2n+1} \\ &= X_f^{even} + e^{2\pi i f/N} X_f^{odd} \quad . \end{aligned} \quad (11.10)$$

Instead of one $N$-point transform we've broken it into two $N/2$-point transforms, one on the even points and one on the odd ones. This requires $\mathcal{O}[(N/2)^2] + \mathcal{O}[(N/2)^2] = \mathcal{O}(N^2/2)$ steps to do the transforms and one final multiplication and addition to combine each element, instead of the original $\mathcal{O}(N^2)$ steps. The even and odd transforms can likewise be split, and so forth, until we've broken the calculation into $N$ single-point transforms. Reassembling each of them through the hierarchical factoring takes $\log_2 N$ adds and multiplies, for a total of $\mathcal{O}(N \log_2 N)$ steps. If $N = 10^6$, doing this requires $\mathcal{O}(10^7)$ steps (about a second on a 10 Mflop workstation), versus $\mathcal{O}(10^{12})$ operations for the DFT (about a day at 10 Mflops). Quite a savings! The modern incarnation of this clever idea is called the *Fast Fourier Transform* (*FFT*) and is associated with Cooley and Tukey [Cooley & Tukey, 1965], but it has a long history dating all the way back to Gauss in 1805. It is an example of the powerful algorithm design principle of *divide-and-conquer*: if you can't solve a difficult problem, split it into successively smaller problems until they can be solved and then recombine them to find the answer [Aho et al., 1974].

The clarity of the FFT implementation hides many subtleties in its application [Oppenheim & Schafer, 1989]. The highest frequency possible in a DFT is $f = 1/2$; beyond that the $2\pi$ periodicity of the exponential will wrap still higher components in $x_n$ onto lower frequencies. This is the phenomenon of *aliasing* and requires that a signal be sampled at more than twice the highest frequency of interest (called the *Nyquist frequency*). And since the transform is done over a finite time it is equivalent to transforming an infinite series multipied by a finite-length pulse. Since multiplication in the time domain is equal to convolution in the frequency domain, and the Fourier transform of a pulse is a *sinc* function $\sin(2\pi f \Delta T)/(\pi f)$, sharp features in the transform get spread out by the finite window and spurious *side-lobes* appear. There are many other ways to *window* data with weighting functions other than a rectangular step, in order to optimize desired attributes such as spectral resolution, sidelobe suppression, or phase uniformity. Finally, remember that the discrete sampling of the spectrum done by the DFT can easily miss important features that lie between the points of the transform.

The FFT is one of the most important algorithms in all of numerical mathematics. Beyond the many applications we've already seen for Fourier transforms it crops up in

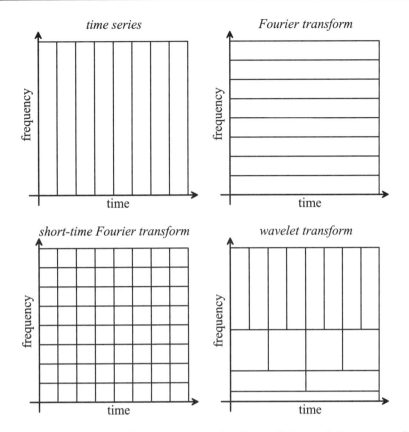

Figure 11.1. Division of time-frequency spaces by the coefficients of discrete transforms.

places where you might not expect it, such as speeding up the multiplication of two long numbers (which is really just a convolution [Knuth, 1981]). When Cooley (then at IBM) first presented the FFT, IBM concluded that it was so significant it should be put in the public domain to prevent anyone from trying to patent it, and so it was published openly. Ironically, its very success has made this kind of behavior much less common now.

## 11.3 WAVELETS

*Wavelets* are families of orthogonal transformations that generalize Fourier transforms in a very important way by introducing locality (Figure 11.1). Trigonometric functions are defined everywhere. This makes them good at describing global properties, such as the frequency of a signal, but very bad at describing locally varying properties. On the other hand, a time series represents a signal as a series of local impulses, which have an infinite spectrum of Fourier coefficients. A sine wave is most conveniently expressed in the frequency domain, and a step function is much more naturally defined in the time domain. In between these extremes lie most signals of interest, for which neither a global nor a local representation is best. A *short-time Fourier transform* (*STFT*) tries to do this by transforming short windows of data. This has the problem that low-frequency

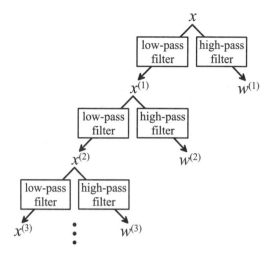

Figure 11.2. Interpretation of the wavelet transform as a hierarchical filter bank.

estimates need big windows to be meaningful, while high-frequency estimates need small windows to be relevant. This is exactly the happy compromise that wavelets provide, retaining a useful notion of both location and frequency.

Wavelets can be understood as a hierarchical *filter bank*, shown in Figure 11.2. A signal is applied to two filters, one passing the high-frequency part of the signal and the other passing the low-frequency part. Then, the low-frequency part goes through a pair of filters, separating it into a new high-frequency component and an even lower-frequency one. This procedure is continued until the signals at the bottom are left with a single point. Since we don't want the transform to throw away information unless we explicitly decide to, each of these steps is done invertibly.

The earliest wavelets were based on expanding a function in terms of rectangular steps, the *Haar wavelets* [Haar, 1910]. This is usually a very poor approximation; we will instead start with the *Daubechies wavelets*, which are among the simplest but still most important families [Daubechies, 1988]. Given a record of $N$ points $x_n$, the first step is to write down a linear filter

$$y_n = \sum_{i=0}^{M-1} b_i x_{n-i} \qquad (11.11)$$

that is zero for "smooth" signals. To design it we certainly want it to vanish for a constant, so that (taking the order $M = 4$ for example)

$$b_0 \cdot 1 + b_1 \cdot 1 + b_2 \cdot 1 + b_3 \cdot 1 = 0 \quad . \qquad (11.12)$$

The next thing that we could ask for is that it vanish for a linear ramp

$$b_0 \cdot 0 + b_1 \cdot 1 + b_2 \cdot 2 + b_3 \cdot 3 = 0 \quad . \qquad (11.13)$$

Since this is a linear filter it will then vanish for any $x = \alpha n + \beta$. It will turn out that for a fourth-order wavelet this is all that we can do; given the other constraints to be included six terms will be needed if we want it to vanish for a quadratic curve, and so forth. Next, we want to define another filter

$$z_n = \sum_{i=0}^{M-1} c_i x_{n-i} \qquad (11.14)$$

that responds exactly oppositely, being large for smooth signals and small for nonsmooth signals. A linear filter is just a convolution of the signal with the filter's coefficients, so the series of the coefficients is the signal that the filter responds maximally to (Chapter 15). Therefore, if the output of our second filter vanishes when the coefficients of the first one are input to it, it will be as unlike the first one as two linear filters can be. This means that we want

$$\sum_{i=0}^{M-1} c_i b_i = 0 \qquad (11.15)$$

(remember that because a linear filter is a convolution, the associated time series flips the order of the coefficients, and so both have the same index in this sum). A pair of filters with this property are called *quadrature mirror* filters. For $M = 4$ the equation to be solved is

$$c_0 b_0 + c_1 b_1 + c_2 b_2 + c_3 b_3 = 0 \quad . \qquad (11.16)$$

By inspection, this can be enforced flipping the order of the coefficients as well as the sign of every other one:

$$b_0 = c_3 \quad b_1 = -c_2 \quad b_2 = c_1 \quad b_3 = -c_0 \quad . \qquad (11.17)$$

We now have two filters: one is large for the smooth parts of the signal, and the other for the nonsmooth parts. To apply them to an input vector we can write it as a matrix problem:

$$\begin{bmatrix} c_0 & c_1 & c_2 & c_3 & & & & & & \\ c_3 & -c_2 & c_1 & -c_0 & & & & & & \\ & & c_0 & c_1 & c_2 & c_3 & & & & \\ & & c_3 & -c_2 & c_1 & -c_0 & & & & \\ & & & & \ddots & & & & & \\ & & & & & & c_0 & c_1 & c_2 & c_3 \\ & & & & & & c_3 & -c_2 & c_1 & -c_0 \\ c_2 & c_3 & & & & & & & c_0 & c_1 \\ c_1 & -c_0 & & & & & & & c_3 & -c_2 \end{bmatrix} \begin{pmatrix} x_0 \\ x_1 \\ \vdots \\ \vdots \\ \vdots \\ \vdots \\ \vdots \\ x_{N-1} \end{pmatrix} = \begin{pmatrix} x_0^{(1)} \\ w_0^{(1)} \\ x_1^{(1)} \\ w_1^{(1)} \\ \vdots \\ x_{N/2-2}^{(1)} \\ w_{N/2-2}^{(1)} \\ x_{N/2-1}^{(1)} \\ w_{N/2-1}^{(1)} \end{pmatrix}$$

(11.18)

(all empty matrix elements are 0). Such a representation of a moving filter is called a *circulant* matrix. Periodic boundary conditions were used to wrap around the coefficients, but it's also possible to define special coefficients for the boundaries to avoid that if necessary. I've called the output of the "smooth" filter $x^{(1)}$, and the output of the "nonsmooth" filter $w^{(1)}$. Each component has half as many points as the original series. The former is a lower resolution description of the signal, and the latter contains the fine structure that was lost in the smoothing.

It is convenient if the transformation is orthogonal so that the inverse is just the transpose. Requiring that the matrix times its transpose results in the identity matrix gives two nontrivial equations

$$c_0^2 + c_1^2 + c_2^2 + c_3^2 = 1$$

$$c_2 c_0 + c_3 c_1 = 0 \quad . \tag{11.19}$$

We also had two equations for the filter

$$c_3 - c_2 + c_1 - c_0 = 0$$

$$-c_2 + 2c_1 - 3c_0 = 0 \tag{11.20}$$

(written in terms of the $c$'s instead of the $b$'s). This is four equations in four unknowns, which can be solved to find

$$c_0 = \frac{1 + \sqrt{3}}{4\sqrt{2}} \qquad c_1 = \frac{3 + \sqrt{3}}{4\sqrt{2}}$$

$$c_2 = \frac{3 - \sqrt{3}}{4\sqrt{2}} \qquad c_3 = \frac{1 - \sqrt{3}}{4\sqrt{2}} \quad . \tag{11.21}$$

Using these coefficients, the transformation can be inverted by using the transpose

$$\begin{bmatrix} c_0 & c_3 & & & & & c_2 & c_1 \\ c_1 & -c_2 & & & & & c_3 & -c_0 \\ c_2 & c_1 & c_0 & c_3 & & & & \\ c_3 & -c_0 & c_1 & -c_2 & & & & \\ & & c_2 & c_1 & & & & \\ & & c_3 & -c_0 & & & & \\ & & & & \ddots & & & \\ & & & & c_0 & c_3 & & \\ & & & & c_1 & -c_2 & & \\ & & & & c_2 & c_1 & c_0 & c_3 \\ & & & & c_3 & -c_0 & c_1 & -c_2 \end{bmatrix} \begin{pmatrix} x_0^{(1)} \\ w_0^{(1)} \\ \vdots \\ \vdots \\ \vdots \\ \vdots \\ \vdots \\ \vdots \\ x_{N/2-1}^{(1)} \\ w_{N/2-1}^{(1)} \end{pmatrix} = \begin{pmatrix} x_0 \\ x_1 \\ \vdots \\ \vdots \\ \vdots \\ \vdots \\ \vdots \\ x_{N-2} \\ x_{N-1} \end{pmatrix}$$

Let's now multiply the output from the filters by another orthonormal matrix that splits the two types of results:

$$\begin{bmatrix} 1 & 0 & & & & & & \\ 0 & 1 & & & & & & \\ & & 1 & 0 & & & & \\ & & 0 & 1 & 0 & & & \\ & & & & \ddots & & & \\ 0 & 1 & 0 & & & & & \\ & 0 & 1 & & & & & \\ & & & & & & 1 & 0 \\ & & & & & & 0 & 1 \end{bmatrix} \begin{pmatrix} x_0^{(1)} \\ w_0^{(1)} \\ x_1^{(1)} \\ w_1^{(1)} \\ \vdots \\ x_{N/2-2}^{(1)} \\ w_{N/2-2}^{(1)} \\ x_{N/2-1}^{(1)} \\ w_{N/2-1}^{(1)} \end{pmatrix} = \begin{pmatrix} x_0^{(1)} \\ x_1^{(1)} \\ \vdots \\ x_{N/2-2}^{(1)} \\ x_{N/2-1}^{(1)} \\ w_0^{(1)} \\ w_1^{(1)} \\ \vdots \\ w_{N/2-1}^{(1)} \end{pmatrix}$$

The first half of the resulting vector is a smoothed version of the original signal at half the time resolution, and the second half contains the details lost in the smoothing. The original series can be recovered by multiplying by the transposes of the two matrices used. We can now go ahead and do the same sequence of operations on the new $x$'s, to give a version at even lower resolution as well as some more "detail" coefficients. Repeating the filtering and shuffling operations until we're left with just two $x$ values and so can go no further gives the following sequence of coefficient vectors:

$$
\begin{array}{ccccccc}
x_0 & x_0^{(1)} & x_0^{(1)} & x_0^{(2)} & x_0^{(2)} & x_0^{(\log_2 N-1)} & x_0^{(\log_2 N-1)} \\
x_1 & w_0^{(1)} & x_1^{(1)} & w_0^{(2)} & x_1^{(2)} & w_0^{(\log_2 N-1)} & x_1^{(\log_2 N-1)} \\
\vdots & x_1^{(1)} & \vdots & x_1^{(2)} & \vdots & x_1^{(\log_2 N-1)} & w_0^{(\log_2 N-1)} \\
\vdots & w_1^{(1)} & \vdots & w_1^{(2)} & x_{N/4-1}^{(2)} & w_1^{(\log_2 N-1)} & w_1^{(\log_2 N-1)} \\
\vdots & \vdots & \vdots & \vdots & w_0^{(2)} & w_0^{(\log_2 N-2)} & w_0^{(\log_2 N-2)} \\
\vdots & \vdots & \vdots & \vdots & w_1^{(2)} & \vdots & \vdots \\
\vdots & \vdots & \vdots & x_{N/4-1}^{(2)} & \vdots & \cdots & \vdots & \vdots \\
\vdots & \vdots & x_{N/2-1}^{(1)} & w_{N/4-1}^{(2)} & w_{N/4-1}^{(2)} & \vdots & \vdots \\
\vdots & \vdots & w_0^{(1)} & w_0^{(1)} & w_0^{(1)} & \vdots & \vdots \\
\vdots & \vdots & w_1^{(1)} & w_1^{(1)} & w_1^{(1)} & \vdots & \vdots \\
\vdots & x_{N/2-1}^{(1)} & \vdots & \vdots & \vdots & \vdots & \vdots \\
x_{N-1} & w_{N/2-1}^{(1)} & w_{N/2-1}^{(1)} & w_{N/2-1}^{(1)} & w_{N/2-1}^{(1)} & w_{N/2-1}^{(1)} & w_{N/2-1}^{(1)}
\end{array}
$$

This defines the *Discrete Wavelet Transformation (DWT)*, and the final $w$'s are the wavelet coefficients. They represent structure at many scales as well as at many locations. If any of the wavelet coefficients are small they can be set to zero to approximate the original series with less information, but the beauty of this kind of compression is that it can find important regions in the time-frequency space rather than projecting all of the information onto the frequency axis (as done by an FFT) or the time axis (by impulses).

A sine wave looks like, well, a sine wave. What does a wavelet look like? We can find out by setting one of the wavelet coefficients to 1 and all the others to 0, and then running the inverse wavelet transform back to find the $x$ series that produces it (just as inverting a Fourier transform of an impulse gives a sinusoidal function). Problem 11.2 shows that this results in quite a curious looking function. To understand it, consider that after one pass of the smoothing filter,

$$x_n^{(1)} = c_0 x_{2n} + c_1 x_{2n+1} + c_2 x_{2n+2} + c_3 x_{2n+3} \quad . \tag{11.22}$$

If a function exists that satisfies

$$X_n = c_0 X_{2n} + c_1 X_{2n+1} + c_2 X_{2n+2} + c_3 X_{2n+3} \tag{11.23}$$

then it will be unchanged by the smoothing (this is a *dilation equation*, instead of a difference or differential equation). The associated wavelet function is

$$W_n = c_3 X_{2n} - c_2 X_{2n+1} + c_1 X_{2n+2} - c_0 X_{2n+3} \quad . \tag{11.24}$$

In the limit of many iterations, so that $n$ approaches a continuous variable, these are the basic functions that are invariant under the transformation. Remarkably, in the continuum limit these apparently innocent and certainly useful functions are very complicated, not even differentiable.

We've been looking at fourth-order wavelets; higher orders are similarly defined. For each two additional coefficients used it's possible to go to one higher derivative of the function that can be matched. Beyond order 6 the coefficients must be found numerically. Our wavelets also have had *compact support* (they are zero everywhere except for where they are defined); this is convenient numerically but can be relaxed in order to get other benefits such as the analytical form and simple spectrum of the *harmonic wavelets* [Newland, 1994].

Just as a high-dimensional Fourier transform can be done by transforming each axis in turn, wavelets can be extended to higher dimensions by transforming each axis separately [Press et al., 1992]. This restricts the wavelets to the axes of the space; it is also possible to define more general multi-dimensional wavelets. The state of the art in wavelets has advanced rapidly since their introduction; see for example [Chui et al., 1994]. Beyond wavelets there are other time-frequency transforms, such as *Wigner functions* [Hlawatsch & Boudreaux-Bartels, 1992], which first arose as a probabilistic representation of quantum mechanics for studying semi-classical systems [Balazs & Jennings, 1984].

## 11.4 PRINCIPAL COMPONENTS

Wavelets were constructed based on the assumption that time and frequency are the interesting axes against which a signal can be viewed. This certainly need not be true, and doesn't even apply to a set of measurements that have no particular temporal or spatial ordering. Rather than designing one transform to apply to all data we might hope to do better by customizing a transform to provide the best representation for a given data set (where "best" of course will reflect some combination of what we hope to achieve and what we know how to accomplish).

Let's once again let $\vec{x}$ be a measurement vector, and $\vec{y} = \mathbf{M} \cdot \vec{x}$ be a transformation to a new set of variables with more desirable properties. The *covariance matrix* of $\vec{y}$ is defined by

$$\mathbf{C}_y \equiv \langle (\vec{y} - \langle \vec{y} \rangle) \cdot (\vec{y} - \langle \vec{y} \rangle)^T \rangle \quad , \tag{11.25}$$

where the *outer product* of two column vectors $\vec{A}$ and $\vec{B}$ is

$$(\vec{A} \cdot \vec{B}^T)_{ij} = A_i B_j \quad , \tag{11.26}$$

and the average is taken over an ensemble of measurements. A reasonable definition of "best" is to ask that the covariance matrix of $\vec{y}$ be diagonal, so that each of its elements is uncorrelated.

To find the required transformation, the covariance matrix of $\vec{y}$ can be related to that of $\vec{x}$:

$$\begin{aligned} \mathbf{C}_y &= \langle (\vec{y} - \langle \vec{y} \rangle) \cdot (\vec{y} - \langle \vec{y} \rangle)^T \rangle \\ &= \langle [\mathbf{M} \cdot (\vec{x} - \langle \vec{x} \rangle)] \cdot [\mathbf{M} \cdot (\vec{x} - \langle \vec{x} \rangle)]^T \rangle \\ &= \langle [\mathbf{M} \cdot (\vec{x} - \langle \vec{x} \rangle)] \cdot [(\vec{x} - \langle \vec{x} \rangle)^T \cdot \mathbf{M}^T] \rangle \end{aligned}$$

$$= \mathbf{M} \cdot \langle (\vec{x} - \langle \vec{x} \rangle) \cdot (\vec{x} - \langle \vec{x} \rangle)^T \rangle \cdot \mathbf{M}^T$$
$$= \mathbf{M} \cdot \mathbf{C}_x \cdot \mathbf{M}^T \quad . \quad (11.27)$$

Because $\mathbf{C}_x$ is a real symmetric matrix it's possible to find an orthonormal set of eigenvectors [Golub & Van Loan, 1996]. Now consider what happens if the columns of $\mathbf{M}^T$ are taken to be these eigenvectors. After multiplication by $\mathbf{C}_x$ each eigenvector is returned multiplied by its corresponding eigenvalue. Then because of the orthonormality, the multiplication of this matrix by $\mathbf{M}$ gives zeros off-diagonal, and returns the values of the eigenvalues on the diagonal. Therefore $\mathbf{C}_y$ is a diagonal matrix as desired. If there are linear correlations among the elements of $\vec{x}$ then some of the eigenvalues will vanish; these components of $\vec{y}$ can be dropped from subsequent analysis. For real data sets the elements might not be exactly equal to zero, but the relative magnitudes of them let the important components be found and the less important ones be ignored. Such *variable subset selection* is frequently the key to successful modeling.

Use of the covariance matrix of a set of measurements to find a transformation to new variables that are uncorrelated is called *Principal Components Analysis (PCA)*. It is such a useful idea that it led to many other related three-letter acronyms (*TLAs*). One is the *Karhunen–Loéve Transform (KLT)* [Fukunaga, 1990]. Here, a measurement vector $\vec{y}$ (such as a time series, or the values of the pixels in an image) is expanded in a sum over orthonormal basis vectors $\vec{\varphi}_i$ with expansion coefficients $x_i$,

$$\vec{y} = \sum_i x_i \vec{\varphi}_i \quad . \quad (11.28)$$

Given an ensemble of measurements of $\vec{y}$, the goal is to choose a set of $\vec{\varphi}_i$ that make the $x_i$'s as independent as possible. Defining $\mathbf{M}$ to be a matrix that has the $\vec{\varphi}_i$ as column vectors, the expansion of $\vec{y}$ can be written as $\vec{y} = \mathbf{M} \cdot \vec{x}$, where $\vec{x}$ is a column vector of the expansion coefficients. We've already seen that the covariance matrices of $\vec{y}$ and $\vec{x}$ are related by

$$\mathbf{C}_y = \mathbf{M} \cdot \mathbf{C}_x \cdot \mathbf{M}^T \quad (11.29)$$

or

$$\mathbf{M}^T \cdot \mathbf{C}_y \cdot \mathbf{M} = \mathbf{C}_x \quad . \quad (11.30)$$

Therefore if we choose the $\vec{\varphi}_i$ to be the eigenvectors of $\mathbf{C}_y$ then $\mathbf{C}_x$ will be diagonal. Since the $\vec{\varphi}_i$ are orthonormal, given a new measurement $\vec{y}$ the expansion coefficients can be found from

$$\vec{y} \cdot \vec{\varphi}_j = \sum_i x_i \vec{\varphi}_i \cdot \vec{\varphi}_j = \sum_i x_i \delta_{ij} = x_j \quad . \quad (11.31)$$

This provides a convenient way to do *lossy compression* for storage or communications, by using only the significant coefficients to partially reconstruct a data vector from the bases.

PCA starts with the covariance matrix of all of the original variables and then throws out the insignificant components. *Factor Analysis* directly seeks a smaller set of variables that can explain the covariance structure of the observations [Hair *et al.*, 1998]. *Independent Components Analysis (ICA)* goes further to use a nonlinear search algorithm to find a linear transformation that makes the new variables independent ($p(y_i, y_j) = p(y_i)p(y_j)$)

rather than just uncorrelated ($\langle y_i y_j \rangle = 0$) [Comon, 1994, Bell & Sejnowski, 1995]. This can be done by minimizing the mutual information among the components.

These are all global transformations; they can also be done locally to capture nonlinear relationships (Section 14.4; [Ghahramani & Hinton, 1998]). From there it is a small step to the hierarchical architectures to be covered in the next chapter, which have been called *Multi-Layer Perceptrons* (*MLP*s) or *Artificial Neural Networks* (*ANN*s).

## 11.5 SELECTED REFERENCES

[Golub & Van Loan, 1996] Golub, G.H., & Van Loan, C.F. (1989). *Matrix Computations*. 2nd edn. Baltimore, MD: Johns Hopkins University Press.

> Everything you always wanted to know about transformations with matrices.

[Fukunaga, 1990] Fukunaga, Keinosuke (1990). *Introduction to Statistical Pattern Recognition*. 2nd edn. Boston, MA: Academic Press.

> Much of the effort in pattern recognition goes into finding good representations.

## 11.6 PROBLEMS

(11.1) Prove that the DFT is unitary.

(11.2) Calculate the inverse wavelet transform, using Daubechies fourth-order coefficients, of a vector of length $2^{12}$, with a 1 in the 5th and 30th places and zeros elsewhere.

(11.3) Consider a measurement of a three-component vector $\vec{x}$, with $x_1$ and $x_2$ being drawn independently from a Gaussian distribution with zero mean and unit variance, and $x_3 = x_1 + x_2$.

  (a) Analytically calculate the covariance matrix of $\vec{x}$.
  (b) What are the eigenvalues?
  (c) Numerically verify these results by drawing a data set from the distribution and computing the covariance matrix and eigenvalues.
  (d) Numerically find the eigenvectors of the covariance matrix, and use them to construct a transformation to a new set of variables $\vec{y}$ that have a diagonal covariance matrix with no zero eigenvalues. Verify this on the data set.

# 12 Architectures

This chapter looks at some of the ways that functions can be connected to *interpolate* among and *extrapolate* beyond observations. The essence of successful generalization lies in finding a form that is just capable enough, but no more. There is a recurring tension between describing too little (using a function that does not have enough flexibility to follow the data) and too much (using a function that is so flexible that it fits noise, or produces artifacts absent in the original data). Along with matching the data, supporting concerns will be how easy a family of functions is to use and to understand. Most anyone who has done any kind of mathematical modeling has made such choices, often without being aware of the problems of old standbys and the power of less well-known alternatives.

Polynomials are often the first, and sometimes the last, functional form encountered for fitting. Familiarity can breed contempt; in this chapter we'll see some of the problems with using polynomials for nontrivial problems, and consider some of the appealing alternatives.

## 12.1 POLYNOMIALS

### 12.1.1 Padé Approximants

As we've already seen many times, a *polynomial* expansion has the form (in 1D)

$$y(x) = \sum_{n=0}^{N} a_n x^n \quad . \tag{12.1}$$

One failing of polynomials is that they have a hard time matching a function that has a pole. Since $(1-x)^{-1} = 1+x+x^2+x^3+\ldots$, a simple pole is equivalent to an infinite-order power series. This is more serious than you might think because even if a function is evaluated only along the real axis, a nearby complex pole can slow down the convergence of a polynomial expansion. An obvious improvement is to use a ratio of polynomials,

$$y(x) = \frac{\sum_{n=1}^{N} a_n x^n}{1 + \sum_{m=1}^{M} b_m x^m} \quad . \tag{12.2}$$

This is called a *Padé approximant* [Baker & Graves-Morris, 1996]. By convention the constant term in the denominator is taken to be 1 since it's always possible to rescale the coefficients by multiplying the numerator and denominator by a constant. The order of the approximation is written $[N/M]$.

One application of Padé approximants is for functional approximation. If a function has a known Taylor expansion

$$f(x) = \sum_{l=0}^{\infty} c_l x^l \quad , \tag{12.3}$$

we can ask that it be matched up to some order by a Padé approximant

$$\frac{\sum_{n=1}^{N} a_n x^n}{1 + \sum_{m=1}^{M} b_m x^m} = \sum_{l=0}^{L} c_l x^l \quad . \tag{12.4}$$

The coefficients can be found by multiplying both sides by the denominator of the left hand side,

$$\sum_{n=0}^{N} a_n x^n = \sum_{l=0}^{L} c_l x^l + \sum_{m=1}^{M} \sum_{l=0}^{L} b_m c_l x^{l+m} \quad , \tag{12.5}$$

and equating powers of $x$. For powers $n \leq N$ this gives

$$a_n = c_n + \sum_{m=1}^{N} b_m c_{n-m} \quad (n = 0, \ldots, N) \tag{12.6}$$

and for higher powers

$$0 = c_l + \sum_{m=1}^{M} b_m c_{l-m} \quad (l = N+1, \ldots, N+M) \quad . \tag{12.7}$$

The latter relationship gives us $M$ equations that can be solved to find the $M$ $b_m$'s in terms of the $c$'s, which can then be plugged into the former equation to find the $a$'s.

Alternatively, given a data set $\{y_i, x_i\}$ we can look for the best agreement by a Padé approximant. Relabeling the coefficients, this can be written as

$$\frac{\sum_{n=1}^{N} a_n x_i^n}{1 + \sum_{n=N+1}^{M+N} a_n x_i^{n-N}} = y_i \quad . \tag{12.8}$$

Once again multiplying both sides by the denominator of the left hand side and rearranging terms gives

$$\sum_{n=0}^{N} x_i^n a_n - \sum_{n=N+1}^{N+M} x_i^{n-N} y_i a_n = y_i \quad , \tag{12.9}$$

or $\mathbf{M} \cdot \vec{a} = \vec{y}$, where the $i$th row of $\mathbf{M}$ is $(x_i^0, \ldots, x_i^N, -x_i^1 y_i, \ldots, -x_i^N y_i)$, $\vec{a} = (a_0, \ldots, a_{N+M})$, and $\vec{y}$ is the vector of observations. This can now be solved by SVD to find $\vec{a} = \mathbf{M}^{-1} \cdot \vec{y}$, the set of coefficients that minimize the sum of the squares of the errors.

When Padé approximants are appropriate they work very well. They can converge remarkably quickly (Problem 12.1), and can have an almost mystical ability to find analytical structure in a function that lets the approximation be useful far beyond where there are any small parameters in an expansion. However, their use requires care. After all they're guaranteed to have poles whether or not the function being approximated does; the domain and rate of convergence can be difficult to anticipate.

### 12.1.2 Splines

Another problem with global polynomials is that they have a hard time following local behavior, resulting in undesirable artifacts such as overshooting and ringing (Problem 12.2). In Chapter 8 we saw one solution to this problem: define polynomials in finite elements, and then patch these local functions together with appropriate matching conditions. *Splines* use cubic polynomials to match their value and first and second derivatives at the boundaries (although the term is sometimes used even if just slopes are being matched). There are almost as many types of splines as there are applications, differing in how the constraints are imposed. Splines are not commonly used for data fitting or functional approximation, because they require finding and keeping track of the boundaries as well as the coefficients, but they are the workhorse of computer graphics, where they are used to represent curves and surfaces.

*Natural splines* pass through a given set of points, matching the derivatives as they cross the points. This is the most obvious way to define a spline, but it has the liability that moving one point affects all of the others. To be useful for an interactive application such as computer graphics it must be possible to make a local change and not have to recompute the entire curve. *Nonuniform B-splines* provide a clever solution to this problem [Foley et al., 1990]. The "B" part refers to the bases the polynomials provide, and the "nonuniform" part refers to the flexibility in defining how they join together.

A nonuniform B-spline curve is defined by

$$\vec{x}_i(t) = \vec{c}_{i-3}\varphi_{i-3}(t) + \vec{c}_{i-2}\varphi_{i-2}(t) \qquad (12.10)$$
$$+ \vec{c}_{i-1}\varphi_{i-1}(t) + \vec{c}_i\varphi_i(t) \quad (t_i \leq t < t_{i+1}) \ .$$

$t$ parameterizes the position along the curve. The $\vec{c}_i$'s are *control points*, defined at given values of $t$ called *knots*, $t_i$. The $\varphi_i$'s are the basis (or *blending*) functions that weight the control points. The basis functions are found recursively:

$$\varphi_i^{(1)}(t) = \begin{cases} 1 & (t_i \leq t < t_{i+1}) \\ 0 & \text{otherwise} \end{cases}$$
$$\varphi_i^{(2)}(t) = \frac{t - t_i}{t_{i+1} - t_i}\varphi_i^{(1)}(t) + \frac{t_{i+2} - t}{t_{i+2} - t_{i+1}}\varphi_{i+1}^{(1)}(t)$$
$$\varphi_i^{(3)}(t) = \frac{t - t_i}{t_{i+2} - t_i}\varphi_i^{(2)}(t) + \frac{t_{i+3} - t}{t_{i+3} - t_{i+1}}\varphi_{i+1}^{(2)}(t)$$
$$\varphi_i(t) \equiv \varphi_i^{(4)}(t) = \frac{t - t_i}{t_{i+3} - t_i}\varphi_i^{(3)}(t) + \frac{t_{i+4} - t}{t_{i+4} - t_{i+1}}\varphi_{i+1}^{(3)}(t) \ . \qquad (12.11)$$

The first line sets the basis function to zero except in intervals where the knots are increasing. The second linearly interpolates between these constants, the third does a linear interpolation between these linear functions to give quadratic polynomials, and the last line does one more round of linear interpolation to provide the final cubics. This hierarchy of linear interpolation preserves the original normalization, so that in each interval the basis functions sum to 1. This means that the curve is bounded by the polygon that has the control points as vertices, with the basis functions controlling how close it comes to them.

Figure 12.1 shows the construction of the basis functions for the knot sequence [1 2 3 4 5 6 7 8]. The curve will approach each control point in turn, but not quite

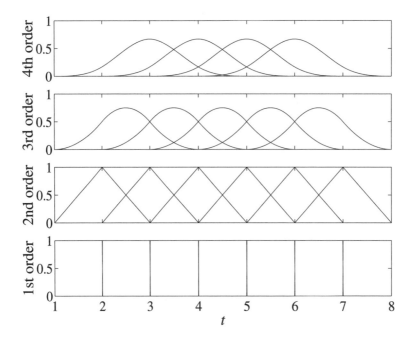

Figure 12.1. Construction of non-uniform B-spline basis functions for the knot sequence [1 2 3 4 5 6 7 8].

reach it because the maximum of the basis functions is less than 1. Figure 12.2 repeats the construction for the knot sequence [0 0 0 0 1 1 1 1]. The repetitions force the curve to go through the first and last control points, and the two intermediate control points set the slopes at the boundaries. In the context of computer graphics the B-spline is called a *Bézier curve*, and the basis functions are called *Bernstein polynomials*.

B-splines can be generalized to higher dimensions, parameterized by more than one degree of freedom. Another important extension is to divide the polynomials for each coordinate by a scale-setting polynomial. These are called *nonuniform rational B-splines*, or *NURBS*. Along with the other benefits of ratios of polynomials described in the last section, this is convenient for implementing transformations such as a change of perspective. Nonuniform nonrational B-splines are a special case with the divisor polynomial equal to 1.

## 12.2 ORTHOGONAL FUNCTIONS

In a polynomial fit the coefficients at each order are intimately connected. Dropping a high-order term gives a completely different function; it's necessary to re-fit to find a lower-order approximation. For many applications it's convenient to have a functional representation that lets successive terms be added to improve the agreement without changing the coefficients that have already been found. For example, if an image is

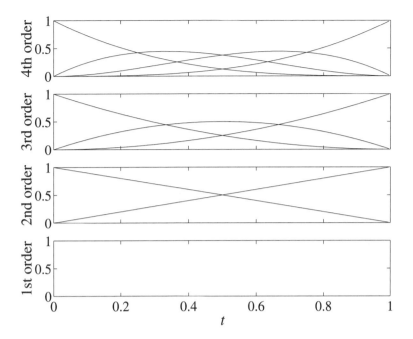

Figure 12.2. Construction of nonuniform B-spline basis functions for the knot sequence [0 0 0 0 1 1 1 1].

stored this way then the fidelity with which it is retrieved can be varied by changing the number of terms used based on the available display, bandwidth, and processor.

Expansions that let successive corrections be added without changing lower-order approximations are done with *orthogonal functions*, the functional analog to the orthogonal transformations that we saw in the last chapter. We have been writing an expansion of a function as a sum of scaled basis terms

$$y(\vec{x}) = \sum_{i=1}^{M} a_i f_i(\vec{x}) \tag{12.12}$$

without worrying about how the $f_i$'s relate to each other. However, if we choose them to be *orthonormal*

$$\int_{-\infty}^{\infty} f_i(\vec{x}) f_j(\vec{x}) \, d\vec{x} = \delta_{ij} \quad , \tag{12.13}$$

then the expansion coefficients can be found by projection:

$$\int_{-\infty}^{\infty} f_i(\vec{x}) y(\vec{x}) \, d\vec{x} = \int_{-\infty}^{\infty} f_i(\vec{x}) \sum_{j=1}^{M} a_j f_j(\vec{x}) \, d\vec{x}$$
$$= \sum_{j=1}^{M} a_j \delta_{ij}$$
$$= a_i \quad . \tag{12.14}$$

Such an expansion is useful if we can analytically or numerically do the integral on the left hand side of equation (12.14).

Assume that we're given an arbitrary *complete* set of basis functions $\varphi_i(\vec{x})$ that can represent any function of interest, for example polynomials up to a specified order. An orthonormal set $f_i$ can be constructed by *Gram–Schmidt orthogonalization*. This is exactly the same procedure used to construct orthogonal vectors, with the dot product between vectors replaced by an integral over the product of functions. The first orthonormal function is found by scaling the first basis function

$$f_1(\vec{x}) = \frac{\varphi_1(\vec{x})}{[\int_{-\infty}^{\infty} \varphi_1(\vec{x})\varphi_1(\vec{x})\, d\vec{x}]^{1/2}} \qquad (12.15)$$

so that it is normalized:

$$\int_{-\infty}^{\infty} f_1(\vec{x})f_1(\vec{x})\, d\vec{x} = \frac{\int_{-\infty}^{\infty} \varphi_1(\vec{x})\varphi_1(\vec{x})\, d\vec{x}}{\int_{-\infty}^{\infty} \varphi_1(\vec{x})\varphi_1(\vec{x})\, d\vec{x}} = 1 \quad . \qquad (12.16)$$

The next member of the set is found by subtracting off the $f_1$ component of $\varphi_2$

$$f_2'(\vec{x}) = \varphi_2(\vec{x}) - f_1(\vec{x}) \int_{-\infty}^{\infty} f_1(\vec{x})\varphi_2(\vec{x})\, d\vec{x} \qquad (12.17)$$

to produce a function that is orthogonal to $f_1$

$$\int_{-\infty}^{\infty} f_1(\vec{x})f_2'(\vec{x})\, d\vec{x} = \int_{-\infty}^{\infty} f_1(\vec{x})\varphi_2(\vec{x})\, d\vec{x}$$
$$- \underbrace{\int_{-\infty}^{\infty} f_1(\vec{x})f_1(\vec{x})\, d\vec{x}}_{1} \int_{-\infty}^{\infty} f_1(\vec{x})\varphi_2(\vec{x})\, d\vec{x}$$
$$= 0 \quad , \qquad (12.18)$$

and then normalizing it

$$f_2(\vec{x}) = \frac{f_2'(\vec{x})}{[\int_{-\infty}^{\infty} f_2'(\vec{x})f_2'(\vec{x})\, d\vec{x}]^{1/2}} \quad . \qquad (12.19)$$

The third one is found by subtracting off these first two components and normalizing:

$$f_3'(\vec{x}) = \varphi_3 - f_1 \int_{-\infty}^{\infty} f_1 \varphi_3\, d\vec{x} - f_2 \int_{-\infty}^{\infty} f_2 \varphi_3\, d\vec{x}$$
$$f_3(\vec{x}) = \frac{f_3'(\vec{x})}{[\int_{-\infty}^{\infty} f_3'(\vec{x})f_3'(\vec{x})\, d\vec{x}]^{1/2}} \quad , \qquad (12.20)$$

and so forth until the basis is constructed.

If a set of experimental observations $\{y_n, \vec{x}_n\}_{n=1}^{N}$ is available instead of a function form $y(\vec{x})$ then it is not possible to directly evaluate the projection integrals in equation (12.14). We can still use orthonormal functions in this case, but they must now be constructed to be orthonormal with respect to the unknown probability density $p(\vec{x})$ of the measurements (assuming that the system is stationary so that the density exists). If we use the same expansion (equation (12.12)) and now choose the $f_i$'s so that

$$\langle f_i(\vec{x})f_j(\vec{x}) \rangle = \int_{-\infty}^{\infty} f_i(\vec{x})f_j(\vec{x})p(\vec{x})\, d\vec{x} = \delta_{ij} \quad , \qquad (12.21)$$

then the coefficients are found by the expectation

$$\langle y(\vec{x})f_i(\vec{x})\rangle = \int_{-\infty}^{\infty} y(\vec{x})f_i(\vec{x})p(\vec{x})\,d\vec{x} \qquad (12.22)$$
$$= a_i \ .$$

By definition an integral over a distribution can be approximated by an average over variables drawn from the distribution, providing an experimental means to find the coefficients

$$a_i = \int_{-\infty}^{\infty} y(\vec{x})f_i(\vec{x})p(\vec{x})\,d\vec{x}$$
$$\approx \frac{1}{N}\sum_{n=1}^{N} y_n f_i(\vec{x}_n) \ . \qquad (12.23)$$

But where do these magical $f_i$'s come from? The previous functions that we constructed were orthogonal with respect to a flat distribution. To orthogonalize with respect to the experimental distribution all we need to do is replace the integrals with sums over the data,

$$f_1(\vec{x}) = \frac{\varphi_1(\vec{x})}{[\int_{-\infty}^{\infty} \varphi_1(\vec{x})\varphi_1(\vec{x})p(\vec{x})]^{1/2}\,d\vec{x}}$$
$$\approx \frac{\varphi_1(\vec{x})}{[\frac{1}{N}\sum_{n=1}^{N} \varphi_1(\vec{x}_n)\varphi_1(\vec{x}_n)]^{1/2}}$$
$$f_2'(\vec{x}) = \varphi_2(\vec{x}) - f_1(\vec{x})\int_{-\infty}^{\infty} f_1(\vec{x})\varphi_2(\vec{x})p(\vec{x})\,d\vec{x}$$
$$\approx \varphi_2(\vec{x}) - f_1(\vec{x})\frac{1}{N}\sum_{n=1}^{N} f_1(\vec{x}_n)\varphi_2(\vec{x}_n) \qquad (12.24)$$

and so forth. This construction lets functions be fit to data by evaluating experimental expectations rather than doing an explicit search for the fit coefficients. Further work can be saved if the starting functions are chosen to satisfy known constraints in the problem, such as using sin functions to force the expansion to vanish at the boundaries of a rectangular region.

Basis functions can of course be orthogonalized with respect to known distributions. For polynomials, many of these families have been named because of their value in mathematical methods, including the *Hermite polynomials* (orthogonal to with respect to $e^{-x^2}$), *Laguerre polynomials* (orthogonal with respect to $e^{-x}$), and the *Chebyshev polynomials* (orthogonal with respect to $(1-x^2)^{-1/2}$) [Arfken & Weber, 1995].

## 12.3 RADIAL BASIS FUNCTIONS

Even dressed up as a family of orthogonal functions, there is a serious problem inherent in the use of any kind of polynomial expansion: the only way to improve a fit is to add higher-order terms that diverge ever more quickly. Matching data that isn't similarly diverging requires a delicate balancing of the coefficients, resulting in a function that is increasingly "wiggly." Even if it passes near the training data it will be useless for interpolation or extrapolation. This is particularly true for functions that have isolated features, such as discontinuities or sharp peaks (Problem 12.2).

*Radial basis functions* (*RBFs*) offer a sensible alternative to the hopeless practice of fighting divergences with still faster divergences. The idea is to use as the basis a set of identical functions that depend only on the radius from the data point to a set of "representative" locations $\vec{c}_i$:

$$y(\vec{x}) = \sum_{i=1}^{M} f(|\vec{x} - \vec{c}_i|; \vec{a}_i) \quad . \tag{12.25}$$

The $\vec{a}_i$ are coefficients associated with the $i$th center $\vec{c}_i$. Now extra terms can be added without changing how quickly the model diverges, and the centers can be placed where they're needed to improve the fit in particular areas. And unlike splines the basis functions are defined everywhere, eliminating the need to keep track of element boundaries.

If the centers are fixed and the coefficients enter linearly then fitting an RBF becomes a linear problem

$$y = \sum_{i=1}^{M} a_i f(|\vec{x} - \vec{c}_i|) \quad . \tag{12.26}$$

Common choices for these $f$ include linear ($f(r) = r$), cubic ($f(r) = r^3$), and fixed-width Gaussian ($f(r) = \exp(-r^2)$). $f(r) = r^2$ is not included in this list because it is equivalent to a single global second-order polynomial ($|\vec{x} - \vec{c}|^2 = \sum_j [x_j - c_j]^2$). For example, fitting $f(r) = r^3$ requires a singular value decomposition to invert

$$\begin{pmatrix} |\vec{x}_1 - \vec{c}_1|^3 & |\vec{x}_1 - \vec{c}_2|^3 & \cdots & |\vec{x}_1 - \vec{c}_M|^3 \\ |\vec{x}_2 - \vec{c}_1|^3 & |\vec{x}_2 - \vec{c}_2|^3 & \cdots & |\vec{x}_2 - \vec{c}_M|^3 \\ |\vec{x}_3 - \vec{c}_1|^3 & |\vec{x}_3 - \vec{c}_2|^3 & \cdots & |\vec{x}_3 - \vec{c}_M|^3 \\ \vdots & \vdots & \cdots & \vdots \\ |\vec{x}_{N-1} - \vec{c}_1|^3 & |\vec{x}_{N-1} - \vec{c}_2|^3 & \cdots & |\vec{x}_{N-1} - \vec{c}_M|^3 \\ |\vec{x}_N - \vec{c}_1|^3 & |\vec{x}_N - \vec{c}_2|^3 & \cdots & |\vec{x}_N - \vec{c}_M|^3 \end{pmatrix} \begin{pmatrix} a_1 \\ a_2 \\ \vdots \\ a_M \end{pmatrix} = \begin{pmatrix} y_1 \\ y_2 \\ y_3 \\ \vdots \\ y_{N-1} \\ y_N \end{pmatrix}$$

to find the solution with the least squares error.

Linear and cubic basis functions appear to violate the spirit of locality of RBFs because they try to expand the unknown function in a set of globally diverging functions, which might seem to require the same kind of delicate cancellation that dooms a high-order polynomial expansion. This intuition misses the fact that all of the terms have the same order and so this can be a sensible thing to do. In fact, a possibly counter-intuitive result is that these diverging basis functions can have better error convergence properties than local ones [Powell, 1992]. They share the crucial attribute that if the data are rescaled so that $r \to \alpha r$, then $r^3 \to (\alpha r)^3 = \alpha^3 r^3$, simply changing the amplitude of the term, which we are going to fit anyway. But for a Gaussian $\exp(-r^2) \to \exp(-\alpha^2 r^2)$, changing the variance of the Gaussian which cannot be modified by the amplitude of the term.

There are many strategies for choosing the $\vec{c}_i$: randomly, uniformly, drawn from the data, uniformly on the support of the data (regions where the probability to see a point is nonzero). If the basis coefficients enter nonlinearly (for example, changing the variance of a Gaussian) then an iterative nonlinear search is needed and RBFs are closely related to the network architectures to be discussed in Section 12.6. If the centers are allowed to

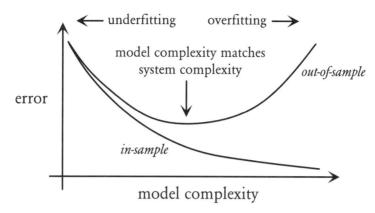

Figure 12.3. Cross-validation.

move (for example, by varying the mean of a Gaussian), RBFs become a particular case of the general theory of approximation by clustering (Chapter 14). The last decision in using an RBF is the number of centers $M$, to be discussed in the next section.

## 12.4 OVERFITTING

Overfitting is a simple idea with deep implications. Any reasonable model with $N$ free parameters should be able to exactly fit $N$ data points, but that is almost always a bad idea. The goal of fitting usually is to be able to interpolate or extrapolate. If the data are at all noisy, a model that passes through every point is carefully fitting idiosyncracies of the noise that do not generalize. On the other hand, if too few parameters are used, the model may be forced to not only ignore the noise but also miss the meaningful behavior of the data. Successful function fitting requires balancing *underfitting*, where there are *model mismatch errors* because the model can not describe the data, with *overfitting*, where there are *model estimation errors* from poor parameter choices in a flexible model. This is another example of a bias–variance tradeoff: you must choose between a reliable estimate of a biased model, or a poor estimate of a model that is capable of a better fit.

Unlike the parameters of a model that are determined by fitting, the *hyper-parameters* that control the architecture of a model must be determined by some kind of procedure above the fitting. *Cross-validation* is a simple quantitative way to do this. The idea is to fit your model on part of the data set, but evaluate the fit on another part of the data set that you have withheld (for example, by randomly choosing a subset of the data). Now repeat this procedure as you vary the complexity of the model, such as the number of RBF centers. The *in-sample* error will continue to decrease if the model is any good, but the *out-of-sample* error will initially decrease but then stop improving and possibly start increasing once the model begins to overfit (Figure 12.3). The best model is the one at the minimum of this curve (although the curve usually won't be this nice). Once this set of hyper-parameters has been found the model can be retrained on the full data set to make predictions for completely new values. The essential observation is that the only way to distinguish between a signal and noise, without any extra information, is by testing the model's generalization ability. Cross-validation is a sensible, useful way to do that, albeit with a great deal of associated debate about when and how to use it properly, e.g., [Goutte, 1997].

A strict Bayesian views (correctly) cross-validation as an *ad hoc* solution to finding hyperparameters. Better to use equation (10.1) and integrate over all possible models [Buntine & Weigend, 1991, Besag *et al.*, 1995]. This is an enormous amount of work, and requires specifying priors on all the variables in the problem. There are some problems where this effort is justified, but for the common case of there being many equally acceptable models cross-validation is a quick and easy way to find one of them.

Cross-validation does assume that there are enough data for both training and testing. While it would be nice to have unlimited amounts of both, that's rarely the case. *Bootstrap* methods provide an alternative for testing models on small data sets [Efron, 1983]. If we could consult an oracle and draw new data points from the unknown distribution that generated our original observations then we could produce as much data as we need. But, we assume that our data set was drawn from this distribution, so why not use it instead? Subsamples can be drawn from the data set and fit as if they were independent observations. This idea appears to be hopelessly circular when first encountered but does have a solid statistical foundation, although bootstrap methods do have the problem that mixing up the training and testing data can easily lead to overconfidence in a model's generalization ability.

Finally, there's an elegant theory of the descriptive power of models that can relate an architecture to the amount of data needed to use it, in advance of any measurements. The *Vapnik–Chervonenkis* (*VC*) dimension is equal to the size of the largest set of points that can be classified in all possible ways by a model, and for various architectures it can be used to find the worst-case scaling of the number of points required to learn a model to a given uncertainty [Vapnik & Chervonenkis, 1971, Kearns & Vazirani, 1994]. Unfortunately, as with so many other grand unifying principles in learning theory, the real-world problems where such guidance is most needed is where it is most difficult to apply.

## 12.5 CURSE OF DIMENSIONALITY

There are two broad classes of basis functions: those with linear coefficients

$$y(\vec{x}) = \sum_{i=1}^{M} a_i f_i(\vec{x}) \quad , \tag{12.27}$$

and those with coefficients that enter nonlinearly

$$y(\vec{x}) = \sum_{i=1}^{M} f_i(\vec{x}; \vec{a}_i) \quad . \tag{12.28}$$

The former has a single global least-squares minimum that can be found by singular value decomposition; in general the latter requires an iterative search with the attendant problems of avoiding local minima and deciding when to stop. It's impossible to find an algorithm that can solve an arbitrary nonlinear search problem in a single step because the function $f$ could be chosen to implement the mapping performed by a general-purpose computer from an input program $\vec{x}$ to an output $y$; if such an omniscient search algorithm existed it would be able to solve computationally intractable problems [Garey & Johnson, 1979].

For a nonparametric fit where we have no idea what the functional form should be, why would we ever want to use nonlinear coefficients? The answer is the *curse of dimensionality*. This ominous sounding problem is invoked by a few different incantations, all relating to serious difficulties that occur as the dimensionality of a problem is increased. Let's say that we wanted to do a second-order polynomial fit. In 1D this is simply

$$y = a_0 + a_1 x + a_2 x^2 \quad . \tag{12.29}$$

In 2D, we have

$$y = a_0 + a_1 x_1 + a_2 x_2 + a_3 x_1 x_2 + a_4 x_1^2 + a_5 x_2^2 \quad . \tag{12.30}$$

We've gone from three to six terms. In 3D we need ten terms:

$$\begin{aligned} y = a_0 + a_1 x_1 + a_2 x_2 + a_3 x_3 + a_4 x_1 x_2 + a_5 x_1 x_3 + a_6 x_2 x_3 \\ + a_7 x_1^2 + a_8 x_2^2 + a_9 x_3^2 \quad , \end{aligned} \tag{12.31}$$

and in $d$ dimensions we need $1 + 2d + d(d-1)/2$ terms. As $d$ is increased the quadratic $d^2$ part quickly dominates. If we try to get around this by leaving off some of the polynomial terms then we do not let those particular variables interact at that order. As we increase the order of the polynomial, the exponent of the number of terms required increases accordingly. A 10th-order fit in 10D will require on the order of $10^{10}$ terms, clearly prohibitive. This rapid increase in the number of terms required is an example of the curse of dimensionality. Because of this many algorithms that give impressive results for a low-dimensional test problem fail miserably for a realistic high-dimensional problem.

The reason to use nonlinear coefficients is to avoid the curse. Given some mild assumptions, a nonlinear function of the form of equation (12.27) will have a typical approximation error of $\mathcal{O}(1/M^{2/d})$ when fitting an arbitrary but reasonable function (one with a bound on the first moment of the Fourier transform), where $M$ is the number of terms and $d$ is the dimension of the space, while a linear function of the form of equation (12.28) will have an error of order $\mathcal{O}(1/M)$ [Barron, 1993]. In low dimensions the difference is small, but in high dimensions it is essential: exponentially more terms are needed if linear coefficients are used. The nonlinear coefficients are much more powerful, able to steer the basis functions where they are needed in a high-dimensional space. Therefore, in low dimensions it's crazy not to use linear coefficients, while in high dimensions it is crazy to use them.

There is another sense of the curse of dimensionality to be mindful of. In a high-dimensional space, everything is surface. To understand this consider the volume of a hyper-sphere of radius $r$ in a $d$-dimensional space:

$$V(r) = \frac{\pi^{d/2} r^d}{(d/2)!} \tag{12.32}$$

(noninteger factorials are defined by the gamma function $\Gamma(n+1) = n!$). Now let's look at the fraction of volume in a shell of thickness $\epsilon$, compared to the total volume of the sphere, in the limit that $d \to \infty$:

$$\frac{V(r) - V(r-\epsilon)}{V(r)} = \frac{r^d - (r-\epsilon)^d}{r^d}$$

$$= 1 - \underbrace{\left(1 - \frac{\epsilon}{r}\right)^d}_{\lim_{d \to \infty} = 0}$$

$$= 1 \quad . \tag{12.33}$$

As $d \to \infty$ all of the volume is in a thin shell near the surface. This is because as $d$ is increased there is more surface available. Now think about what this implies for data collection. "Typical" points come from the interior of a distribution, not near the edges, but in a high-dimensional space distributions are essentially all edges. This means that in a huge data set there might not be a single point in the interior of the distribution, and so any analysis will be dominated by edge effects. Most obviously, the data will appear to be drawn from a lower-dimensional distribution [Gershenfeld, 1992]. Even if your basis functions work in high dimensions you may not be able to collect enough data to use them.

## 12.6 NEURAL NETWORKS

The study of *neural networks* started as an attempt to build mathematical models that worked in the same way that brains do. While biology is so complex that such explicit connections between computation *in vivo* and *in silico* have been hard to make outside of specialized areas [Richards, 1988], the effort to do so has lead to a powerful language for using large flexible nonlinear models. The spirit of neural network or *connectionist* modeling is to use fully nonlinear functions (to handle the curse of dimensionality), and use a large number of terms (so that model mismatch errors are not a concern). Instead of matching the architecture of the model to a problem, a model is used that can describe almost anything, and careful training of the model is used to constrain it to describe the data.

The input to a typical neural network is a vector of elements $\{x_k\}$ (Figure 12.4). These are combined by a series of linear filters with weights $w_{jk}$ to give the inputs to the *hidden units*

$$h_j = \sum_k w_{jk} x_k \quad . \tag{12.34}$$

Next comes the nonlinearity: these inputs are passed through a layer of *activation functions* $g(h_j)$ to give the output

$$X_j = g(h_j) = g\left(\sum_k w_{jk} x_k\right) \quad . \tag{12.35}$$

The activation or "squashing" functions are usually chosen to clip for large magnitudes to keep the response bounded; common choices are

$$g(h) = \frac{1}{1 + e^{-2\beta h}} \tag{12.36}$$

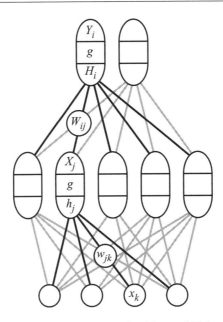

Figure 12.4. A neural network with one hidden layer.

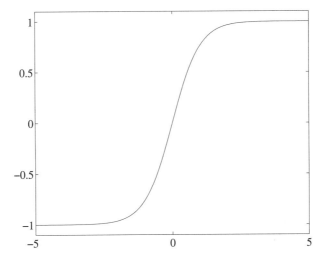

Figure 12.5. tanh function.

or

$$g(h) = \tanh(\beta h) = \frac{e^{\beta h} - e^{-\beta h}}{e^{\beta h} + e^{-\beta h}} \tag{12.37}$$

(shown in Figure 12.5). $\beta$ controls the steepness; typical values are 1/2 or 1. If it is too steep the neural network becomes a binary function and loses its ability to do continuous optimizations; if it is too flat then the network is linear. It's also possible to use other nonlinear functions such as Gaussians.

The outputs from the hidden units can then go through another layer of filters,

$$H_i = \sum_j W_{ij} X_j = \sum_j W_{ij}\, g\left(\sum_k w_{jk} x_k\right) \quad, \tag{12.38}$$

and be fed through another layer of squashing functions to finally produce the outputs

$$Y_i = g(H_i) = g\left[\sum_j W_{ij}\, g\left(\sum_k w_{jk} x_k\right)\right] \quad. \tag{12.39}$$

If the network is being used for regressing variables instead of classifying features, linear output functions should be used instead [Rumelhart et al., 1996].

This model is very general. With no hidden units such a network can classify only linearly separable problems (ones for which the possible output values can be separated by global hyperplanes). These are called *perceptrons*, and this negative result led people to assume that nonlinear networks are not useful for general tasks [Minsky & Papert, 1988]. However, it has since been shown that with one hidden layer a network can describe any continuous function (if there are enough hidden units), and that with two hidden layers it can describe any function at all [Kolmogorov, 1957, Cybenko, 1989]. For local nonlinear basis functions such as Gaussian RBFs a single layer suffices [Hartman et al., 1990].

### 12.6.1 Back Propagation

The purpose of training a neural network is to find coefficients that reduce the error between the set of outputs and given test data $y_i(\vec{x}_n)$. This is usually done by minimizing the least squares error

$$\chi^2 = \frac{1}{2}\sum_n \sum_i [y_i(\vec{x}_n) - Y_i(\vec{x}_n)]^2$$

$$= \frac{1}{2}\sum_n \sum_i \left[y_i(\vec{x}_n) - g\left(\sum_j W_{ij}\, g\left(\sum_k w_{jk} x_{n,k}\right)\right)\right]^2 \quad. \tag{12.40}$$

One way to reduce $\chi^2$ is to use gradient descent. The update step in the output weights can be found by differentiating ($\eta$ is a scale factor that controls how big the step is):

$$\Delta W_{ij} = -\eta \frac{\partial \chi^2}{\partial W_{ij}}$$

$$= \eta \sum_n [y_i(\vec{x}_n) - Y_i(\vec{x}_n)] g'(H_i) X_j$$

$$\equiv \eta \sum_n \Delta_i X_j \quad, \tag{12.41}$$

with the definition

$$\Delta_i = [y_i(\vec{x}_n) - Y_i(\vec{x}_n)] g'(H_i) \quad. \tag{12.42}$$

The update in the input weights can be found from the chain rule:

$$\Delta w_{jk} = -\eta \frac{\partial \chi^2}{\partial w_{jk}}$$
$$= \eta \sum_n \sum_i [y_i(\vec{x}_n) - Y_i(\vec{x}_n)] \, g'(H_i) W_{ij} \, g'(h_j) x_{n,k}$$
$$= \eta \sum_n \sum_i \Delta_i W_{ij} \, g'(h_j) x_k$$
$$\equiv \eta \sum_x \delta_j x_k \quad , \tag{12.43}$$

defining

$$\delta_j = g'(h_j) \sum_i W_{ij} \Delta_i \quad . \tag{12.44}$$

The deltas for the input layer are found in terms of the deltas for the output layer by running them backwards through the network $W_{ij}$'s. This is straightforward to generalize to networks with more than one hidden layer. Training a network by gradient descent, feeding the errors backwards through the network like this, is called *back propagation*. In the last chapter we saw that gradient descent slows down near minima, so the Levenberg–Marquardt method can be used to improve the convergence. The search can easily get stuck in local minima; some easy improvements are to add momentum so that it rolls out of small wells, or add some random fluctuations to help kick it out of minima. These ideas will be discussed in the next chapter, as well as why such a simple-minded search in the high-dimensional space of weights works so surprisingly well.

We have been discussing *feed-forward* networks, in which the inputs control the outputs. Just as a finite impulse response filter can be generalized to an infinite impulse response filter by letting the outputs influence the inputs, in a *recurrent network* the outputs are fed back to the inputs. This is a good thing to do if the net should learn to model such feedback behavior. Much of the art in using neural networks lies in choosing an architecture that reflects the structure in a problem. For example, for *multistationary* data that switch among different regimes it's advantageous to train a collection of networks that are forced to specialize in particular regimes [Weigend *et al.*, 1995].

## 12.7 REGULARIZATION

In neural networks, model complexity was originally controlled by *early stopping*. This uses a large network, and continues training it through back propagation until the out-of-sample predictions degrade. In effect, degrees of freedom are introduced by the training. Rather than match the form of the model to the data *a priori*, an enormous model is used that can describe anything, and it is constrained by the training. In fact, it is common for neural networks to have more parameters than the size of the training data set! Unlike a polynomial fit, the network can choose from a much larger space of possible models. The order of a polynomial restricts how many variables can interact; in a neural network all of the variables can always interact (although some of the connections will decay during training).

Early stopping had some early successes, but does not provide a way to guide the training with advance knowledge. A more general framework for constraining flexible models is provided by the powerful concept of *regularization* [Weigend et al., 1990, Girosi et al., 1995]. This is the name for the knob that any good algorithm should have that controls a trade-off between matching the data and enforcing your prior beliefs. As we saw in Section 10.1, maximum *a posteriori* estimation requires finding

$$\max_{\text{model}} p(\text{model}|\text{data}) = \max_{\text{model}} \frac{p(\text{data}|\text{model}) \, p(\text{model})}{p(\text{data})} \quad , \tag{12.45}$$

or taking logarithms

$$\max_{\text{model}} \log p(\text{model}|\text{data}) = \log p(\text{data}|\text{model})$$
$$+ \log p(\text{model}) - \log p(\text{data}) \quad . \tag{12.46}$$

The first term on the right hand side measures how well the model matches the data. This could be a Gaussian error model, or it might be defined to be the profit in a stock trading model. The second term, called a *prior*, expresses advance beliefs about what kind of model is reasonable. The final term comes in only if training is done over multiple data sets. Taken together, these components define a *cost function* or *penalty function* to be searched to find the best model (using methods to be covered in the next chapter).

If there truly is no prior knowledge at all, then setting $p(\text{model}) = 1$ reduces to maximum likelihood estimation. But that's rarely the case, and even simple priors have great value. Although priors can be formally expressed as a probability distribution, they're usually just added to the cost function with a Lagrange multiplier. A common prior is the belief that the model should be locally linear unless the data forces it to do otherwise; this is measured by the integral of the square of the curvature of the model. If a least squares (Gaussian) error model is used then the quantity to be minimzed for a 1D problem is

$$I = \sum_{n=1}^{N}[y_n - f(x_n, \vec{a})]^2 + \lambda \int \left[\frac{d^2 f}{dx^2}\right]^2 dx \quad , \tag{12.47}$$

where $\vec{a}$ is the set of coefficients to be determined and the bounds of integration are over the interval being fit. The left hand term gives the agreement between the data and the model, and the term on the right depends on the model alone. Making the sum of both terms extremal by solving $\partial I/\vec{a} = 0$ (Chapter 4) for a particular value of $\lambda$ gives a set of equations to be solved to find the corresponding $\vec{a}$. The best value for $\lambda$ can then be found iteratively by a procedure such as cross-validation. Problem 12.2 gives an example of the use of this regularizer.

Other common regularizers include the entropy of a distribution, expressing the belief that the distribution should be flat, and the total magnitude of a model's coefficients, seeking the most parsimonious model. Regularization plays a central role in the theory of *inverse problems* [Parker, 1977, Engl et al., 1996], the essential task that arises throughout science and engineering of deducing the state of a system from a set of measurements that underdetermine it.

Flexible models such as neural networks do have an impressive ability to find a useful model from a huge space of possibilities, but they aren't magic, and they don't replace the need to think [Weigend & Gershenfeld, 1993]. Clever training alone is not a cure for bad data. In practical applications much of the progress comes from collecting good training data sets, finding appropriate features to use for the model inputs and outputs, and imposing domain-specific constraints on the model to improve its ability to generalize from limited observations. Salvation lies in these details. Given unlimited data and unlimited resolution almost any reasonable model should work with data in whatever form you present it, but in the real world of imperfect measurements, insight into model architectures must be enhaced by a combination of advance analysis of a system and experimentation with the data.

## 12.8 SELECTED REFERENCES

[Hertz et al., 1991] Hertz, John A., Krogh, Anders S., & Palmer, Richard G. (1991). *Introduction to the Theory of Neural Computation.* Redwood City, CA: Addison-Wesley.

[Bishop, 1996] Bishop, Christopher M. (1996). *Neural Networks for Pattern Recognition.* Oxford: Oxford University Press.

Excellent introductions to neural networks.

## 12.9 PROBLEMS

(12.1) Find the first five *diagonal* Padé approximants $[1/1], \ldots, [5/5]$ to $e^x$ around the origin. Remember that the numerator and denominator can be multiplied by a constant to make the numbers as convenient as possible. Evaluate the approximations at $x = 1$ and compare with the correct value of $e = 2.718281828459045$. How is the error improving with the order?

(12.2) Take as a data set $x = \{-10, -9, \ldots, 9, 10\}$, and $y(x) = 0$ if $x \leq 0$ and $y(x) = 1$ if $x > 0$.

(a) Fit the data with a polynomial with 5, 10, and 15 terms, using a pseudo-inverse of the Vandermonde matrix (such as Matlab's *pinv* function).

(b) Fit the data with 5, 10, and 15 $r^3$ RBFs uniformly distributed between $x = -10$ and $x = 10$.

(c) Using the coefficients found for these six fits, evaluate the total out-of-sample error at $x = \{-10.5, -9.5, \ldots, 9.5, 10.5\}$.

(d) Using a 10th-order polynomial, fit the data with the curvature regularizer in equation (12.47), and plot the fits for $\lambda = 0, 0.01, 0.1, 1$ (this part is harder than the others).

# 13 Optimization and Search

Once you've gathered your data, selected a representation to work with, chosen an architecture for functional approximation, specified an error metric, and expressed your prior beliefs about the model, then comes the essential step of choosing the best parameters. If they enter linearly, the best global values can be found in one step with a singular value decomposition, but as we saw in the last chapter coping with the curse of dimensionality usually requires parameters to be inside nonlinearities. This entails an iterative search starting from an initial guess. Such exploration is similar to the challenge faced by a mountaineer in picking a route up a demanding peak, but with two essential complications: the search might be in a 200-dimensional space instead of just 2D, and because of the cost of function evaluation you must do the equivalent of climbing while looking down at your feet, using only information available in a local neighborhood.

The need to search for parameters to make a function extremal occurs in many kinds of optimization. We already saw one nice way to do nonlinear search, the Levenberg–Marquardt method (Section 10.4.1). But this is far from the end of the story. Levenberg–Marquardt assumes that it is possible to calculate both the first and second derivatives of the function to be minimized, that the starting parameter values are near the desired extremum of the cost function, and that the function is reasonably smooth. In practice these assumptions often do not apply, and so in this chapter we will look at ways to relax them.

Throughout, we will assume that the goal is to find an acceptable solution rather than a provably optimal one. The latter is of course much more work, extra effort that is pointless if there are many solutions that are equally good, particularly if other sources of uncertainty such as measurement noise are larger than the differences among the solutions. If an optimal answer really is needed then all possible ones much be checked, although in many problems as the search proceeds it's possible to prune parts of the space that can be shown to be worse than the current best solution (the *Branch-and-Bound* algorithm, [Lawler & Wood, 1966]).

Perhaps because nature must solve these kinds of optimization problems so frequently, the algorithms in this chapter share a kind of natural familiarity missing in most numerical methods. We will start with blobs (for the downhill simplex search), add mass (momentum for avoiding local minima), then temperature (via simulated annealing), and finally reproduction (with genetic algorithms). The cost of this kind of intuition is rigor. While there is some supporting theory, the real justification for this collection of algorithms is their empirical performance. Consequently it's important to view them not as fixed received wisdom, but rather as a framework to guide further exploration.

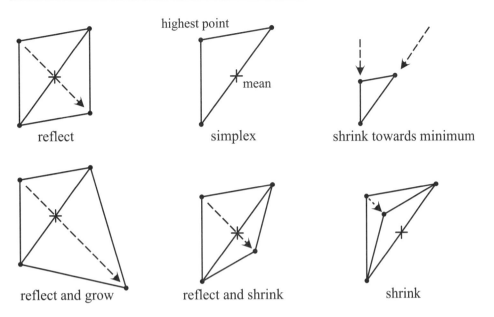

Figure 13.1. Downhill simplex update moves.

## 13.1 MULTIDIMENSIONAL SEARCH

As features are added to a model it quickly becomes inconvenient, if not impossible, to analytically calculate the partial derivatives with respect to the parameters (impossible because the function being searched might itself be the result of a numerical simulation). A simple-minded solution is to evaluate the function at a constellation of points around the current location and use a finite difference approximation to find the partials, but the number of function evaluations required at each step rapidly becomes prohibitive in a high-dimensional space. After each function evaluation there should be some kind of update of the search to make sure that the next function evaluation is as useful as possible. Each function evaluation requires choosing not only the most promising direction but also the length scale to be checked. In Levenberg–Marquardt we used the Hessian to set the scale, but doing that numerically would require still more function evaluations. The *Downhill Simplex* method, also called the *Nelder–Mead* method after its original authors [Nelder & Mead, 1965], is a delightful (although far from optimal) solution to these problems. Of all algorithms it arguably provides the most functionality for the least amount of code, along with the most entertainment.

A *simplex* is a geometrical figure that has one more vertex than dimension: a triangle in 2D, a tetrahedtron in 3D, and so forth. Nelder–Mead starts by choosing an initial simplex, for example by picking a random location and taking unit vectors for the edges, and evaluating the function being searched at the $D + 1$ vertices of the simplex. Then an iterative procedure attempts to improve the vertex with the highest value of the function at each step (assuming that the goal is minimization; for maximization the vertex with the smallest value is updated).

The moves are shown in Figure 13.1. Let $\vec{x}_i$ by the location of the $i$th vertex, ordered so that $f(\vec{x}_1) > f(\vec{x}_2) > \ldots > f(\vec{x}_{D+1})$. The first step is to calculate the center of the face of the simplex defined by all of the vertices other than the one we're trying to improve:

$$\vec{x}_{mean} = \frac{1}{D} \sum_{i=2}^{D+1} \vec{x}_i \quad . \tag{13.1}$$

Since all of the other vertices have a better function value it's a reasonable guess that they give a good direction to move in. Therefore the next step is to reflect the point across the face:

$$\begin{aligned} \vec{x}_1 \to \vec{x}_1^{new} &= \vec{x}_{mean} + (\vec{x}_{mean} - \vec{x}_1) \\ &= 2\vec{x}_{mean} - \vec{x}_1 \quad . \end{aligned} \tag{13.2}$$

If $f(\vec{x}_1^{new}) < f(x_{D+1})$ the new point is now the best vertex in the simplex and the move is clearly a good one. Therefore it's worth checking to see if it's even better to double the size of the step:

$$\begin{aligned} \vec{x}_1 \to \vec{x}_1^{new} &= \vec{x}_{mean} + 2(\vec{x}_{mean} - \vec{x}_1) \\ &= 3\vec{x}_{mean} - 2\vec{x}_1 \quad . \end{aligned} \tag{13.3}$$

If growing like this gives a better function value than just reflecting we keep the move, otherwise we go back to the point found by reflecting alone. If growing does succeed it's possible to try moving the point further still in the same direction, but this isn't done because it would result in a long skinny simplex. For the simplex to be most effective its size in each direction should be appropriate for the scale of variation of the function, therefore after growing it's better to go back and improve the new worse point (the one that had been $\vec{x}_2$).

If after reflecting $\vec{x}_1^{new}$ is still the worst point it means that we overshot the minimum of the function. Therefore instead of reflecting and growing we can try reflecting and shrinking:

$$\begin{aligned} \vec{x}_1 \to \vec{x}_1^{new} &= \vec{x}_{mean} + \frac{1}{2}(\vec{x}_{mean} - \vec{x}_1) \\ &= \frac{3}{2}\vec{x}_{mean} - \frac{1}{2}\vec{x}_1 \quad . \end{aligned} \tag{13.4}$$

If $f(\vec{x}_1^{new}) < f(\vec{x}_2)$ we accept the move and try to improve $f(\vec{x}_2)$; if after reflecting and shrinking $f(\vec{x}_1^{new})$ is still worse we can try just shrinking:

$$\begin{aligned} \vec{x}_1 \to \vec{x}_1^{new} &= \vec{x}_{mean} - \frac{1}{2}(\vec{x}_{mean} - \vec{x}_1) \\ &= \frac{1}{2}(\vec{x}_{mean} + \vec{x}_1) \quad . \end{aligned} \tag{13.5}$$

If after shrinking $f(\vec{x}_1^{new})$ is still worse the conclusion is that the moves we're making are too big to find the minimum, so we give up and shrink all of the vertices towards the best one:

$$\begin{aligned} \vec{x}_i \to \vec{x}_i^{new} &= \vec{x}_i - \frac{1}{2}(\vec{x}_i - \vec{x}_{D+1}) \quad (i = 1, \ldots, D) \\ &= \frac{1}{2}(\vec{x}_i + \vec{x}_{D+1}) \quad . \end{aligned} \tag{13.6}$$

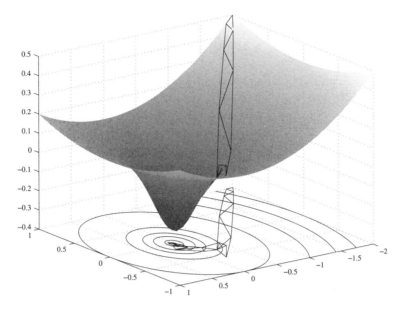

Figure 13.2. Downhill simplex function minimization. Above is the evolution of the simplex shown on the surface; below the 2D projection on a contour map.

Taken together these moves result in a surprisingly adept simplex. As it tumbles downhill it grows and shrinks to find the most advantageous length scale for searching in each direction, squeezing through narrow valleys where needed and racing down smooth regions where possible. When it reaches a minimum it will give up and shrink down around it, triggering a stopping decision when the values are no longer improving. The beauty of this algorithm is that other than the stopping criterion there are no adjustable algorithm parameters to set. Figure 13.2 shows what a simplex looks like searching for the minimum of a function.

A simplex search takes on a particularly simple form in 1D, where the simplex is just a pair of points. Each step takes the higher point and moves it around the lower point (in 1D this can be optimized by using triples of points with scale factors based on the golden mean [Press et al., 1992]). A 1D search is useful in more than 1D, because it can be used to find the minimum of a function in a given direction, $\min_\alpha f(\vec{x}_0 + \alpha \hat{x})$. This is called *line minimization*.

Performing a series of line minimizations provides another way to do multi-dimensional search. A naive way to do this is to cycle through the axes of the space, minimizing along each direction in turn. This does not work well, because each succeeding minimization can influence the optimizations that came before it, leading to a large number of zigs and zags to follow the function along directions that might not be lined up with the axes of the space. *Powell's method* improves on this idea by updating the directions that are searched to try to find a set of directions that don't interfere with each other [Acton, 1990]. Starting from an initial guess $\vec{x}_0$, $D$ line minimizations are performed, initially along each of the $D$ axes of the space. The resulting point $\vec{x}_1$ defines a change vector, $\Delta \vec{x} = \vec{x}_1 - \vec{x}_0$. This is a good direction to keep for future minimizations, assuming that moving in that direction again is advantageous, which is easily checked.

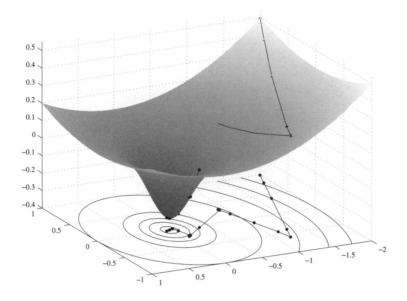

Figure 13.3. Minimization by Powell's method in 2D, illustrated by starting with random search directions. After each pair of line minimizations the net change is taken as a new search direction, replacing the most similar previous direction.

If it is, $\Delta \vec{x}$ is added to the set of directions used for minimization. Since we need to keep a set of $D$ directions to span the $D$-dimensional space, the new direction should replace the one most similar to it. This could be determined by computing all the dot products among the vectors, or more simply estimated by throwing out the direction in which there was the largest change in $f$ (and hence the direction which most likely made the most important contribution to $\Delta \vec{x}$). An example is shown in Figure 13.3. Notice how each line minimization ends when it is tangent to the function contours, and that the algorithm quickly reduces the overlap between the search directions. If the gradient of the function is available, *conjugate gradient* algorithms improve on Powell's method by explicitly constructing a set of directions that will be locally noninterfering.

Multidimensional search has a surprising use when applied to the simple quadratic function $f(\vec{x}) = \vec{x}^T \cdot \mathbf{M} \cdot \vec{x}/2 - \vec{x}^T \cdot \vec{v}$. At the minimum the gradient vanishes,

$$0 = \nabla f(\vec{x}) = \mathbf{M} \cdot \vec{x} - \vec{v}$$
$$\Rightarrow \vec{x} = \mathbf{M}^{-1} \cdot \vec{v} \quad . \tag{13.7}$$

We've inverted the matrix! For a dense matrix the effort to perform the search does not offer an improvement over more conventional techniques, but for a very large sparse matrix this can be the most efficient way to find a solution [Golub & Van Loan, 1996].

Finally, if the function being minimized is associated with a physical system, an analog circuit can do a multidimensional search by filtering a noisy evaluation [Kirk *et al.*, 1993].

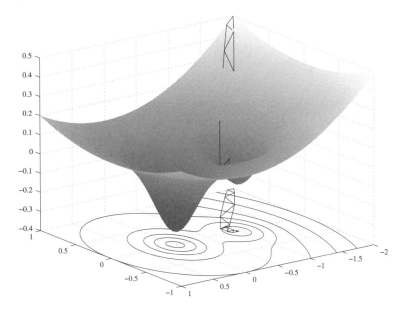

Figure 13.4. Downhill simplex function minimization getting caught in a local mininum.

## 13.2 LOCAL MINIMA

In functional optimization, like life, sometimes taking short-term gains does not lead to the best long-term return. Most nontrivial functions have local minima that will trap an unwary algorithm that always accepts a step that improves the function. Consider Figure 13.4, which shows the behavior of downhill simplex search on a function with two minima. Since by definition it always makes moves that improve the local value of the function it necessarily gets caught in the basin around the starting point.

All techniques for coping with local minima must in some way force the search to sometimes make moves that are not locally desirable, a numerical analog of learning from mistakes. A crude but eminently usable way to do this is by adding *momentum* to the search. Just as momentum causes a particle to keep moving in an old direction as a new force is applied, a fraction of whatever update was last done in a search can be added to whatever new update the algorithm deems best. This is usually done by loose analogy rather than by formally mapping the problem onto classical mechanics.

For downhill simplex, let $\Delta \vec{x}(n)$ be the total change in the $n$th move to the worst point in the simplex, $\vec{x}_1^{new}(n) = \vec{x}_1^{old}(n) + \Delta \vec{x}(n)$. Momentum adds to this some of the change from the previous move

$$\vec{x}_1^{new}(n) = \vec{x}_1^{old}(n) + \Delta \vec{x}(n) + \alpha [\vec{x}_1^{new}(n-1) - \vec{x}_1^{old}(n-1)] \quad . \tag{13.8}$$

As Figure 13.5 shows, this lets the simplex "roll" out of a local minima and find a better one. As $\alpha$ is increased the simplex is able to climb out of deeper minima, but takes longer to converge around the final minimum.

The attraction of using momentum is that it is a trivial addition to almost any other

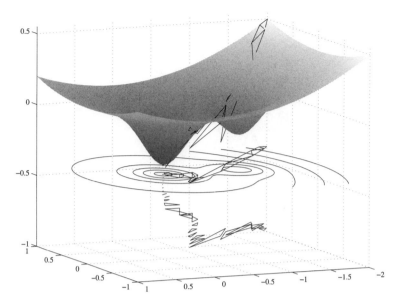

Figure 13.5. Downhill simplex function minimization with momenum added, avoiding a local minimum. At the bottom the updates are shown displaced vertically as a function of the iteration for clarity.

algorithm. Of course including momentum does not guarantee that local minima will be avoided; that depends on the details of how much is used, what the configuration of the search space is, and where the initial condition is located. As more momentum is included it is possible to climb out of deeper minima, and the reasonable hope is that the resulting directions in which the search proceeds will be sufficiently randomized to have a good chance of finding better minima. This is making an implicit statistical assumption about the dynamics of the search, in effect asking that there be enough redirection of the search for it to appear ergodic and hence be able to reach the whole search space. In the rest of this chapter we will turn to explicit statistical assumptions to randomize searches.

## 13.3 SIMULATED ANNEALING

The growth of a crystal from a liquid solves a difficult optimization problem. If the liquid was instantaneously frozen, the atoms would be trapped in the configuration they had in the liquid state. This arrangement has a higher energy than the crystalline ordering, but there is an energy barrier to be surmounted to reach the crystalline state from a glassy one. If instead the liquid was slowly cooled, the atoms would start at a high temperature exploring many local rearrangements, and then at a lower temperature become trapped in the lowest energy configuration that was reached. The slower the cooling rate, the more likely it is that they will find the ordering that has the lowest global energy. This process of slowly cooling a system to eliminate defects is called *annealing*.

Annealing was introduced to numerical methods in the 1950s by Metropolis and coworkers [Metropolis et al., 1953]. They were studying statistical mechanical systems and wanted to include thermodynamic fluctuations. In thermodynamics, the relative probability of seeing a system in a state with an energy $E$ is proportional to an exponential of the energy divided by the temperature times Boltzmann's constant $k$ [Balian, 1991]:

$$p(E) \propto e^{-E/kT} \equiv e^{-\beta E} \quad . \tag{13.9}$$

This is called a *Boltzmann factor*. At $T = 0$ this means that the system will be in the lowest energy state, but at $T > 0$ there is some chance to see it in any of the states. Metropolis proposed that to update a simulation, a new state reachable from the current state be randomly selected and have its energy evaluated. If the energy is lower, the state should always be accepted. But if the energy is higher, it is accepted based on the probability to see a fluctuation of that size, which is proportional to the exponential of the change in energy $\exp(-\beta \Delta E)$. This is easily implemented by accepting the move if a random number drawn from a uniform distribution between 0 and 1 is less than a threshold value equal to the desired probability.

In the 1980s Scott Kirkpatrick realized that the same idea could be used to search for solutions to other hard problems, such as finding a good routing of the wires on a printed circuit board [Kirkpatrick et al., 1983, Kirkpatrick, 1984]. This idea is called by analogy *simulated annealing*. Now the energy is replaced by a cost function, such as the total length of wire on the circuit board, or a MAP likelihood estimate. Trial moves that update the state are evaluated. Moves that improve the search are always accepted, and moves that make it worse are taken with a probability given by a Boltzmann factor in the change in the cost function. At a high temperature this means that the search indiscriminately accepts any move offered, and hence pays little attention to the function being searched and tumbles around the space. As the temperature is lowered the function becomes more and more significant, until as $T \rightarrow 0$ the search becomes trapped in the lowest minima that it has reached. Figure 13.6 shows simulated annealing working on a function with multiple minima.

There are two key elements in implementing simulated annealing. The first is the selection of the trial moves to be evaluated. This could be done randomly, but that makes little sense since in a high-dimensional space the chance of guessing a good direction is very small. Better is to use an algorithm such as downhill simplex or conjugate gradient that makes intelligent choices for the updates, but include the Boltzmann factor to allow it to make mistakes. The second part of implementing simulated annealing is choosing the cooling schedule. Cool too fast and you freeze the system in a bad solution; cool too slowly and you waste computer time. There is a large literature on techniques motivated by thermodynamics that analyze the fluctuations to determine if the system is being kept near equilibrium. A more modest empirical approach is to try runs at successively slower cooling rates until the answer stops improving.

Simulated annealing requires repeatedly making a probabilistic decision to accept a move proportional to its Boltzmann factor. If $\alpha \in [0, 1]$ is the probability with which we want to accept a move, this can be done simply by drawing a random number uniformly distributed between 0 and 1. If the number is less than $\alpha$ we accept the move, and if the number is greater than $\alpha$ we reject it. By definition, this gives a probability of success of $\alpha$, and for failure of $1 - \alpha$.

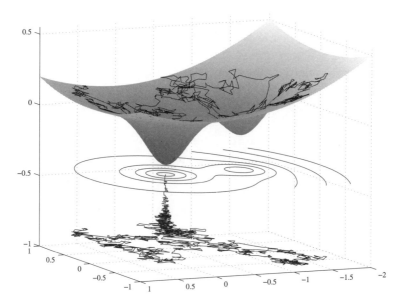

Figure 13.6. Optimization by simulated annealing. As the temperature is decreased, the random walker gets caught in the mininum. In the lower trace the iterations are displaced vertically.

## 13.4 GENETIC ALGORITHMS

There is a natural system that solves even harder problems than crystal growth: evolution. Atoms in a crystal have a very limited repertoire of possible moves compared to the options of, say, a bird in choosing how to fly. The bird must determine the size and shape of the wings, their structural and aerodynamic properties, the control strategy for flight, as well as all of the other aspects of its life-cycle that support its ability to fly.

Evolution works by using a large population to explore many options in parallel, rather than concentrating on trying many changes around a single design. The same can be true of numerical methods. Simulated annealing keeps one set of search parameters that are repeatedly updated. An alternative is to keep an ensemble of sets of parameters, spending less time on any one member of the ensemble. Techniques that do this have come to be called *genetic algorithms*, or *GAs*, by analogy with the evolutionary update of the genome [Forrest, 1993].

The state of a GA is given by a population, with each member of the population being a complete set of parameters for the function being searched. The whole population is updated in generations. There are four steps to the update:

- *Fitness*
  The fitness step evaluates the function being searched for the set of parameters for each member of the population.
- *Reproduction*
  The members of the new population are selected based on their fitness. The total size of the population is fixed, and the probability for a member of the previous generation to appear in the new one is proportional to its fitness. A set of parameters with a

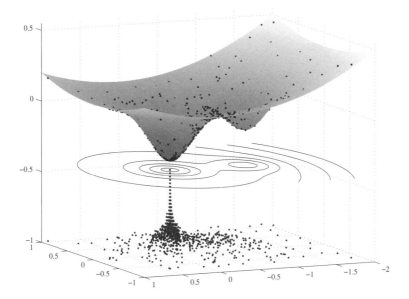

Figure 13.7. Optimization by a genetic algorithm, with populations displaced vertically at the bottom.

low fitness might disappear, and one with a high fitness can be duplicated many times. The weighting of how strongly the fitness determines the relative rates of reproduction is analogous to the temperature of simulated annealing. Low selectivity accepts any solution; high selectivity forces one solution to dominate.

- *Crossover*

  In crossover, members of the ensemble can share parameters. After two parents are randomly chosen based on their fitness, the offspring gets its parameter values based on some kind of random selection from the parents. The usefulness of crossover depends on the nature of the function being searched. If it naturally decouples into subproblems, then one parent may be good at one part of the problem and the other at another, so taking a block of parameters from each can be advantageous. If on the other hand the parameter values are all intimately linked then crossover has little value and can be skipped. Crossover introduces to the search process a notion of collaboration among members of the ensemble, making it possible to jump to solutions in new parts of the space without having to make a series of discrete moves to get there.

- *Mutation*

  This step introduces changes into the parameters. Like simulated annealing it could be done randomly, but preferably takes advantage of whatever is known about how to generate good moves for the problem.

Figure 13.7 shows the evolution of the population in a GA searching the same function used in the previous examples. Clearly this is a long way from downhill simplex search. There the only choice was when to stop; for a GA there are many decisions to be made about how to implement each of the steps. For simple problems it can be possible to do this optimally [Prügel-Bennett & Shapiro, 1994], but the most important use of GAs is

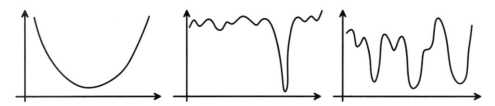

Figure 13.8. Types of search problems. Left, a global minimum that can be found by gradient descent. Right, a function with many equally good nearby minima that can be found with a local stochastic search. And center, a function with a large difference between typical local minima and the global minimum, requiring a global search.

for problems where such calculations are not possible. Even so, plausible choices for each of the options can lead to surprisingly satisfactory performance. After all, evolution has come a long way with many suboptimal choices.

## 13.5 THE BLESSING OF DIMENSIONALITY

This chapter has presented an escalating series of search algorithm enhancements to handle successively more difficult problems. Figure 13.8 illustrates the types of functions to which they apply. On the left is a smooth function with a broad minimum. As long as you have reasonable confidence that you can guess starting parameter values that put you within the basin of the desired minimum, variants of gradient descent will get you there (Nelder–Mead or Powell's method if only function evaluation is possible, conjugate gradient if first derivatives are available, Levenberg–Marquardt if first and second derivatives are available). At the opposite extreme is a function with very many equally good local minima. Wherever you start out, a bit of randomness can get you out of the small local minima and into one of the good ones, making this an appropriate problem for simulated annealing. But simulated annealing is less suitable for the problem in the middle, which has one global minimum that is much better than all the other local ones. Here it's very important that you find that basin, and so it's better to spend less time improving any one set of parameters and more time working with an ensemble to examine more parts of the space. This is where GAs are best.

It's not possible to decide which of these applies to a given nontrivial problem, because only a small part of the search space can ever be glimpsed. But a good clue is provided by the statistics of a number of local searches starting from random initial conditions. If the same answer keeps being found the case on the left applies, if different answers are always found but they have similar costs then the case on the right applies, and if there is a large range in the best solutions found then it's the one in the middle.

It should be clear that simulated annealing and genetic algorithms are not disjoint alternatives. Both start with a good generator of local moves. Simulated annealing introduces stochasticity to cope with local minima; genetic algorithms introduce an ensemble to cope with still rougher search landscapes. One idea came from observing how thermodynamic systems solve problems, the other from biology, but divorced from the legacy of their origins both have sensible lessons for numerical search.

Perhaps the biggest surprise of all about high-dimensional nonlinear search is just how

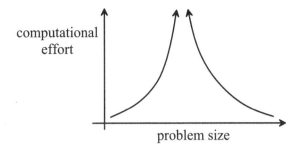

Figure 13.9. Characteristic computational effort as a function of the problem size.

well it can work. Variants of simulated annealing are now used routinely for such hard problems as routing fleets of airplanes. The explanation for this success can be called the *blessing of dimensionality*. The kind of landscape shown on the right of Figure 13.8 has come to be well understood through the study of *spin glasses* [Mezard, 1987]. These are physical systems in which atomic spin degrees of freedom have random interaction strengths; Problems 13.1 and 13.2 work through an example. For a large number of spins it is extremely difficult to find the global minimum, but there is an enormous number of local minima that are all almost equally good (remember that the number of configurations is exponential in the number of spins). Almost anywhere you look you will be near a reasonably low-energy solution.

In fact, the really hard problems are not the big ones. If there are enough planes in a fleet, there are many different routings that are comparable. Small problems by definition are not hard; for a small fleet the routing options can be exhaustively checked. What's difficult is the intermediate case, where there are too many planes for a simple search to find the global best answer, but there are too few planes for there to be many alternative acceptable routings. Many other problems have been observed to have this kind of behavior, shown in Figure 13.9. The effort to find an answer goes down for small problems because there are so few possible answers to check, and it goes down for big problems because there are so many different ways to solve a problem. In between is a transition between these regimes that can be very difficult to handle. This crossover has been shown to have many of the characteristics of a phase transition [Cheeseman *et al.*, 1991, Kirkpatrick & Selman, 1994], which is usually where the most complex behavior in a system occurs.

Figure 13.9 explains a great deal of the confusion about the relative claims for search algorithms. Many different techniques work equally well on toy problems, because they are so easy. And many techniques can work on what look like hard problems, if they are given a big enough space to work in. But the really hard problems in between can trip up techniques with stellar records at either extreme. This figure also helps explain the success of neural networks. If a model is not going to have a small number of meaningful parameters, then the best thing to do is to give it so many adjustable parameters that there's no trouble finding a good solution, and prevent overfitting by imposing priors.

## 13.6 SELECTED REFERENCES

[Acton, 1990] Acton, Forman S. (1990). *Numerical Methods That Work*. Washington, DC: Mathematical Association of America.

This should be required reading for anyone who uses any search algorithm, to understand why the author feels that "They are the first refuge of the computational scoundrel, and one feels at times that the world would be a better place if they were quietly abandoned."

[Press et al., 1992] Press, William H., Teukolsky, Saul A., Vetterling, William T., & Flannery, Brian P. (1992). *Numerical Recipes in C: The Art of Scientific Computing*. 2nd edn. New York, NY: Cambridge University Press.

*Numerical Recipes* has a good collection of search algorithms.

## 13.7 PROBLEMS

(13.1) Consider a 1D spin glass defined by a set of spins $S_1, S_2, \ldots, S_N$, where each spin can be either $+1$ or $-1$. The energy of the spins is defined by

$$E = -\sum_{i=1}^{N} J_i S_i S_{i+1} \quad , \tag{13.10}$$

where the $J_i$'s are random variables drawn from a Gaussian distribution with zero mean and unit variance, and $S_{N+1} = S_1$ (periodic boundary conditions). Find a low-energy configuration of the spins by simulated annealing. The minimum possible energy is bounded by

$$E_{min} = -\sum_{i=1}^{N} |J_i| \quad ; \tag{13.11}$$

compare your result with this. At each iteration flip a single randomly chosen spin, and if the energy increases by $\Delta E$ accept the move with a probability

$$p = e^{-\beta \Delta E} \tag{13.12}$$

(always accept a move that decreases the energy). Take $\beta$ to be proportional to time, $\beta = \alpha t$ (where $t$ is the number of iterations), and repeat the problem for $\alpha = 0.1, 0.01$, and $0.001$ for $N = 100$. Choose a single set of values for the $J$'s and the starting values for the spins and use these for each of the cooling rates.

(13.2) Now solve the same problem with a genetic algorithm (keep the same values for the $J$'s as the previous problem). Start with a population of 100 randomly drawn sets of spins. At each time step evaluate the energy of each member of the population, and then assign it a probability for reproduction of

$$p \propto e^{-\beta(E - E_{min})} \quad . \tag{13.13}$$

Generate 100 new strings by, for each string, choosing two of the strings from the previous population by drawing from this probability distribution, choosing a random crossover point, taking the bits to the left of the crossover point in the first string and the bits to the right in the second, and then mutating by randomly flipping a bit in the resulting string. Plot the minimum energy in the population as a function of time step for $\beta = 10, 1, 0.1, 0.01$.

# 14 Clustering and Density Estimation

We've learned more and more about how to describe and fit functions, but the decision to fit a function can itself be unduly restrictive. Instead of assuming a noise model for the data (such as Gaussianity), why not go ahead and deduce the underlying probability distribution from which the data were drawn? Given the distribution (or *density*) any other quantity of interest, such as a conditional forecast of a new observation, can be derived. Perhaps because this is such an ambitious goal it is surprisingly poorly covered in the literature and in a typical education, but in fact it is not only possible but extremely useful.

This chapter will cover *density estimation* at three levels of generality and complexity. First will be methods based on binning the data that are easy to implement but that can require impractical amounts of data. This is solved by casting density estimation as a problem in functional approximation, but that approach can have trouble representing all of the things that a density can do. The final algorithms are based on clustering, merging the desirable features of the preceeding ones while avoiding the attendant liabilities.

## 14.1 HISTOGRAMMING, SORTING, AND TREES

Given a set of measurements $\{x_n\}_{n=1}^N$ that were drawn from an unknown distribution $p(x)$, the simplest way to model the density is by histogramming $x$. If $N_i$ is the number of observations between $x_i$ and $x_{i+1}$, then the probability $p(x_i)$ to see a new point in the bin between $x_i$ and $x_{i+1}$ is approximately $N_i/N$ (by the Law of Large Numbers).

This idea is simple but not very useful. The first problem is the amount of storage required. Let's say that we're looking at a 15-dimensional data set sampled at a 16-bit resolution. Binning it would require an array of $2^{16 \times 15} \approx 10^{72}$ elements, more elements than there are atoms in the universe! But this is an unreasonable approach because most of those bins would be empty. The storage requirement becomes feasible if only occupied bins are allocated. This might appear to be circular reasoning, but in fact can be done quite simply.

The trick is to perform a *lexicographic sort*. The digits of each vector are appended to make one long number, in the preceeding example $15 \times 16 = 240$ bits long. Then the one-dimensional string of numbers is sorted, requiring $\mathcal{O}(N \log N)$ operations. A final pass through the sorted numbers counts the number of times each number appears, giving the occupancy of the multi-dimensional bin uniquely indexed by that number.

Since the resolution is known in advance this can further be reduced to a linear time algorithm. Sorting takes $\mathcal{O}(N \log N)$ when the primitive operation is a comparison of

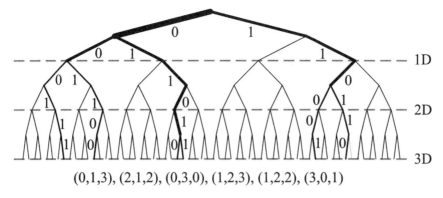

(0,1,3), (2,1,2), (0,3,0), (1,2,3), (1,2,2), (3,0,1)

Figure 14.1. Multi-dimensional binning on a binary tree.

two elements, but it can be done in linear time at a fixed resolution by sorting on a tree. This is shown schematically in Figure 14.1. Each bit of the appended number is used to decide to branch right or left in the tree. If a node already exists then the count of points passing through it is incremented by one, and if the node does not exist then it is created. Descending through the tree takes a time proportional to the number of bits, and hence the dimension, and the total running time is proportional to the number of points. The storage depends on the amount of space the data set occupies (an idea that will be revisited in Chapter 16 to characterize a time series).

Sorting on a tree makes it tractable to accumulate a multi-dimensional histogram but it leaves another problem: the number of points in each bin is variable. Assuming Poisson counting errors, this means that the relative uncertainty of the count of each bin ranges from 100% for a bin with a single point in it to a small fraction for a bin with a large occupancy. Many operations (such as taking a logarithm in calculating an entropy) will magnify the influence of the large errors associated with the low probability bins.

The solution to this problem is to use bins that have a fixed occupancy rather than a fixed size. These can be constructed by using a $k$-$D$ tree [Preparata & Shamos, 1985], shown for 2D in Figure 14.2. A vertical cut is initially chosen at a location $x_1^1$ that has half of the data on the left and half on the right. Then each of these halves is cut horizontally at $y_1^1$ and $y_2^1$ to leave half of the data on top and half on bottom. These four bins are next cut horizontally at $x_{1-4}^2$, and so forth. In a higher-dimensional space the cuts cycle over the directions of the axes. By construction each bin has the same number of points, but the volume of the bin varies. Since the probability is estimated by the number of points divided by the volume, instead of varying the number of points for a given volume as was done before we can divide the fixed number of points by the varying volumes. This gives a density estimate with constant error per bin.

Building a $k$-$D$ tree is more work than fixed-mass binning on a tree, particularly if new points need to be added later (which can require moving many of the $k$-$D$ tree partitions). There is a simple trick that approximates fixed-volume binning while retaining the convenience of a linear-time tree. The idea is to interleave the bits of each coordinate of the point to be sorted rather than appending them, so that each succeeding bit is associated with a different direction. Then the path that the sorted sequence of numbers takes through the space is a fractal curve [Mandelbrot, 1983], shown in 2D in Figure 14.3. This

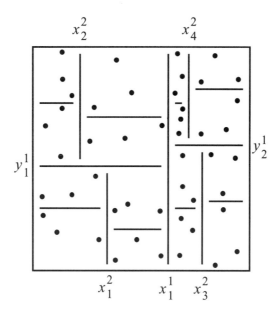

Figure 14.2. $k$-D tree binning in 2D.

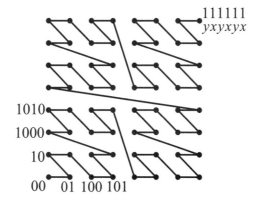

Figure 14.3. Fractal binning in 2D.

has the property that the average distance between two points in the space is proportional to the linear distance along the curve connecting them. Therefore the difference between the indices of two bins estimates the volume that covers them.

Even histogramming with constant error per bin still has a serious problem. Let's say that we're modestly looking at three-dimensional data sampled at an 8-bit resolution. If it is random with a uniform probability distribution and is histogrammed in $M$ bins then the probability of each bin will be roughly $1/M$. The entropy of this distribution is

$$H = -\sum_{i=1}^{M} p_i \log_2 p_i$$
$$= \log_2 M$$
$$\Rightarrow 2^H = M \quad .$$

(14.1)

But we know that the entropy is $H = 8 \times 3 = 24$ bits, therefore there must be $2^{24} \approx 10^7$ occupied bins or the entropy will be underestimated. Since the smallest possible data set would have one point per bin this means that at least 10 million points must be used, an unrealistic amount in most experimental settings.

The real problem is that histogramming has no notion of neighborhood. Two adjacent bins could represent different letters in an alphabet just as well as nearby parts of a space, ignoring the strong constraint that neighboring points might be expected to behave similarly. What's needed is some kind of functional approximation that can make predictions about the density in locations where points haven't been observed.

## 14.2 FITTING DENSITIES

Let's say that we're given a set of data points $\{x_n\}_{n=1}^N$ and want to infer the density $p(x)$ from which they were drawn. If we're willing to add some kind of prior belief, for example that nearby points behave similarly, then we can hope to find a functional representation for $p(x)$. This must of course be normalized to be a valid density, and a possible further constraint is *compact support*: the density is restricted to be nonzero on a bounded set. In this section we'll assume that the density vanishes outside the interval [0,1]; the generalization is straightforward.

One way to relate this problem to what we've already studied is to introduce the *cumulative* distribution

$$P(x) \equiv \int_0^x p(x)\,dx \quad . \tag{14.2}$$

If we could find the cumulative distribution then the density follows by differentiation,

$$p(x) = \frac{dP}{dx} \quad . \tag{14.3}$$

There is in fact an easy way to find a starting guess for $P(x)$: sort the data set. $P(x_i)$ is defined to be the fraction of points below $x_i$, therefore if $i$ is the index of where point $x_i$ appears in the sorted set, then

$$y_i \equiv \frac{i}{N+1} \approx P(x_i) \quad . \tag{14.4}$$

The denominator is taken to be $N + 1$ instead of $N$ because the normalization constraint is effectively an extra point $P(1) = 1$.

To estimate the error, recognize that $P(x_i)$ is the probability to draw a new point below $x_i$, and that $NP(x_i)$ is the expected number of such points in a data set of $N$ points. If we assume Gaussian errors then the variance around this mean is $NP(x_i)[1 - P(x_i)]$ [Gershenfeld, 1999a]. Since $y_i = i/(N+1)$, the standard deviation in $y_i$ is the standard deviation in the number of points below it, $i$, divided by $N+1$,

$$\sigma_i = \frac{1}{N+1}\{NP(x_i)[1-P(x_i)]\}^{1/2}$$
$$= \frac{1}{N+1}\left[N\frac{i}{N+1}\left(1 - \frac{i}{N+1}\right)\right]^{1/2} \quad . \tag{14.5}$$

This properly vanishes at the boundaries where we know that $P(0) = 0$ and $P(1) = 1$.

We can now fit a function to the data set $\{x_i, y_i, \sigma_i\}$. A convenient parameterization is

$$P(x) = x + \sum_{m=1}^{M} a_m \sin(m\pi x)$$

$$\Rightarrow p(x) = 1 + \sum_{m=1}^{M} a_m m\pi \cos(m\pi x) \quad, \tag{14.6}$$

which enforces the boundary conditions $P(0) = 0$ and $P(1) = 1$ regardless of the choice of coefficients (this representation also imposes the constraint that $dp/dx = 0$ at $x = 0$ and 1, which may or may not be appropriate).

One way to impose the prior that nearby points behave similarly is to use the integral square curvature of $P(x)$ as a regularizer (Section 12.7), which seeks to minimize the quantity

$$I_0 = \int_0^1 \left(\frac{d^2 P}{dx^2}\right)^2 dx \quad. \tag{14.7}$$

The model mismatch is given by

$$I_1 = \frac{1}{N} \sum_{i=1}^{N} \left[\frac{P(x_i) - y_i}{\sigma_i}\right]^2 \quad. \tag{14.8}$$

The expected value of $I_1$ is 1; we don't want to minimize it because passing exactly through the data is a very unlikely event given the uncertainty. Therefore we want to impose this constraint by introducing a Lagrange multiplier in a variational sum

$$I = I_0 + \lambda I_1 \quad. \tag{14.9}$$

Solving

$$\frac{\partial I}{\partial a_m} = 0 \tag{14.10}$$

for all $m$ gives the set of coefficients $\{a_m\}$ as a function of $\lambda$, and then given the coefficients we can do a one-dimensional search to find the value of $\lambda$ that results in $I_1 = 1$.

Plugging in the expansion (14.5) and doing the derivatives and integrals shows that if the matrix $\mathbf{A}$ is defined by

$$A_{lm} = 2\pi^4 l^2 m^2 \begin{cases} \frac{1}{2(l^2 - m^2)\pi} \{(l+m)\sin[(l-m)\pi] \\ \qquad - (l-m)\sin[(l+m)\pi]\} & (l \neq m) \\ \frac{1}{2} - \sin(2m\pi)/(4m\pi) & (l = m) \end{cases} \tag{14.11}$$

the matrix $\mathbf{B}$ by

$$B_{lm} = \frac{2}{N} \sum_{i=1}^{N} \sin(l\pi x_i)\sin(m\pi x_i)/\sigma_i^2 \quad, \tag{14.12}$$

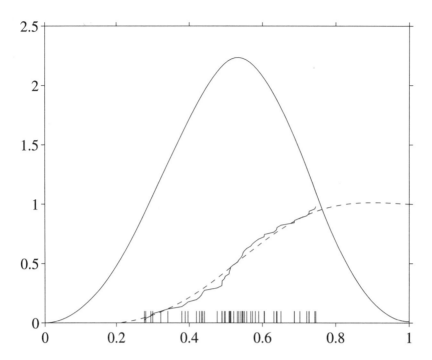

Figure 14.4. Cumulative histogram (solid) and regularized fit (dashed), for the points shown along the axis drawn from a Gaussian distribution. Differentiating gives the resulting density estimate.

and the vector $\vec{c}$ by

$$c_m = \frac{2}{N} \sum_{i=1}^{N} (y_i - x_i) \sin(m\pi x_i)/\sigma_i^2 \quad , \tag{14.13}$$

then the vector of expansion coefficients $\vec{a} = (a_1, \ldots, a_M)$ is given by

$$\vec{a} = \lambda(\mathbf{A} + \lambda\mathbf{B})^{-1} \cdot \vec{c} \quad . \tag{14.14}$$

An example is shown in Figure 14.4. Fifty points were drawn from a Gaussian distribution and the cumulative density was fit using 50 basis functions. Even though there are as many coefficients as data points, the regularization insures that the resulting density does not overfit the data. This figure also demonstrates that the presence of the regularizer biases the estimate of the variance upwards, because the curvature is minimized if the fitted curve starts above the cumulative histogram and ends up below it.

## 14.3 MIXTURE DENSITY ESTIMATION AND EXPECTATION-MAXIMIZATION

Fitting a function lets us generalize our density estimate away from measured points, but it unfortunately does not handle many other kinds of generalization very well. The first problem is that it can be hard to express local beliefs with a global prior. For

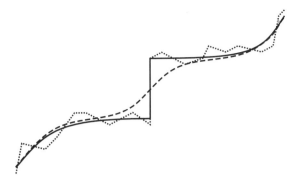

Figure 14.5. The problem with global regularization. A smoothness prior (dashed line) misses the discontinuty in the data (solid line); a maximum entropy prior (dotted) misses the smoothness.

example, if a smooth curve has some discontinuities, then a smoothness prior will round out the discontinuities, and a maximum entropy prior will fit the discontinuity but miss the smoothness (Figure 14.5). What's needed is a way to express a statement like "the density is smooth everywhere, except for where it isn't."

Another problem is with the kinds of functions that need to be represented. Consider points distributed on a low-dimensional surface in a high-dimensional space. Transverse to the surface the distribution is very narrow, something that is hard to expand with trigonometric or polynomial basis functions.

These problems with capturing local behavior suggest that density estimation should be done using local rather than global functions. *Kernel density estimation* [Silverman, 1986] does this by placing some kind of smooth bump, such as a Gaussian, on each data point. An obvious disadvantage of this approach is that the resulting model requires retaining all of the data. A better approach is to find interesting places to put a smaller number of local functions that can model larger neighborhoods. This is done by *mixture models* [McLachlan & Basford, 1988], which are closely connected to the problem of splitting up a data set by *clustering*, and are an example of *unsupervised learning*. Unlike function fitting with a known target, the algorithm must learn for itself where the interesting places in the data set are.

In $D$ dimensions a mixture model can be written by factoring the density over multivariate Gaussians

$$p(\vec{x}) = \sum_{m=1}^{M} p(\vec{x}, c_m)$$

$$= \sum_{m=1}^{M} p(\vec{x}|c_m)\, p(c_m)$$

$$= \sum_{m=1}^{M} \frac{|\mathbf{C}_m^{-1}|^{1/2}}{(2\pi)^{D/2}} e^{-(\vec{x}-\vec{\mu}_m)^T \cdot \mathbf{C}_M^{-1} \cdot (\vec{x}-\vec{\mu}_m)/2}\, p(c_m) \ , \qquad (14.15)$$

where $|\cdot|^{1/2}$ is the square root of the determinant, and $c_m$ refers to the $m$th Gaussian with mean $\vec{\mu}_m$ and covariance matrix $\mathbf{C}_m$. The challenge of course is to find these parameters.

If we had a single Gaussian, the mean value $\vec{\mu}$ could be estimated simply by averaging the data,

$$\vec{\mu} = \int_{-\infty}^{\infty} \vec{x}\, p(\vec{x})\, d\vec{x}$$

$$\approx \frac{1}{N} \sum_{n=1}^{N} \vec{x}_n \quad . \tag{14.16}$$

The second line follows because an integral over a density can be approximated by a sum over variables drawn from the density; we don't know the density but by definition it is the one our data set was taken from. This idea can be extended to more Gaussians by recognizing that the $m$th mean is the integral with respect to the conditional distribution,

$$\begin{aligned}
\vec{\mu}_m &= \int \vec{x}\, p(\vec{x}|c_m)\, d\vec{x} \\
&= \int \vec{x}\, \frac{p(c_m|\vec{x})}{p(c_m)} p(\vec{x})\, d\vec{x} \\
&\approx \frac{1}{Np(c_m)} \sum_{n=1}^{N} \vec{x}_n\, p(c_m|\vec{x}_n) \quad .
\end{aligned} \tag{14.17}$$

Similarly, the covariance matrix could be found from

$$\mathbf{C}_m \approx \frac{1}{Np(c_m)} \sum_{n=1}^{N} (\vec{x}_n - \vec{\mu}_m)(\vec{x}_n - \vec{\mu}_m)^T\, p(c_m|x_n) \quad , \tag{14.18}$$

and the expansion weights by

$$\begin{aligned}
p(c_m) &= \int_{-\infty}^{\infty} p(\vec{x}, c_m)\, d\vec{x} \\
&= \int_{-\infty}^{\infty} p(c_m|\vec{x})\, p(\vec{x})\, d\vec{x} \\
&\approx \frac{1}{N} \sum_{n=1}^{N} p(c_m|\vec{x}_n) \quad .
\end{aligned} \tag{14.19}$$

But how do we find the posterior probability $p(c_m|\vec{x})$ used in these sums? By definition it is

$$\begin{aligned}
p(c_m|\vec{x}) &= \frac{p(\vec{x}, c_m)}{p(\vec{x})} \\
&= \frac{p(\vec{x}|c_m)\, p(c_m)}{\sum_{m=1}^{M} p(\vec{x}|c_m)\, p(c_m)} \quad ,
\end{aligned} \tag{14.20}$$

which we can calculate using the definition (equation 14.15). This might appear to be circular reasoning, and it is! The probabilities of the points can be calculated if we know the parameters of the distributions (means, variances, weights), and the parameters can be found if we know the probabilities. Since we start knowing neither, we can start with a random guess for the parameters and go back and forth, iteratively updating the probabilities and then the parameters. Calculating an expected distribution given parameters, then finding the most likely parameters given a distribution, is called *Expectation-*

*Maximization (EM)*, and it converges to the maximum likelihood distribution starting from the initial guess [Dempster et al., 1977].

To see where this magical property comes from let's take the log-likelihood $L$ of the data set (the log of the product of the probabilities to see each point) and differentiate with respect to the mean of the $m$th Gaussian:

$$\nabla_{\vec{\mu}_m} L = \nabla_{\vec{\mu}_m} \log \prod_{n=1}^{N} p(\vec{x}_n)$$

$$= \sum_{n=1}^{N} \nabla_{\vec{\mu}_m} \log p(\vec{x}_n)$$

$$= \sum_{n=1}^{N} \frac{1}{p(\vec{x}_n)} \nabla_{\vec{\mu}_m} p(\vec{x}_n)$$

$$= \sum_{n=1}^{N} \frac{1}{p(\vec{x}_n)} p(\vec{x}_n, c_m) \mathbf{C}_m^{-1} \cdot (\vec{x}_n - \vec{\mu}_m)$$

$$= \sum_{n=1}^{N} p(c_m|\vec{x}_n) \mathbf{C}_m^{-1} \cdot \vec{x}_n - \sum_{n=1}^{N} p(c_m|\vec{x}_n) \mathbf{C}_m^{-1} \cdot \vec{\mu}_m$$

$$= N p(c_m) \mathbf{C}_m^{-1} \cdot \left[ \frac{1}{N p(c_m)} \sum_{n=1}^{N} \vec{x}_n \, p(c_m|\vec{x}_n) - \vec{\mu}_m \right] . \qquad (14.21)$$

Writing the change in the mean after one EM iteration as $\delta \vec{\mu}_m$ and recognizing that the term on the left in the bracket is the update rule for the mean, we see that

$$\delta \vec{\mu}_m = \frac{\mathbf{C}_m}{N p(c_m)} \cdot \nabla_{\vec{\mu}_m} L \quad . \qquad (14.22)$$

Now look back at Section 10.4, where we wrote a gradient descent update step in terms of the gradient of a cost function suitably scaled. In one EM update the mean moves in the direction of the gradient of the log-likelihood, scaled by the Gaussian's covariance matrix divided by the weight. The weight is a positive number, and the covariance matrix is *positive definite* (it has positive eigenvalues [Strang, 1986]), therefore the result is to increase the likelihood. The changes in the mean stop at the maximum when the gradient vanishes. Similar equations apply to the variances and weights [Lei & Jordan, 1996].

This is as close as data analysis algorithms come to a free lunch. We're maximizing a quantity of interest (the log-likelihood) merely by repeated function evaluations. The secret is in equation (14.20). The numerator measures how strongly one Gaussian predicts a point, and the denominator measures how strongly all the Gaussians predict the point. The ratio gives the fraction of the point to be associated with each Gaussian. Each point has one unit of explanatory power, and it gives it out based on how well it is predicted. As a Gaussian begins to have a high probability at a point, that point effectively disappears from the other Gaussians. The Gaussians therefore start exploring the data set in parallel, "eating" points that they explain well and forcing the other Gaussians towards ones they don't. The interaction among the Gaussians happens in the denominator of the posterior term, collectively performing a high-dimensional search.

The EM cycle finds the local maximum of the likelihood that can be reached from

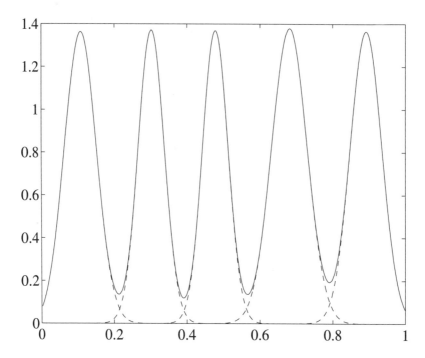

Figure 14.6. Mixture density estimation with Gaussians for data uniformly distributed between 0 and 1.

the starting condition, but that usually is not a significant limitation because there are so many equivalent arrangements that are equally good. The Gaussians can for example be started with random means and variances large enough to "feel" the entire data set. If local maxima are a problem, the techniques of the last chapter can be used to climb out of them.

## 14.4 CLUSTER-WEIGHTED MODELING

As appealing as mixture density estimation is, there is still a final problem. Figure 14.6 shows the result of doing EM with a mixture of Gaussians on random data uniformly distributed over the interval [0, 1]. The individual distributions are shown as dashed lines, and the sum by a solid line. The correct answer is a constant, but because that is hard to represent in this basis we get a very bumpy distribution that depends on the precise details of how the Gaussians overlap. In gaining locality we've lost the ability to model simple functional dependence.

The problem is that a Gaussian captures only proximity; anything nontrivial must come from the overlap of multiple Gaussians. A better alternative is to base the expansion of a density around models that can locally describe more complex behavior. This powerful idea has re-emerged in many flavors under many names, including *Bayesian networks* [Buntine, 1994], *mixtures of experts* [Jordan & Jacobs, 1994], and *gated experts* [Weigend et al., 1995]. The version that we will cover, *cluster-weighted modeling* [Gershenfeld

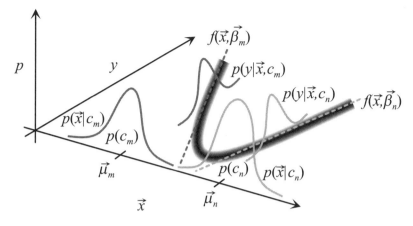

Figure 14.7. The spaces in cluster-weighted modeling.

*et al.*, 1998], is just general enough to be able to describe many situations, but not so general that it presents difficult architectural decisions.

The goal now is to capture the functional dependence in a system as part of a density estimate. Let's start with a set of $N$ observations $\{y_n, \vec{x}_n\}_{n=1}^N$, where the $\vec{x}_n$ are known inputs and the $y_n$ are measured outputs. $y$ might be the exchange rate between the dollar and the mark, and $\vec{x}$ a group of currency indicators. Or $y$ could be a future value of a signal and $\vec{x}$ a vector of past lagged values. For simplicity we'll take $y$ to be a scalar, but the generalization to vector $\vec{y}$ is straightforward.

Given the joint density $p(y, \vec{x})$ we could find any derived quantity of interest, such as a conditional forecast $\langle y | \vec{x} \rangle$. We'll proceed by expanding the density again, but now in terms of explanatory clusters that contain three terms: a weight $p(c_m)$, a domain of influence in the input space $p(\vec{x}|c_m)$, and a dependence in the output space $p(y|\vec{x}, c_m)$:

$$p(y, \vec{x}) = \sum_{m=1}^{M} p(y, \vec{x}, c_m)$$

$$= \sum_{m=1}^{M} p(y, \vec{x}|c_m)\, p(c_m)$$

$$= \sum_{m=1}^{M} p(y|\vec{x}, c_m)\, p(\vec{x}|c_m)\, p(c_m) \quad . \tag{14.23}$$

These terms are shown in Figure 14.7. As before, $p(c_m)$ is a number that measures the fraction of the data set explained by the cluster. The input term could be taken to be a $D$-dimensional separable Gaussian,

$$p(\vec{x}|c_m) = \prod_{d=1}^{D} \frac{1}{\sqrt{2\pi\sigma_{m,d}^2}} e^{-(x_d - \mu_{m,d})^2 / 2\sigma_{m,d}^2} \quad , \tag{14.24}$$

or one using the full covariance matrix,

$$p(\vec{x}|c_m) = \frac{|\mathbf{C}_m^{-1}|^{1/2}}{(2\pi)^{D/2}} e^{-(\vec{x} - \vec{\mu}_m)^T \cdot \mathbf{C}_m^{-1} \cdot (\vec{x} - \vec{\mu}_m)/2} \quad . \tag{14.25}$$

Separable Gaussians require storage and computation linear in the dimension of the space while nonseparable ones require quadratic storage and a matrix inverse at each step. Conversely, using the covariance matrix lets one cluster capture a linear relationship that would require many separable clusters to describe. Therefore covariance clusters are preferred in low-dimensional spaces, but cannot be used in high-dimensional spaces.

The output term will also be taken to be a Gaussian, but with a new twist:

$$p(y|\vec{x}, c_m) = \frac{1}{\sqrt{2\pi\sigma_{m,y}^2}} e^{-[y-f(\vec{x},\vec{\beta}_m)]^2/2\sigma_{m,y}^2} \quad . \tag{14.26}$$

The mean of the Gaussian is now a function $f$ that depends on $\vec{x}$ and a set of parameters $\vec{\beta}_m$. The reason for this can be seen by calculating the conditional forecast

$$\begin{aligned}
\langle y|\vec{x}\rangle &= \int y \, p(y|\vec{x}) \, dy \\
&= \int y \, \frac{p(y,\vec{x})}{p(\vec{x})} \, dy \\
&= \frac{\sum_{m=1}^{M} \int y \, p(y|\vec{x}, c_m) \, dy \, p(\vec{x}|c_m) \, p(c_m)}{\sum_{m=1}^{M} p(\vec{x}|c_m) \, p(c_m)} \\
&= \frac{\sum_{m=1}^{M} f(\vec{x}, \vec{\beta}_m) \, p(\vec{x}|c_m) \, p(c_m)}{\sum_{m=1}^{M} p(\vec{x}|c_m) \, p(c_m)} \quad .
\end{aligned} \tag{14.27}$$

We see that in the forecast the Gaussians are used to control the interpolation among the local functions, rather than directly serving as the basis for functional approximation. This means that $f$ can be chosen to reflect a prior belief on the local relationship between $\vec{x}$ and $y$, for example locally linear, and even one cluster is capable of modeling this behavior. Using locally linear models says nothing about the global smoothness because there is no constraint that the clusters need to be near each other, so this architecture can handle the combination of smoothness and discontinuity posed in Figure 14.5.

Equation (14.27) looks like a kind of network model, but it is a derived property of a more general density estimate rather than an assumed form. We can also predict other quantities of interest such as the error in our forecast of $y$,

$$\begin{aligned}
\langle \sigma_y^2|\vec{x}\rangle &= \int (y - \langle y|\vec{x}\rangle)^2 \, p(y|\vec{x}) \, dy \\
&= \int (y^2 - \langle y|\vec{x}\rangle^2) \, p(y|\vec{x}) \, dy \\
&= \frac{\sum_{m=1}^{M}[\sigma_{m,y}^2 + f(\vec{x}, \vec{\beta}_m)^2] \, p(\vec{x}|c_m) \, p(c_m)}{\sum_{m=1}^{M} p(\vec{x}|c_m) \, p(c_m)} - \langle y|\vec{x}\rangle^2.
\end{aligned} \tag{14.28}$$

Note that the use of Gaussian clusters does not assume a Gaussian error model; there can be multiple output clusters associated with one input value to capture multimodal or other kinds of non-Gaussian distributions.

The estimation of the parameters will follow the EM algorithm we used in the last section. From the forward probabilities we can calculate the posteriors

$$p(c_m|y,\vec{x}) = \frac{p(y,\vec{x},c_m)}{p(y,\vec{x})}$$

$$= \frac{p(y|\vec{x},c_m)\,p(\vec{x}|c_m)\,p(c_m)}{\sum_{m=1}^{M} p(y,\vec{x},c_m)}$$

$$= \frac{p(y|\vec{x},c_m)\,p(\vec{x}|c_m)\,p(c_m)}{\sum_{m=1}^{M} p(y|\vec{x},c_m)\,p(\vec{x}|c_m)\,p(c_m)} \qquad (14.29)$$

which are used to update the cluster weights

$$p(c_m) = \int p(y,\vec{x},c_m)\,dy\,d\vec{x}$$

$$= \int p(c_m|y,\vec{x})\,p(y,\vec{x})\,dy\,d\vec{x}$$

$$\approx \frac{1}{N}\sum_{n=1}^{N} p(c_m|y_n,\vec{x}_n) \qquad . \qquad (14.30)$$

A similar calculation is used to find the new means,

$$\vec{\mu}_m^{new} = \int \vec{x}\,p(\vec{x}|c_m)\,d\vec{x}$$

$$= \int \vec{x}\,p(y,\vec{x}|c_m)\,dy\,d\vec{x}$$

$$= \int \vec{x}\,\frac{p(c_m|y,\vec{x})}{p(c_m)}\,p(y,\vec{x})\,dy\,d\vec{x}$$

$$\approx \frac{1}{N\,p(c_m)}\sum_{n=1}^{N}\vec{x}_n\,p(c_m|y_n,\vec{x}_n)$$

$$= \frac{\sum_{n=1}^{N}\vec{x}_n\,p(c_m|y_n,\vec{x}_n)}{\sum_{n=1}^{N} p(c_m|y_n,\vec{x}_n)}$$

$$\equiv \langle\vec{x}\rangle_m \qquad (14.31)$$

(where the last line defines the cluster-weighted expectation value). The second line introduces $y$ as a variable that is immediately integrated over, an apparently meaningless step that in fact is essential. Because we are using the stochastic sampling trick to evaluate the integrals and guide the cluster updates, this permits the clusters to respond to both where data are in the input space and how well their models work in the output space. A cluster won't move to explain nearby data if they are better explained by another cluster's model, and if two clusters' models work equally well then they will separate to better explain where the data are.

The cluster-weighted expectations are also used to update the variances

$$\sigma_{m,d}^{2,new} = \langle(x_d - \mu_{m,d})^2\rangle_m \qquad (14.32)$$

or covariances

$$[\mathbf{C}_m]_{ij}^{new} = \langle(x_i - \mu_i)(x_j - \mu_j)\rangle_m \qquad . \qquad (14.33)$$

The model parameters are found by choosing the values that maximize the cluster-weighted log-likelihood:

$$0 = \frac{\partial}{\partial \vec{\beta}_m} \log \prod_{n=1}^{N} p(y_n, \vec{x}_n)$$

$$= \sum_{n=1}^{N} \frac{\partial}{\partial \vec{\beta}_m} \log p(y_n, \vec{x}_n)$$

$$= \sum_{n=1}^{N} \frac{1}{p(y_n, \vec{x}_n)} \frac{\partial p(y_n, \vec{x}_n)}{\partial \vec{\beta}_m}$$

$$= \sum_{n=1}^{N} \frac{1}{p(y_n, \vec{x}_n)} p(y_n, \vec{x}_n, c_m) \frac{y_n - f(\vec{x}_n, \vec{\beta}_m)}{\sigma_{m,y}^2} \frac{\partial f(\vec{x}_n, \vec{\beta}_m)}{\partial \vec{\beta}_m}$$

$$= \frac{1}{\sigma_{m,y}^2} \sum_{n=1}^{N} p(c_m | y_n, \vec{x}_n)[y_n - f(\vec{x}_n, \vec{\beta}_m)] \frac{\partial f(\vec{x}_n, \vec{\beta}_m)}{\partial \vec{\beta}_m}$$

$$= \frac{1}{Np(c_m)} \sum_{n=1}^{N} p(c_m | y_n, \vec{x}_n)[y_n - f(\vec{x}_n, \vec{\beta}_m)] \frac{\partial f(\vec{x}_n, \vec{\beta}_m)}{\partial \vec{\beta}_m}$$

$$= \left\langle [y - f(\vec{x}, \vec{\beta}_m)] \frac{\partial f(\vec{x}, \vec{\beta}_m)}{\partial \vec{\beta}_m} \right\rangle_m \quad . \tag{14.34}$$

In the last few lines remember that since the left hand side is zero we can multiply or divide the right hand side by a constant.

If we choose a local model that has linear coefficients,

$$f(\vec{x}, \vec{\beta}_m) = \sum_{i=1}^{I} \beta_{m,i} f_i(\vec{x}) \quad , \tag{14.35}$$

then this gives for the coefficients of the $m$th cluster

$$0 = \langle [y - f(\vec{x}, \vec{\beta}_m)] f_j(\vec{x}) \rangle_m$$

$$= \underbrace{\langle y f_j(\vec{x}) \rangle_m}_{a_j} - \sum_{i=1}^{I} \beta_{m,i} \underbrace{\langle f_j(\vec{x}) f_i(\vec{x}) \rangle_m}_{B_{ji}}$$

$$\Rightarrow \vec{\beta}_m = B^{-1} \cdot \vec{a} \quad . \tag{14.36}$$

Finally, once we've found the new model parameters then the new output width is

$$\sigma_{m,y}^{2,new} = \langle [y - f(\vec{x}, \vec{\beta}_m)]^2 \rangle_m \quad . \tag{14.37}$$

It's a good idea to add a small constant to the input and output variance estimates to prevent a cluster from shrinking down to zero width for data with no functional dependence; this constant reflects the underlying resolution of the data.

Iterating this sequence of evaluating the forward and posterior probabilities at the data points, then maximizing the likelihood of the parameters, finds the most probable set of parameters that can be reached from the starting configuration. The only algorithm parameter is $M$, the number of clusters. This controls overfitting, and can be determined by cross-validation. The only other choice to make is $f$, the form of the local model. This can be based on past practice for a domain, and should be simple enough that the local model cannot overfit. In the absence of any prior information a local linear or quadratic model is a reasonable choice. It's even possible to include more than one type of local model and let the clustering find where they are most relevant.

Cluster-weighted modeling then assembles the local models into a global model that can handle nonlinearity (by the nonlinear interpolation among local models), discontinuity (since nothing forces clusters to stay near each other), non-Gaussianity (by the overlap of multiple clusters), nonstationarity (by giving absolute time as an input variable so that clusters can grow or shrink as needed to find the locally stationary time scales), find low-dimensional structure in high-dimensional spaces (models are allocated only where there are data to explain), predict not only errors from the width of the output distribution but also errors that come from forecasting away from where training observations were made (by using the input density estimate), build on experience (by reducing to a familiar global version of the local model in the limit of one cluster), and the resulting model has a transparent architecture (the parameters are easy to interpret, and by definition the clusters find "interesting" places to represent the data) [Gershenfeld et al., 1998].

Taken together those specifications are a long way from the simple models we used at the outset of the study of function fitting. Looking beyond cluster-weighted modeling, it's possible to improve on the EM search by doing a full Bayesian integration over prior distributions on all the parameters [Richardson & Green, 1997], but this generality is not needed for the common case of weak priors on the parameters and many equally acceptable solutions. And the model can be further generalized to arbitrary probabilistic networks [Buntine, 1996], a step that is useful if there is some *a priori* insight into the architecture.

Chapter 16 will develop the application to time series analysis; another important special case is observables that are restricted to a discrete set of values such as events, patterns, or conditions. Let's now let $\{s_n, \vec{x}_n\}_{n=1}^{N}$ be the training set of pairs of input points $\vec{x}$ and discrete output states $s$. Once again we'll factor the joint density over clusters

$$p(s, \vec{x}) = \sum_{m=1}^{M} p(s|\vec{x}, c_m)\, p(\vec{x}|c_m)\, p(c_m) \quad , \tag{14.38}$$

but now the output term is simply a histogram of the probability to see each state for each cluster (with no explicit $\vec{x}$ dependence). The histogram is found by generalizing the simple binning we used at the beginning of the chapter to the cluster-weighted expectation of a delta function that equals one on points with the correct state and zero otherwise:

$$p(s|\vec{x}, c_m) = \frac{1}{Np(c_m)} \sum_{n=1}^{N} \delta_{s_n=s} p(c_m|s_n, \vec{x}_n)$$

$$= \frac{1}{Np(c_m)} \sum_{n|s_n=s} p(c_m|s_n, \vec{x}_n) \quad . \tag{14.39}$$

Everything else is the same as the real-valued output case. An example is shown in Figure 14.8, for two states $s$ (labelled "1" and "2"). The clusters all started equally likely to predict either state, but after convergence they have specialized in one of the data types as well as dividing the space.

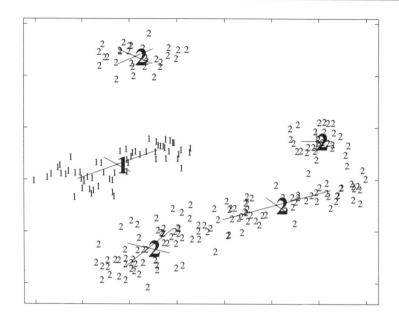

Figure 14.8. Cluster-weighted modeling with discrete output states. The small numbers are the events associated with each point; the large numbers are the most likely event associated with each cluster, and the lines are the cluster covariances.

This model can predict the probability to see each state given a new point by

$$p(s|\vec{x}) = \frac{p(s,\vec{x})}{\sum_s p(s,\vec{x})} \quad . \tag{14.40}$$

For applications where computational complexity or speed is a concern there are easy approximations that can be made. It is possible to use just the cluster variances instead of the full covariance matrix, and to restrict the output models to one state per cluster rather than keeping the complete histogram. If the goal is simply to predict the most likely state rather than the distribution of outcomes, then evaluating the density can be approximated by comparing the square distances to the clusters divided by the variances. Finally, keeping only the cluster means reduces to a linear-time search to find the nearest cluster, $\min_m |\vec{x} - \vec{\mu}_m|$, which defines a *Voronoi tesselation* of the space by the clusters [Preparata & Shamos, 1985].

The final approximation of simply looking up the state of the cluster nearest a new point might appear to not need such a general clustering to start with. A much simpler algorithm would be to iteratively assign points to the nearest cluster and then move the cluster to the mean of the points associated with it. This is called the *k-means* algorithm [MacQueen, 1967, Hartigan, 1975]. It doesn't work very well, though, because in the iteration a cluster cannot be influenced by a large number of points that are just over the boundary with another cluster. By embedding the cluster states in Gaussian kernels as we have done, this is a *soft clustering* unlike the *hard clustering* done by k-means. The result is usually a faster search and a better model.

Predicting states is part of the domain of *pattern recognition*. State prediction can also be used to approximate a set of inputs by a common output; *vector quantization* applies this to a feature vector describing a signal in order to compress it for transmission and storage [Nasrabadi & King, 1988, Gersho & Gray, 1992]). Alternatively, it can be used to retrieve information associated with related inputs stored in a *content-addressable memory* [Kohonen, 1989, Bechtel & Abrahamsen, 1991]. The essence of these problems lies in choosing good places to put the clusters, and the combination of EM with local models provides a good solution.

## 14.5 SELECTED REFERENCES

[Duda & Hart, 1973] Duda, Richard O., & Hart, Peter E. (1973). *Pattern Classification and Scene Analysis*. New York, NY: Wiley.

[Therrien, 1989] Therrien, Charles W. (1989). *Decision, Estimation, and Classification: An Introduction to Pattern Recognition and Related Topics*. New York, NY: Wiley.

[Fukunaga, 1990] Fukunaga, Keinosuke (1990). *Introduction to Statistical Pattern Recognition*. 2nd edn. Boston, MA: Academic Press.

These are classic pattern recognition texts.

[Silverman, 1986] Silverman, B.W. (1986). *Density Estimation for Statistics and Data Analysis*. New York, NY: Chapman and Hall.

Traditional density estimation.

[Bishop, 1996] Bishop, Christopher M. (1996). *Neural Networks for Pattern Recognition*. Oxford: Oxford University Press.

Bishop has a nice treatment of density estimation in a connectionist context.

## 14.6 PROBLEMS

(14.1) Revisit the fitting problem in Chapter 12 with cluster-weighted modeling. Once again take as a data set $x = \{-10, -9, \ldots, 9, 10\}$, and $y(x) = 0$ if $x \leq 0$ and $y(x) = 1$ if $x > 0$. Take the simplest local model for the clusters, a constant. Plot the resulting forecast $\langle y|x \rangle$ and uncertainty $\langle \sigma_y^2|x \rangle$ for analyses with 1, 2, 3, and 4 clusters.

# 15 Filtering and State Estimation

Our study of estimating parameters from observations has presumed that there are unchanging parameters to be estimated. For many (if not most) applications this is not so: not only are the parameters varying, but finding their variation in time may be the goal of the data analysis. This chapter and the next bring time back into the picture. Here we will look at the problem of estimating a time-dependent set of parameters that describe the state of a system, given measurements of observable quantities along with some kind of model for the relationship between the observations and the underlying state. For example, in order to navigate, an airplane must know where it is. Many relevant signals arrive at the airplane, such as radar echoes, GPS messages, and gyroscopic measurements. The first task is to reduce these raw signals to position estimates, and then these estimates must be combined along with any relevant past information to provide the best overall estimate of the plane's position. Closely related tasks are *smoothing*, *noise reduction*, or *signal separation*, using the collected data set to provide the best estimate of previous states (given new measurements, where do we think the airplane was?), and *prediction*, using the data to forecast a future state (where is the airplane going?). These tasks are often described as filtering problems, even though they really are general algorithm questions, because they evolved from early implementations in analog filters.

We will start with the simple example of *matched filters* to detect a known signal, extend that to *Wiener filters* to separate a signal from noise, and then turn to the much more general, useful, and important *Kalman filters*. Much of estimation theory is based on linear techniques. Since the world is not always obligingly linear, we will look at how nonlinearity makes estimation more difficult, and simpler. The chapter closes with the use of *Hidden Markov Models* to help find models as well as states.

## 15.1 MATCHED FILTERS

Consider a signal $x(t)$ passed through a linear filter with impulse response $f(t)$ (go back to Chapter 2 if you need a review of linear systems theory). The frequency domain response of the output $Y(\omega)$ will be the product of the Fourier transforms of the input and the filter

$$Y(\omega) = X(\omega)F(\omega) \quad , \tag{15.1}$$

and the time domain response will be the convolution

$$y(t) = x(t) * f(t) = \int_0^T x(t-u)f(u)\,du \quad , \tag{15.2}$$

where the bounds of the integral are the interval during which the signal has been applied to the filter. The magnitude of the output can be bounded by *Schwarz's inequality*:

$$y^2(t) = \left| \int_0^T x(t-u)f(u)\,du \right|^2$$

$$\leq \int_0^T |x(t-u)|^2\,du \int_0^T |f(u)|^2\,du \quad . \tag{15.3}$$

By inspection, this bound will be saturated (reach its maximum value) if

$$f(u) = A\, x^*(t-u) \tag{15.4}$$

for any constant $A$. The filter will produce the maximum output for a given input signal if the impulse response of the filter is proportional to the complex conjugate of the signal reversed in time. This is called a *matched filter*, and is used routinely to detect and time known signals. For example, to measure the arrival time of radar echoes, the output from a filter matched to the transmitted pulses goes to a comparator, and the time when the output exceeds a preset threshold is used to determine when a pulse has arrived.

## 15.2 WIENER FILTERS

Next, consider a time-invariant filter with impulse response $f(t)$ that receives an input $x(t) + \eta(t)$ and produces an output $y(t)$, with $x(t)$ a desired signal and $\eta(t)$ noise added to the signal (such as from the front end of an amplifier). In the time domain the output is the convolution

$$y(t) = \int_{-\infty}^{\infty} f(u)[x(t-u) + \eta(t-u)]\,du \quad , \tag{15.5}$$

for now assuming that the signals are defined for all time. How should the filter be designed to make $y(t)$ as close as possible to $x(t)$? One way to do this is by minimizing the mean square error between them (in Chapter 10 we saw that this implicitly assumes Gaussian statistics, but is an assumption that is commonly and relatively reliably used more broadly). This problem was solved for a linear filter by Norbert Wiener at MIT's Radiation Laboratory in the 1940s, therefore the solution is called a *Wiener filter*.

The expected value of the error at time $t$ is

$$\langle E^2 \rangle = \langle [x(t+\alpha) - y(t)]^2 \rangle \quad , \tag{15.6}$$

where the average is over an ensemble of realizations of the noise process. An offset $\alpha$ has been added to cover the three cases of:

- $\alpha < 0$ : *smoothing* the past;
- $\alpha = 0$ : *filtering* the present;
- $\alpha > 0$ : *predicting* the future.

Substituting in equation (15.5),

$$\langle E^2 \rangle = \langle x^2(t+\alpha) \rangle - 2 \int_{-\infty}^{\infty} f(u) \underbrace{\langle x(t+\alpha)[x(t-u) + \eta(t-u)] \rangle}_{\equiv C_{x,x+\eta}(\alpha+u)} du \qquad (15.7)$$

$$+ \int_{-\infty}^{\infty} \int_{-\infty}^{\infty} f(u)f(v) \underbrace{\langle [x(t-u) + \eta(t-u)][x(t-v) + \eta(t-v)] \rangle}_{\equiv C_{x+\eta,x+\eta}(u-v)} du\, dv \ .$$

We must find the $f(t)$ that minimizes the sum of these integrals over the correlation functions. Since the first term does not depend on the filter function $f(t)$ it can't contribute to the minimization and we will drop it. Because of the double integral we can't use the Euler–Lagrange equation derived in Chapter 4, but we can use a similar argument. Assume that $f(t)$ is the optimal filter that we are looking for, and let $g(t)$ be any arbitrary filter added to it, giving a new filter $f(t) + \epsilon g(t)$. In terms of this the new error is

$$\langle E^2 \rangle = \int_{-\infty}^{\infty} \int_{-\infty}^{\infty} [f(u) + \epsilon g(u)][f(v) + \epsilon g(v)] C_{x+\eta,x+\eta}(u-v)\, du\, dv$$
$$-2 \int_{-\infty}^{\infty} [f(u) + \epsilon g(u)] C_{x,x+\eta}(\alpha + u)\, du \quad . \qquad (15.8)$$

We can now differentiate with respect to $\epsilon$ and look for the minimum at $\epsilon = 0$:

$$\left. \frac{\partial \langle E^2 \rangle}{\partial \epsilon} \right|_{\epsilon=0} = 0$$

$$= \int_{-\infty}^{\infty} \int_{-\infty}^{\infty} g(u) f(v) C_{x+\eta,x+\eta}(u-v)\, du\, dv$$
$$+ \int_{-\infty}^{\infty} \int_{-\infty}^{\infty} f(u) g(v) C_{x+\eta,x+\eta}(u-v)\, du\, dv$$
$$+ 2\epsilon \int_{-\infty}^{\infty} \int_{-\infty}^{\infty} g(u) g(v) C_{x+\eta,x+\eta}(u-v)\, du\, dv$$
$$- 2 \int_{-\infty}^{\infty} g(u) C_{x,x+\eta}(\alpha + u)\, du \quad . \qquad (15.9)$$

The first two terms are the same (interchanging dummy integration variables and using the symmetry of the correlation functions), and the third one vanishes at $\epsilon = 0$, so we're left with

$$\int_{-\infty}^{\infty} g(\tau) \left[ -C_{x,x+\eta}(\alpha + \tau) + \int_{-\infty}^{\infty} f(u) C_{x+\eta,x+\eta}(u - \tau)\, du \right] d\tau = 0 \qquad (15.10)$$

Since $g(\tau)$ is arbitrary, the only way this can be equal to zero for all choices of $g$ is if the term in brackets vanishes

$$\int_{-\infty}^{\infty} f(u) C_{x+\eta,x+\eta}(u-\tau)\, du = C_{x,x+\eta}(\alpha + \tau) \quad . \qquad (15.11)$$

This is now a simpler integral equation to be solved for $f$, but there is a crucial subtlety in equation (15.11). If we solve it for all $\tau$, positive and negative, then it will require that the filter function $f$ be defined for both positive and negative times. This is a *noncausal* filter. The only way that the filter can have access to the signal at all times (other than

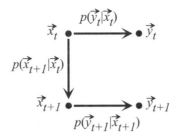

Figure 15.1. Update of an internal state and an external observable.

being psychic) is if it is applied after the full time record has been recorded, which is fine for off-line applications. If we don't mind a noncausal filter, the convolution in equation (15.11) is easily solved by taking (two-sided) Laplace transforms

$$F(s)C_{x+\eta,x+\eta}(s) = C_{x,x+\eta}(s)e^{\alpha s} \tag{15.12}$$

and so

$$F(s) = \frac{C_{x,x+\eta}(s)e^{\alpha s}}{C_{x+\eta,x+\eta}(s)} \quad . \tag{15.13}$$

This has a simple interpretation: the Wiener filter rolls the response off when the signal is much smaller than the noise, and sets the gain to unity if the signal is much larger than the noise. Prediction or smoothing is done simply by the complex phase shift of a linear system.

If a causal filter is needed so that the Wiener filter can be used in real-time then $f(\tau)$ must vanish for negative $\tau$, and equation (15.11) must be solved only for $\tau \geq 0$. In this case it is called the *Wiener–Hopf* equation, and is much more difficult to solve, although there are many special techniques available for doing so because it is so important [Brown & Hwang, 1997]. Beyond Wiener filters, signal separation for nonlinear systems requires the more general time series techniques to be introduced in Chapter 16.

## 15.3 KALMAN FILTERS

Wiener filters are impressively optimal, but practically not very useful. It is important to remember that everything is optimal with respect to something. In the case of Wiener filters we found the "best" linear time-invariant filter, but by design it is therefore linear time-invariant. The result is an almost trivial kind of signal separation, simply cutting off the response where the signal is small compared to the noise. Furthermore, it does not easily generalize to more complex problems with multiple degrees of freedom.

Now consider the general system shown in Figure 15.1. There is an internal state $\vec{x}$, for example, the position, velocity, and acceleration of an airplane as well as the orientations of the control surfaces. Its state is updated in discrete time according to a distribution function $p(\vec{x}_{t+1}|\vec{x}_t)$, which includes both the deterministic and random influences. For the airplane, the deterministic part is the aerodynamics, and the random part includes factors such as turbulence and control errors. The internal state is not directly accessible, but rather must be inferred from measurements of observables $\vec{y}$ (such as the airplane's pitot tube, GPS receiver, and radar returns), which are related to $\vec{x}$ by a relation $p(\vec{y}|\vec{x})$ that can include a random component due to errors in the measurement process. How should the measurements be combined to estimate the system's state? Further, is it possible to iteratively update the state estimate given new measurements without recalculating it from scratch? Kalman filters provide a general solution to this important problem.

To start, let's assume that there are just two random variables, $x$ and $y$, that have a joint probability distribution $p(x, y)$. Given a measurement of $y$, what function $\hat{x}(y)$ should we use to estimate $x$? Once again, we will do this by picking the estimate the minimizes the mean square error over the distribution. This means that we want to minimize

$$\langle [x - \hat{x}(y)]^2 \rangle = \int\int [x - \hat{x}(y)]^2 \, p(x,y) \, dx \, dy$$

$$= \int\int [x^2 - 2x\hat{x}(y) + \hat{x}^2(y)] \, \underbrace{p(x,y)}_{p(x|y)p(y)} \, dx \, dy$$

$$= \int x^2 \underbrace{\int p(x,y) \, dy}_{p(x)} \, dx - 2 \int \hat{x}(y) \underbrace{\int x \, p(x|y) dx}_{\equiv \langle x|y \rangle} \, p(y) \, dy$$

$$+ \int \hat{x}^2(y) \underbrace{\int p(x,y) \, dx}_{p(y)} \, dy$$

$$= \int x^2 \, p(x) \, dx + \int [\hat{x}^2(y) - 2\hat{x}(y)\langle x|y\rangle] \, p(y) \, dy$$

$$= \int x^2 \, p(x) \, dx$$

$$+ \int [\hat{x}^2(y) - 2\hat{x}(y)\langle x|y\rangle + \langle x|y\rangle^2 - \langle x|y\rangle^2] \, p(y) \, dy$$

$$= \int x^2 \, p(x) \, dx$$

$$- \int \langle x|y\rangle^2 \, p(y) \, dy + \int [\hat{x}(y) - \langle x|y\rangle]^2 \, p(y) \, dy \ . \quad (15.14)$$

All integrals are over the limits of the distribution, and in the last line we completed the square. The first two terms don't depend on the unknown estimator $\hat{x}(y)$, and so are irrelevant to the minimization. The last term is the product of two non-negative functions, which will be minimized if the left hand one vanishes:

$$\hat{x}(y) = \langle x|y\rangle = \int x \, p(x|y) \, dx \quad . \quad (15.15)$$

In retrospect, this is perhaps an obvious result: the minimum mean square estimator simply is the most probable value. This result easily generalizes to multi-dimensional distributions.

Now let's assume that the system's update rule is linear, with additive noise $\vec{\eta}$

$$\vec{x}_t = \mathbf{A}_{t-1} \cdot \vec{x}_{t-1} + \vec{\eta}_t \quad (15.16)$$

(we will later relax the assumption of linear updates), and assume a linear relationship

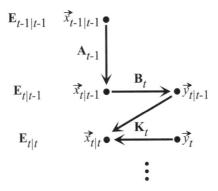

Figure 15.2. Steps in Kalman filtering.

between the state and the observable with additive noise $\vec{\epsilon}$

$$\vec{y}_t = \mathbf{B}_t \cdot \vec{x}_t + \vec{\epsilon}_t \quad . \tag{15.17}$$

The noise sources are assumed to be uncorrelated in time, but can have correlations among the components, as measured by the noise covariance matrices $\mathbf{N}^x$ and $\mathbf{N}^y$

$$\mathbf{N}^x = \langle \vec{\eta} \vec{\eta}^T \rangle \quad \langle \eta_i(t) \eta_j(t') \rangle = N^x_{ij} \delta_{tt'}$$

$$\mathbf{N}^y = \langle \vec{\epsilon} \vec{\epsilon}^T \rangle \quad \langle \epsilon_i(t) \epsilon_j(t') \rangle = N^y_{ij} \delta_{tt'} \tag{15.18}$$

(where as usual $\vec{\epsilon}^T$ is the transpose of $\vec{\epsilon}$). The two noise sources are taken to be uncorrelated with each other

$$\langle \vec{\eta} \vec{\epsilon}^T \rangle = \vec{0} \tag{15.19}$$

and to have zero mean

$$\langle \vec{\eta} \rangle = \langle \vec{\epsilon} \rangle = \vec{0} \quad . \tag{15.20}$$

The elements of Kalman filtering are shown in Figure 15.2. $\vec{x}_t$ is the true (but inaccessible) state of the system at time $t$, $\vec{y}_t$ the observable, and $\mathbf{E}_t$ is the covariance matrix of the error in the estimate of $\vec{x}$. The notation $\vec{x}_{n|m}$ represents the best estimate for $\vec{x}_n$ given the record of measurements up to time $m$

$$\vec{x}_{n|m} = \langle \vec{x}_n | \vec{x}_m, \vec{x}_{m-1}, \ldots \rangle = \int \vec{x}_n \, p(\vec{x}_n | \vec{x}_m, \vec{x}_{m-1}, \ldots) \, d\vec{x}_n \quad . \tag{15.21}$$

The first step in Kalman filtering is to use the best estimate of the previous system state, $\vec{x}_{t-1|t-1}$, to predict the new state $\vec{x}_{t|t-1}$. This is then used to predict the observable $\vec{y}_{t|t-1}$. Then, when the true new observable $\vec{y}_t$ is measured, it and the estimate $\vec{y}_{t|t-1}$ are combined to estimate the new internal state $\vec{x}_{t|t}$. There are two very important and perhaps nonobvious elements of this figure. First, the state estimate updates are done on just the previous state, without needing the full record, but (given the assumptions of the model) this provides just as good an estimate. Second, this estimate will be much better than if the new observable alone was used to estimate the internal state. Kalman filtering is an example of *recursive estimation*: to determine the present estimate it is necessary to know the previous one, which in turn depends on the one before that, and so forth back to the initial conditions.

To do the first prediction step, recall that if two variables $a$ and $b$ with probabilities $p_a(a)$ and $p_b(b)$ are added, then the distribution for their sum $c = a+b$ is the convolution

$$p(c) = \int p_b(b) p_a(c-b) \, db \quad . \tag{15.22}$$

Since $\vec{x}_t$ depends only on the previous value $\vec{x}_{t-1}$ plus the noise term, the expected value will depend only on the previous expected value $\vec{x}_{t-1|t-1}$:

$$\vec{x}_{t|t-1} = \int \vec{x}_t \, p(\vec{x}_t | \vec{x}_{t-1}) \, d\vec{x}_t \quad . \tag{15.23}$$

The conditional distribution $p(\vec{x}_t | \vec{x}_{t-1})$ consists of the deterministic distribution $\delta(\vec{x}_t - \mathbf{A}_t \cdot \vec{x}_{t-1})$ convolved by the (zero mean) noise distribution $p_\eta$, so

$$\vec{x}_{t|t-1} = \int \vec{x}_t \, p_\eta(\vec{x}_t - \mathbf{A}_{t-1} \cdot \vec{x}_{t-1}) \, d\vec{x}_t \quad , \tag{15.24}$$

and since the noise distribution is zero mean

$$\vec{x}_{t|t-1} = \mathbf{A}_{t-1} \cdot \vec{x}_{t-1|t-1} \quad . \tag{15.25}$$

Similarly,

$$\vec{y}_{t|t-1} = \mathbf{B}_t \cdot \vec{x}_{t|t-1} \quad . \tag{15.26}$$

This gives us the estimates for the new internal state and observable. To update the internal state estimate, these can be linearly combined with the new observation $\vec{y}_t$

$$\vec{x}_{t|t} = \vec{x}_{t|t-1} + \mathbf{K}_t \cdot (\vec{y}_t - \vec{y}_{t|t-1}) \quad . \tag{15.27}$$

The matrix $\mathbf{K}_t$ is called the *Kalman gain matrix*; we will derive the optimal form for it. Given this estimate we can define the error covariance matrix in terms of the (inaccessible) true state $\vec{x}_t$ by

$$\mathbf{E}_{t|t} = \langle (\vec{x}_t - \vec{x}_{t|t})(\vec{x}_t - \vec{x}_{t|t})^T \rangle \quad . \tag{15.28}$$

The difference between the true state and the estimate is

$$\vec{x}_t - \vec{x}_{t|t} = \vec{x}_t - \vec{x}_{t|t-1} - \mathbf{K}_t \cdot (\vec{y}_t - \vec{y}_{t|t-1}) \quad , \tag{15.29}$$

and the difference between the predicted and true observation is

$$\vec{y}_t - \vec{y}_{t|t-1} = \mathbf{B}_t \cdot \vec{x}_t + \vec{\epsilon}_t - \mathbf{B}_t \cdot \vec{x}_{t|t-1} \quad . \tag{15.30}$$

Combining these,

$$\begin{aligned} \vec{x}_t - \vec{x}_{t|t} &= \vec{x}_t - \vec{x}_{t|t-1} - \mathbf{K}_t \mathbf{B}_t \cdot (\vec{x}_t - \vec{x}_{t|t-1}) - \mathbf{K}_t \cdot \vec{\epsilon}_t \\ &= (1 - \mathbf{K}_t \mathbf{B}_t) \cdot (\vec{x}_t - \vec{x}_{t|t-1}) - \mathbf{K}_t \cdot \vec{\epsilon}_t \quad . \end{aligned} \tag{15.31}$$

Therefore the error matrix is updated by

$$\begin{aligned} \mathbf{E}_{t|t} &= (1 - \mathbf{K}_t \mathbf{B}_t) \langle (\vec{x}_t - \vec{x}_{t|t-1})(\vec{x}_t - \vec{x}_{t|t-1})^T \rangle \\ &\quad \times (1 - \mathbf{K}_t \mathbf{B}_t)^T + \mathbf{K}_t \langle \vec{\epsilon}_t \vec{\epsilon}_t^T \rangle \mathbf{K}_t^T \\ &= (1 - \mathbf{K}_t \mathbf{B}_t) \, \mathbf{E}_{t|t-1} \, (1 - \mathbf{K}_t \mathbf{B}_t)^T + \mathbf{K}_t \mathbf{N}_t^y \mathbf{K}_t^T \end{aligned} \tag{15.32}$$

(there are no cross terms because the measurement noise $\vec{\epsilon}_t$ is independent of the state estimation error $\vec{x}_t - \vec{x}_{t|t-1}$). The diagonal terms of the error covariance matrix are the state errors; we want to choose the Kalman gain matrix $\mathbf{K}$ to minimize the sum of the diagonal terms of the matrix, i.e., minimize the trace

$$\text{Tr}(\mathbf{E}_{t|t}) = \langle |\vec{x}_t - \vec{x}_{t|t}|^2 \rangle \quad . \tag{15.33}$$

To do this minimization, we will use two matrix identities

$$\frac{d\,\text{Tr}(\mathbf{AB})}{d\mathbf{A}} = \mathbf{B}^T \quad \text{(if } \mathbf{AB} \text{ is square)} \tag{15.34}$$

and

$$\frac{d\,\text{Tr}(\mathbf{ACA}^T)}{d\mathbf{A}} = 2\mathbf{AC} \quad \text{(if } \mathbf{C} \text{ is symmetric)} \quad , \tag{15.35}$$

where

$$\left(\frac{df}{d\mathbf{A}}\right)_{ij} \equiv \frac{df}{dA_{ij}} \tag{15.36}$$

(these can be proved by writing out the components). Equation (15.32) can be expanded out as

$$\begin{aligned}\mathbf{E}_{t|t} = &\,\mathbf{E}_{t|t-1} - \mathbf{K}_t\mathbf{B}_t\mathbf{E}_{t|t-1} - \mathbf{E}_{t|t-1}\mathbf{B}_t^T\mathbf{K}_t^T \\ &+ \mathbf{K}_t(\mathbf{B}_t\mathbf{E}_{t|t-1}\mathbf{B}_t^T + \mathbf{R}_t)\mathbf{K}_t^T\end{aligned} \tag{15.37}$$

(recalling that $(\mathbf{AB})^T = \mathbf{B}^T\mathbf{A}^T$). Applying the two matrix identities to take the derivative of the trace of this equation with respect to $\mathbf{K}$, and using the fact that the trace is unchanged by taking the transpose $\text{Tr}(\mathbf{E}_{t|t-1}\mathbf{B}_t^T\mathbf{K}_t^T) = \text{Tr}([\mathbf{E}_{t|t-1}\mathbf{B}_t^T\mathbf{K}_t^T]^T) = \text{Tr}(\mathbf{K}_t\mathbf{B}_t\mathbf{E}_{t|t-1})$, gives

$$\frac{d\,\text{Tr}(\mathbf{P}_{t|t})}{d\,\mathbf{K}_t} = -2(\mathbf{B}_t\mathbf{P}_{t|t-1})^T + 2\mathbf{K}_t(\mathbf{B}_t\mathbf{P}_{t|t-1}\mathbf{B}_t^T + \mathbf{N}_t^y) = 0 \quad . \tag{15.38}$$

This equation defines the Kalman gain matrix that makes the error extremal; checking the second derivative shows that this is a minimum. Solving for the optimal gain matrix,

$$\mathbf{K}_t = \mathbf{E}_{t|t-1}\mathbf{B}_t^T(\mathbf{B}_t\mathbf{E}_{t|t-1}\mathbf{B}_t^T + \mathbf{N}_t^y)^{-1} \quad . \tag{15.39}$$

Substituting the gain matrix back into equation (15.37), the third and fourth terms cancel, leaving

$$\mathbf{E}_{t|t} = \mathbf{E}_{t|t-1} - \mathbf{E}_{t|t-1}\mathbf{B}_t^T(\mathbf{B}_t\mathbf{E}_{t|t-1}\mathbf{B}_t^T + \mathbf{R}_t)^{-1}\mathbf{B}_t\mathbf{E}_{t|t-1} \tag{15.40}$$

or

$$\mathbf{E}_{t|t} = (1 - \mathbf{K}_t\mathbf{B}_t)\mathbf{E}_{t|t-1} \quad . \tag{15.41}$$

This gives the update rule for the error matrix given a new measurement of the observable.

The last piece that we need is the predicted error after the state prediction step

$\vec{x}_{t+1|t} = \mathbf{A}_t \cdot \vec{x}_{t|t}$, which will be

$$\begin{aligned}
\mathbf{E}_{t+1|t} &= \langle (\vec{x}_{t+1} - \vec{x}_{t+1|t})(\vec{x}_{t+1} - \vec{x}_{t+1|t})^T \rangle \\
&= \langle (\mathbf{A}_t \cdot \vec{x}_t + \vec{\eta}_t - \mathbf{A}_t \cdot \vec{x}_{t|t})(\mathbf{A}_t \cdot \vec{x}_t + \vec{\eta}_t - \mathbf{A}_t \cdot \vec{x}_{t|t})^T \rangle \\
&= \langle (\mathbf{A}_t \cdot (\vec{x}_t - \vec{x}_{t|t}) + \vec{\eta}_t)(\mathbf{A}_t \cdot (\vec{x}_t - \vec{x}_{t|t}) + \vec{\eta}_t)^T \rangle \\
&= \langle \mathbf{A}_t \cdot (\vec{x}_t - \vec{x}_{t|t})(\vec{x}_t - \vec{x}_{t|t})^T \cdot \mathbf{A}^T \rangle + \langle \vec{\eta}_t \vec{\eta}_t^T \rangle \\
&= \mathbf{A}_t \mathbf{E}_{t|t} \mathbf{A}_t^T + \mathbf{N}_t^x \quad .
\end{aligned} \qquad (15.42)$$

This completes the derivation of the Kalman filter, the linear estimator with the minimum square error. Recapping, the procedure starts with an initial estimate for the state $\vec{x}_{t|t-1}$ and error $\mathbf{E}_{t|t-1}$, and then the steps are:

- Estimate the new observable

$$\vec{y}_{t|t-1} = \mathbf{B}_t \cdot \vec{x}_{t|t-1} \quad .$$

- Measure a new value for the observable

$$y_t \quad .$$

- Compute the Kalman gain matrix

$$\mathbf{K}_t = \mathbf{E}_{t|t-1} \mathbf{B}_t^T (\mathbf{B}_t \mathbf{E}_{t|t-1} \mathbf{B}_t^T + \mathbf{N}_t^y)^{-1} \quad .$$

- Estimate the new state

$$\vec{x}_{t|t} = \vec{x}_{t|t-1} + \mathbf{K}_t \cdot (\vec{y}_t - \vec{y}_{t|t-1}) \quad .$$

- Update the error matrix

$$\mathbf{E}_{t|t} = (1 - \mathbf{K}_t \mathbf{B}_t) \mathbf{E}_{t|t-1} \quad .$$

- Predict the new state

$$\vec{x}_{t+1|t} = \mathbf{A}_t \cdot \vec{x}_{t|t} \quad .$$

- Predict the new error

$$\mathbf{E}_{t+1|t} = \mathbf{A}_t \mathbf{E}_{t|t} \mathbf{A}_t^T + \mathbf{N}_t^x \quad .$$

Stepping back from the details of the derivation, these equations have very natural limits. If $\mathbf{B} \to 0$ (the observable $\vec{y}$ does not depend on the internal state $\vec{x}$) or $\mathbf{N}^y \to \infty$ (the observable is dominated by measurement noise) then $\mathbf{K} \to 0$ and the measurements are not used in the state estimate. Conversely, if $\mathbf{N}^y \to 0$ and $\mathbf{B} \to \mathbf{I}$ (there is no noise in the observable, and the transformation from the internal state reduces to the identity matrix) then the update replaces the internal state with the new measurement. Remarkably, the iterative application of this recursive algorithm gives the best estimate of $\vec{x}(t)$ from the history of $\vec{y}(t)$ that can be made by a linear estimator; it cannot be improved by analyzing the entire data set off-line [Catlin, 1989].

## 15.4 NONLINEARITY AND ENTRAINMENT

The derivation of the Kalman filter has assumed linearity in two places: linear observables and dynamics, and linear updates of the state following a new measurement. The former can be relaxed by local linearization; we'll return to the latter in the next chapter.

The nonlinear governing equations now are

$$\vec{x}_t = \vec{f}(\vec{x}_{t-1}) + \vec{\eta}_t \qquad \vec{y}_t = \vec{g}(\vec{x}_t) + \vec{\epsilon}_t \quad . \tag{15.43}$$

The system governing equation is needed to predict the new state $\vec{x}_t = \vec{f}(\vec{x}_{t-1})$, and to predict the new error

$$\begin{aligned}
\mathbf{E}_{t+1|t} &= \langle (\vec{x}_{t+1} - \vec{x}_{t+1|t})(\vec{x}_{t+1} - \vec{x}_{t+1|t})^T \rangle \\
&= \langle [\vec{f}(\vec{x}_t) + \vec{\eta}_t - \vec{f}(\vec{x}_{t|t})][\vec{f}(\vec{x}_t) + \vec{\eta}_t - \vec{f}(x_{t|t})]^T \rangle \quad .
\end{aligned} \tag{15.44}$$

If the prediction error is not large (i.e., the noise $\vec{\eta}$ is small), then $\vec{f}$ can be replaced by its local linearization

$$\begin{aligned}
\vec{f}(\vec{x}_t) - \vec{f}(\vec{x}_{t|t}) &= \left. \frac{\partial \vec{f}}{\partial \vec{x}} \right|_{\vec{x}_{t|t}} \cdot (\vec{x}_t - \vec{x}_{t|t}) \\
&\equiv \mathbf{A}'_t \cdot (\vec{x}_t - \vec{x}_{t|t}) \quad .
\end{aligned} \tag{15.45}$$

With this revised definition for $\mathbf{A}$ this equation is then the same as before (equation (15.42)). Similarly, the observable equation occurs in predicting the new observable $\vec{y}_{t|t-1} = \vec{f}(\vec{x}_{t|t-1})$ and in the derivation of the Kalman gain matrix as

$$\begin{aligned}
\vec{y}_t - \vec{y}_{t|t-1} &= \vec{g}(\vec{x}_t) + \vec{\epsilon}_t - \vec{g}(\vec{x}_{t|t-1}) \\
&= \left. \frac{\partial \vec{g}}{\partial \vec{x}} \right|_{\vec{x}_{t|t-1}} \cdot (\vec{x}_t - \vec{x}_{t|t-1}) + \vec{\epsilon}_t \\
&\equiv \mathbf{B}'_t \cdot (\vec{x}_t - \vec{x}_{t|t-1}) + \vec{\epsilon}_t \quad .
\end{aligned} \tag{15.46}$$

Once again, by taking $\mathbf{B}$ to be the local linearization this is the same as equation (15.30). Redefining the Kalman filter to use local linearizations of nonlinear observables and dynamics in the state and error updates gives the *extended Kalman filter*. As with most things nonlinear it is no longer possible to prove the same kind of optimality results about an extended Kalman filter, a liability that is more than made up for by its broader applicability.

The magic of Kalman filtering happens in the step

$$\vec{x}_{t|t} = \vec{x}_{t|t-1} + \mathbf{K}_t \cdot (\vec{y}_t - \vec{y}_{t|t-1}) \quad . \tag{15.47}$$

A correction is added to the internal state based on the difference between what you predicted and what you observed, scaled by how much you trust your predictions versus the observations. Officially, to be able to apply this you must know enough about the system to be able to calculate the noise covariances in both the dynamics and the measurements. In practice this is often not the case, particularly since the "noise" represents all aspects of the system not covered by the model. Then the noise terms become adjustable parameters that are selected to give satisfactory performance (Problem 15.1 provides an example of this trade-off).

The success of Kalman filtering even when it is not formally justified hints at the power of equation (15.47). Most nonlinear systems share the property that a small interaction with another system can cause their states to become synchronized. This process is called *entrainment*. For example, let $d\vec{x}/dt = \vec{f}(\vec{x})$, and take $d\vec{x}'/dt = \vec{f}(\vec{x}')$ to obey the same governing equation but have different initial conditions. Then if we couple one of the degrees of the freedom of the two systems with a linear correction that seeks to drive those variables to the same value,

$$\frac{dx_i}{dt} = f_i(x_i) + \epsilon(x'_i - x_i) \quad , \tag{15.48}$$

then for most choices of $f$, $\epsilon$, and $i$, $\vec{x}$ will approach $\vec{x}'$ as long as $x_i$ interacts with the other components of $\vec{x}$ and there is dissipation to damp out errors. Because dissipation reduces the dimension of the subspace of a system's configuration space that it actually uses [Temam, 1988], it's needed to separate the tugs from the coupling between systems from the internal evolution of a system. $\epsilon$ is a small parameter that controls the trade-off between responding quickly and ignoring noise.

Entrainment requires that the largest Lyapunov exponent associated with the coupling between the systems is negative [Pecora et al., 1997]; this does not even require the systems to be identical [Parlitz et al., 1997]. A formerly familiar example is provided by mechanical clocks on the wall of a clock shop; the vibrations coupled through the wall could entrain the clock mechanisms so that they would tick in synchrony.

Entrainment can be used to design systems whose simple dynamics replaces complicated algorithms for the job of state estimation. An example is *spread spectrum acquisition*, a very important task in engineering practice. A transmitter that seeks to make optimal use of a communications channel uses a linear feedback shift register (LFSR, Chapter 5) to generate ideal pseudo-random noise as a modulation source. To decode the message, the receiver must maintain a copy of the transmitter's shift register that remains faithful even if there is noise in the transmission or the two system's clocks drift apart. This is conventionally done by a coding search for the best setting of the receiver's shift register [Simon et al., 1994].

An LFSR uses a binary recursion

$$x_n = \sum_{i=1}^{N} a_i x_{n-i} \quad (\text{mod } 2) \quad , \tag{15.49}$$

with the $a_i$'s chosen to make the $z$-transform irreducible. It's possible to add small perturbations to this discrete function if the LFSR is replaced by an *analog feedback shift register* ($AFSR$) [Gershenfeld & Grinstein, 1995],

$$x_n = \frac{1}{2}\left[1 - \cos\left(\pi \sum_{i=1}^{N} a_i x_{n-i}\right)\right] \quad . \tag{15.50}$$

This analog function matches the value of the LFSR for binary arguments. Because the magnitude of the slope of the map is less than 1 at the digital values, these are stable fixed points [Guckenheimer & Holmes, 1983] that attract an arbitrary initial condition of

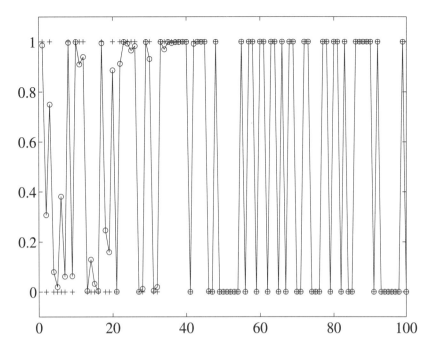

Figure 15.3. Entrainment of an analog feedback shift register (○) with a linear feedback shift register (+).

the register onto the LFSR sequence. If we now add to this a small correction

$$x_n = \frac{1}{2}\left[1 - \cos\left(\pi \sum_{i=1}^{N} a_i x_{n-i}\right)\right] + \epsilon(x'_n - x_n) \quad , \tag{15.51}$$

where $x'_n$ is a signal received from another shift register, then the two systems can entrain. This is shown in Figure 15.3. As usual, if $\epsilon$ is large the receiver locks quickly but it will also try to follow any modulation and noise in the signal; if $\epsilon$ is small it will take longer to lock but will result in a cleaner estimate of the transmitter's state.

## 15.5 HIDDEN MARKOV MODELS

The job of a Kalman filter is to provide an estimate of the internal state given a history of measurements of an external observable. It presumes, however, that you already know how to calculate the transition probabilities, and further that you're not interested in the probability distribution for the internal state. A *Hidden Markov Model* (*HMM*) addresses these limitations.

For example, consider coin flipping done by a corrupt referee who has two coins, one biased and the other fair, with the biased coin occasionally switched in surreptitiously. The observable is whether the outcome is heads or tails; the hidden variable is which coin is being used. Figure 15.4 shows this situation. There are transition probabilities between the hidden states $A$ and $B$, and emission probabilities associated with the observables 0

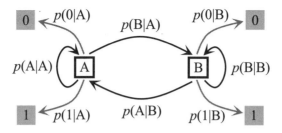

Figure 15.4. A Hidden Markov Model.

and 1. Given this architecture, and a set of measurements of the observable, our task is to deduce both the fixed transition probabilities and the changing probabilities to see the internal states. This is a discrete HMM; it's also possible to use HMMs with continuous time models [Rabiner, 1989].

Just as with the relationship between AR and and MA models (Section 16.1), an HMM can be approximated by an ordinary Markov model, but the latter might require an enormous order to capture the behavior of the former because the rules for dynamics in the present can depend on a change of models that occurred a long time ago.

We'll assume that the internal state $x$ can take on $N$ discrete values, for convenience taken to be $x = \{1, \ldots, N\}$. Call $x_t$ the internal state at time $t$, and let $\{y_1, \ldots, y_T\}$ be a set of measurements of the observable. An HMM is specified by three sets of probabilities: $p(x_{t+1}|x_t)$, the internal transitions, $p(y_t|x_t)$, the emission of an observable given the internal state, and $p(x_1)$, the initial distribution of the internal states.

A key quantity to estimate is $p(x_t, x_{t+1}, y_1, \ldots, y_T)$, the probability to see a pair of internal states along with the observations. From this we can find the probability to see a transition given the observations,

$$p(x_{t+1}|x_t, y_1, \ldots, y_T) = \frac{p(x_t, x_{t+1}, y_1, \ldots, y_T)}{p(x_t, y_1, \ldots, y_T)}$$
$$= \frac{p(x_t, x_{t+1}, y_1, \ldots, y_T)}{\sum_{x_{t+1}=1}^{N} p(x_t, x_{t+1}, y_1, \ldots, y_T)} \quad , \quad (15.52)$$

and then the absolute transition probability can be estimated by averaging over the record

$$p(x_{t+1} = j | x_t = i) \approx \frac{1}{T} \sum_{t'=1}^{T} p(x_{t'+1} = j | x_{t'} = i, y_1, \ldots, y_T) \quad . \quad (15.53)$$

Similarly, the probability of seeing an internal state is

$$p(x_t|y_1, \ldots, y_T) = \frac{p(x_t, y_1, \ldots, y_T)}{p(y_1, \ldots, y_T)}$$
$$= \frac{\sum_{x_{t+1}=1}^{N} p(x_t, x_{t+1}, y_1, \ldots, y_T)}{\sum_{x_t=1}^{N} \sum_{x_{t+1}=1}^{N} p(x_t, x_{t+1}, y_1, \ldots, y_T)} \quad , \quad (15.54)$$

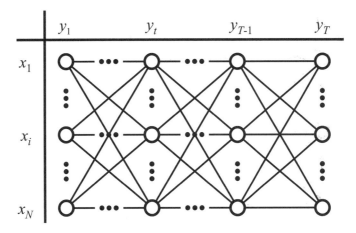

Figure 15.5. Hidden Markov Model trellis.

which can be used to estimate the observable probability by another sum over the data

$$p(y_t = j | x_t = i) \approx \frac{\sum_{t' | y_{t'} = j} p(x_{t'} = i | y_1, \ldots, y_T)}{\sum_{t'=1}^{T} p(x_{t'} = i | y_1, \ldots, y_T)} \quad , \tag{15.55}$$

as well as the absolute probability of an internal state

$$p(x = i) \approx \frac{1}{T} \sum_{t=1}^{T} p(x_t = i | y_1, \ldots, y_T) \quad . \tag{15.56}$$

There is a problem lurking in the estimation of these quantities. Consider the probability of the model to produce the observations $p(y_1, \ldots, y_T)$. Since we don't know the sequence of internal states we have to sum over all of the possibilities

$$p(y_1, \ldots, y_T) = \sum_{x_1=1}^{N} \cdots \sum_{x_T=1}^{N} p(x_1, \ldots, x_T, y_1, \ldots, y_T) \quad . \tag{15.57}$$

This is a set of $T$ sums over $N$ terms, requiring $N^T$ operations. That's a a big number! A model with 10 internal states and an observed sequence of 100 points requires adding $10^{100}$ terms, which is larger than the number of atoms in the universe ($\sim 10^{70}$). The problem may be seen in Figure 15.5. The observed outputs are written across the top, with the possible internal states under them. The exponential explosion comes in the number of different paths through this *trellis*.

The trellis also points to a solution: each column depends only on the previous column, and so we are doing far too much work by recalculating each column over and over for every path that passes through it. Let's start with the last step. Notice that it can be written as a sum over the internal states,

$$p(y_1, \ldots, y_T) = \sum_{x_T=1}^{N} p(x_T, y_1, \ldots, y_T) \tag{15.58}$$

$$= \sum_{x_T=1}^{N} p(y_T | x_T, y_1, \ldots, y_{T-1}) \, p(x_T, y_1, \ldots, y_{T-1}) \quad .$$

The output probability depends only on the internal state, and so this can be simplified to

$$p(y_1,\ldots,y_T) = \sum_{x_T=1}^{N} p(y_T|x_T)\, p(x_T, y_1, \ldots, y_{T-1}) \quad . \tag{15.59}$$

Factored again over the previous step,

$$p(y_1,\ldots,y_T) = \sum_{x_T=1}^{N} p(y_T|x_T) \sum_{x_{T-1}=1}^{N} p(x_T, x_{T-1}, y_1, \ldots, y_{T-1})$$

$$= \sum_{x_T=1}^{N} p(y_T|x_T) \sum_{x_{T-1}=1}^{N} p(x_T|x_{T-1}, y_1, \ldots, y_{T-1})\, p(x_{T-1}, y_1, \ldots, y_{T-1})$$

$$= \sum_{x_T=1}^{N} p(y_T|x_T) \sum_{x_{T-1}=1}^{N} p(x_T|x_{T-1})\, p(x_{T-1}, y_1, \ldots, y_{T-1}) \quad , \tag{15.60}$$

dropping the dependence of the internal transition probability on anything but the previous state. Continuing in this fashion back to the beginning we find that

$$p(y_1,\ldots,y_T) = \sum_{x_T=1}^{N} p(y_T|x_T) \sum_{x_{T-1}=1}^{N} p(x_T|x_{T-1})\, p(y_{T-1}|x_{T-1}) \tag{15.61}$$

$$\cdots \sum_{x_2=1}^{N} p(x_3|x_2)\, p(y_2|x_2) \sum_{x_1=1}^{N} p(x_2|x_1)\, p(y_1|x_1)\, p(x_1) \quad .$$

The $x_1$ sum has $N$ terms and must be done for all values of $x_2$, a total of $N^2$ operations. Since there are $T$ of these, the cost of the calculation drops to $\mathcal{O}(N^2 T)$ – quite a saving over $N^T$. As in so many other areas, a hard problem becomes easy if it is written recursively. For an HMM this is called the *forward algorithm*.

The same idea works in reverse. Start with the probability to see a sequence of observables given a starting initial state, and factor it over the first step:

$$p(y_t,\ldots,y_T|x_t) = \sum_{x_{t+1}=1}^{N} p(x_{t+1}, y_t, \ldots, y_T|x_t)$$

$$= \sum_{x_{t+1}=1}^{N} p(y_t|x_t, x_{t+1}, y_{t+1}, \ldots, y_T)\, p(x_{t+1}, y_{t+1}, \ldots, y_T|x_t)$$

$$= p(y_t|x_t) \sum_{x_{t+1}=1}^{N} p(y_{t+1}, \ldots, y_T|x_t, x_{t+1})\, p(x_{t+1}|x_t)$$

$$= p(y_t|x_t) \sum_{x_{t+1}=1}^{N} p(x_{t+1}|x_t)\, p(y_{t+1}, \ldots, y_T|x_{t+1}) \quad . \tag{15.62}$$

Continuing on to the end,

$$p(y_t, \ldots, y_T | x_t) = p(y_t | x_t) \sum_{x_{t+1}=1}^{N} p(x_{t+1} | x_t) \, p(y_{t+1} | x_{t+1})$$

$$\times \sum_{x_{t+2}=1}^{N} p(x_{t+2} | x_{t+1}) \, p(y_{t+2} | x_{t+2}) \cdots \sum_{x_{T-1}=1}^{N} p(x_{T-1} | x_{T-2}) \, p(y_{T-1} | x_{T-1})$$

$$\times \sum_{x_T=1}^{N} p(x_T | x_{T-1}) \, p(y_T | x_T) \quad . \tag{15.63}$$

This is called (can you guess?) the *backwards algorithm*.

Now return to the probability to see a pair of internal states and the observations. This can be factored as

$$p(x_t, x_{t+1}, y_1, \ldots, y_T) = p(x_t, y_1, \ldots, y_t) \, p(x_{t+1} | x_t, y_1, \ldots, y_t)$$
$$p(y_{t+1}, \ldots, y_T | x_t, x_{t+1}, y_t, \ldots, y_T) \, , \tag{15.64}$$

or dropping irrelevant variables,

$$p(x_t, x_{t+1}, y_1, \ldots, y_T) = p(x_t, y_1, \ldots, y_t) \, p(x_{t+1} | x_t) \, p(y_{t+1}, \ldots, y_T | x_{t+1}) \quad .$$

There are three factors on the right. The first is what we find from the forward algorithm, the middle one is the transition probability specified by the HMM, and the last is the result of the backward algorithm. Therefore this quantity can be calculated for all points by a linear-time pass through the data.

Once that's been done the resulting distributions can be used to update the transition probabilities according to equations (15.53) and (15.55). This procedure can then be iterated, first using the transition probabilities and the observables to update the estimate of the internal probabilities, then using the internal probabilities to find new values of the transition probabilities. Going back and forth between finding probabilities given parameters and finding the most likely parameters given probabilities is just the *Expectation-Maximization* (*EM*) algorithm that we saw in Section 14.3, which in the context of HMMs is called the *Baum–Welch* algorithm. It finds the maximum likelihood parameters starting from initial guesses for them. For a model with continuous parameters the M step becomes a maximization with respect to the parameterized distribution of internal states.

The combination of the forward-backward algorithm and EM finds the parameters of an HMM but it provides no guidance into choosing the architecture. The need to specify the architecture is the weakness, and strength, of using HMM's. In applications where there is some *a priori* insight into the internal states it is straightforward to build that in. A classic example, which helped drive the development of HMMs, is in speech recognition. Here the outputs can be parameters for a sound synthesis model, say ARMA coefficients (Section 16.1), and the internal states are phonemes and then words. It's hard to recognize these primitives from just a short stretch of sound, but the possible utterances are a strong function of what has preceeded them. The same thing applies to many other recognition tasks, such as reading handwriting, where a scrawled letter can be interpreted based on its context. An HMM provides the means to express these ideas.

The most important use of an HMM comes not on the training data but in applying the resulting model to deduce the hidden states given new data. To do this we want to find the most likely sequence of states given the data,

$$\underset{x_1...x_T}{\operatorname{argmax}}\, p(x_1,\ldots,x_T|y_1,\ldots,y_T) = \underset{x_1...x_T}{\operatorname{argmax}}\, \frac{p(x_1,\ldots,x_T,y_1,\ldots,y_T)}{p(y_1,\ldots,y_T)}$$

$$= \underset{x_1...x_T}{\operatorname{argmax}}\, p(x_1,\ldots,x_T,y_1,\ldots,y_T) \quad (15.65)$$

(the denominator can be dropped because it doesn't affect the maximization). $\operatorname{argmax}_x f(x)$ is defined to be the argument $x$ that gives $f$ the maximum value, as compared to $\max_x f(x)$ which is the value of $f$ at the maximum.

Naively this requires checking the likelihood of every path through the trellis, an $\mathcal{O}(N^T)$ calculation. Not surprisingly, the same recursive trick that we used before also works here. Start by factoring out the final step and dropping terms that are irrelevant to the distribution,

$$\underset{x_1...x_T}{\max}\, p(x_1,\ldots,x_T,y_1,\ldots,y_T)$$
$$= \underset{x_1...x_T}{\max}\, p(x_T,y_T|x_1,\ldots,x_{T-1},y_1,\ldots,y_{T-1})\, p(x_1,\ldots,x_{T-1},y_1,\ldots,y_{T-1})$$
$$= \underset{x_1...x_T}{\max}\, p(x_T,y_T|x_{T-1})\, p(x_1,\ldots,x_{T-1},y_1,\ldots,y_{T-1}) \quad (15.66)$$
$$= \underset{x_1...x_T}{\max}\, p(y_T|x_T,x_{T-1})\, p(x_T|x_{T-1})\, p(x_1,\ldots,x_{T-1},y_1,\ldots,y_{T-1})$$
$$= \underset{x_T}{\max}\, p(y_T|x_T)\, \underset{x_1...x_{T-1}}{\max}\, p(x_T|x_{T-1})\, p(x_1,\ldots,x_{T-1},y_1,\ldots,y_{T-1})\ .$$

Continuing in this fashion back to the beginning,

$$\underset{x_1...x_T}{\max}\, p(x_1,\ldots,x_T,y_1,\ldots,y_T)$$
$$= \underset{x_T}{\max}\, p(y_T|x_T)\, \underset{x_{T-1}}{\max}\, p(x_T|x_{T-1})\, p(y_{T-1}|x_{T-1}) \quad (15.67)$$
$$\cdots \underset{x_2}{\max}\, p(x_3|x_2)\, p(y_2|x_2)\, \underset{x_1}{\max}\, p(x_2|x_1)\, p(y_1|x_1)\, p(x_1)\ .$$

This is now once again an $\mathcal{O}(N^2 T)$ calculation. It is called the *Viterbi* algorithm, and is very important beyond HMMs in decoding signals sent through noisy channels that have had correlations introduced by a convolutional coder [Sklar, 1988]. There is one subtlety in implementing it: each maximization has a set of outcomes based on the unknown value of the following step. This is handled by using the maximum value for each outcome and keeping track of which one was used at each step, then backtracking from the end of the calculation once the final maximization is known.

Figures 15.1 and 15.5 were used to help explain Kalman filtering and HMMs by drawing the connections among the variables. It's possible to go much further with such diagrams, using them to write down probabilistic models with more complex dependencies than what we've covered, and applying graphical techniques to arrive at the kinds of simplifications we found to make the estimation problems tractable [Smyth et al., 1997]. Such architectural complexity is useful when there is advance knowledge to guide it; the next chapter turns to the opposite limit.

## 15.6 SELECTED REFERENCES

[Brown & Hwang, 1997] Brown, Robert Grover, & Hwang, Patrick Y.C.(1997). *Introduction to Random Signals and Applied Kalman Filtering*. 3rd edn. New York, NY: Wiley.

A good practical introduction to estimation theory.

[Catlin, 1989] Catlin, Donald E. (1989). *Estimation, Control, and the Discrete Kalman Filter*. Applied Mathematical Sciences, vol. 71. New York, NY: Springer Verlag.

The rigorous mathematical background of Kalman filtering.

[Honerkamp, 1994] Honerkamp, Josef (1994). *Stochastic Dynamical Systems: Concepts, Numerical Methods, Data Analysis*. New York, NY: VCH. Translated by Katja Lindenberg.

The connection between estimation theory and the theory of stochastic dynamical systems.

## 15.7 PROBLEMS

(15.1) Take as a test signal a periodically modulated sinusoid with noise added,

$$y_n = \sin[0.1t + 4\sin(0.01t)] + \eta \equiv \sin(\theta_n) + \eta \quad , \tag{15.68}$$

where $\eta$ is a Gaussian noise process with $\sigma = 0.1$. Design a Kalman filter to estimate the noise-free signal. Use a two-component state vector $\vec{x}_n = (\theta_n, \theta_{n-1})$, and assume for the internal model a linear extrapolation $\theta_{n+1} = \theta_n + (\theta_n - \theta_{n-1})$. Take the system noise matrix $\mathbf{N}^x$ to be diagonal, and plot the predicted value of $y$ versus the measured value of $y$ if the standard deviation of the system noise is chosen to be $10^{-1}$, $10^{-3}$, and $10^{-5}$. Use the identity matrix for the initial error estimate.

# 16 Linear and Nonlinear Time Series

A Kalman filter, or a Hidden Markov Model, starts with some notion of the dynamics of a system and then seeks to match it to observations. As powerful as these ideas are, what if you're given a signal without *a priori* insight into the system that produced it? What if your goal is to learn more about the nature of the system, not just what its state is? This is the domain of *time series analysis*. The field is as broad as time itself; it is defined not by any particular tools (it draws on many of the preceeding chapters), but rather by the intent of their use.

Time series problems arise in almost all disciplines, ranging from studying variations in currency exchange rates to variations in heart-rates. Wherever they occur there are three recurring tasks:

- *Characterize:* What kind of system produced the signal? How many degrees of freedom does it have? How random is it? Is it linear? How does noise influence the system?
- *Forecast:* Based on an estimate of the current state, what will the system do next?
- *Model:* What are the governing equations for the system? What is their long-term behavior?

These are closely related but not identical. For example, a model with good long-term properties may not be the best way to make short-term forecasts and *vice versa*. And although it's possible to characterize a system without explicitly writing down a model, some of the most powerful characterization techniques are based on first building a model.

This chapter will assume that the analyst is an observer, not a participant. Beyond modeling comes manipulation. If it is possible to influence a system then these kinds of descriptions can be used to choose informative inputs (by selecting them where the model uncertainty is large), and to drive the system to a desired state (by reversing the model to predict inputs based on outputs, the domain of *control theory* [Doyle *et al.*, 1992, Auerbach *et al.*, 1992]).

Time series originally were analyzed, not surprisingly, in the time domain. Characterization consisted of looking at the series, and the only kind of forecasting or modeling was simple extrapolation. A major step was Yule's 1927 analysis of the sunspot cycle [Yule, 1927]. This was perhaps the first time that a model with internal degrees of freedom (what we would now call a linear autoregressive model) was inferred from measurements of an external observable (the sunspot series). This rapidly bloomed into the theory of linear time series, which is mature, successful, ubiquitous, and applicable only to linear systems. It arises in two very different limits: deterministic systems that are so simple they can be described by linear governing equations, or systems which are so stochastic that their deviation from ideal randomness is governed by linear random variable equations. In between these two extremes lies the rest of the world, for which nonlinearity does matter. The theories of nonlinear differential equations or stochastic processes in general have no general results, but rather there are many particular tractable cases. However, there is a powerful theory emerging for the characterization and modeling of nonlinear systems without making any linear assumptions. This chapter starts with the linear canon and closes with these newer ideas.

## 16.1 LINEAR TIME SERIES

The most general linear system produces an output $y$ that is a linear function of external inputs $x$ (sometimes called *innovations*) and its previous outputs:

$$y_t = a_t + \underbrace{\sum_{m=1}^{M} b_m y_{t-m}}_{\text{AR, IIR}} + \underbrace{\sum_{n=0}^{N} c_n x_{t-n}}_{\text{MA, FIR}} \quad . \tag{16.1}$$

Typically the $a_t$ term is nonzero only for an initial transient, which imposes the initial conditions on the system. Depending on the side of campus that you are on, the two parts of this equation are called:

- Statistics
  - *Auto-Regressive* (*AR*): The output is a linear regression of its $M$ previous values.
  - *Moving Average* (*MA*): The output is an $N$-point moving average of the input.
  - Taken together, they define an $ARMA(M, N)$ model.
- Engineering
  - *Infinite Impulse Response* (*IIR*): The output can continue after the input stops.
  - *Finite Impulse Response* (*FIR*): The output stops after the input stops.

For the statistician these are random variables, while the engineer usually tries to make sure that they are not. *Trending* in a nonstationary signal can be removed by differencing it to some order before model building (first-order time differences remove a linear drift, second-order removes a polynomial trend, and so forth), giving an *ARIMA* (Auto-Regressive Integrated Moving Average) model.

The $z$-transform of the output

$$Y(z) \equiv \sum_{n=-\infty}^{n=\infty} y_n z^n \tag{16.2}$$

provides a complete analysis of the system (Chapter 2). Since convolution in the time domain equals multiplication in the $z$ domain, the $z$-transform can easily be solved:

$$y_t = a_t + \sum_{m=1}^{M} b_m y_{t-m} + \sum_{n=0}^{N} c_n x_{t-n}$$

$$Y(z) = A(z) + B(z)Y(z) + C(z)X(z)$$
$$= \frac{A(z)}{1 - B(z)} + \frac{C(z)}{1 - B(z)} X(z) \quad . \quad (16.3)$$

The $z$-transform of the output consists of two terms. The first depends on the initial transient, and the second term is equal to the $z$-transform of the input multiplied by a system *transfer function* that is independent of the input. The output $Y(z)$ consists of a ratio of polynomials $A(z)$ and $C(z)$ divided by $1 - B(z)$ reflecting the system's structure, and a possible non-polynomial part from the input $X(z)$. The numerators, due to the inputs, can have *zeros*, and the denominator, due to the memory of the output, can have *poles*. As we've seen, the location of these poles and zeros determines the system's characteristics (such as stability and oscillation frequencies).

The AR and MA coefficients can be determined from the correlation coefficients. Taking $\langle y \rangle$ to denote the time average expectation value of $y$ (written in the statistics literature as $E[y]$), the *autocorrelation function* is defined to be

$$\kappa_\tau \equiv \frac{\langle (y_t - \langle y_t \rangle)(y_{t-\tau} - \langle y_t \rangle) \rangle}{\langle (y_t - \langle y_t \rangle)^2 \rangle}$$
$$= \frac{\langle (y_t - \mu_y)(y_{t-\tau} - \mu_y) \rangle}{\sigma_y^2} \quad . \quad (16.4)$$

$\mu_y$ is the mean and $\sigma_y^2$ is the variance. This can also be written as

$$\kappa_\tau = \frac{\langle (y_t - \mu_y)(y_{t-\tau} - \mu_y) \rangle}{\langle (y_t - \mu_y)(y_t - \mu_y) \rangle}$$
$$= \frac{\langle y_t y_{t-\tau} \rangle - \mu_y \langle y_{t-\tau} \rangle - \mu_y \langle y_t \rangle + \mu_y \mu_y}{\langle y_t y_t \rangle - \mu_y \langle y_t \rangle - \mu_y \langle y_t \rangle + \mu_y \mu_y}$$
$$= \frac{\langle y_t y_{t-\tau} \rangle - \mu_y^2}{\langle y_t y_t \rangle - \mu_y^2} \quad (16.5)$$

since time averages are independent of the time origin for a stationary process. The autocorrelation function ranges from 1 for perfect correlation between two times, to 0 for uncorrelation, and to $-1$ for anticorrelation.

For an MA model ($a = b = 0$), if the input is assumed to be zero mean ($\langle x_t \rangle = \mu_x = 0$) then $\mu_y = 0$, and the autocorrelation function becomes

$$\kappa_\tau = \frac{\left\langle \left( \sum_{n=0}^{N} c_n x_{t-n} \right) \left( \sum_{n'=0}^{N} c_{n'} x_{t-\tau-n'} \right) \right\rangle}{\left\langle \left( \sum_{n=0}^{N} c_n x_{t-n} \right) \left( \sum_{n'=0}^{N} c_{n'} x_{t-n'} \right) \right\rangle}$$
$$= \frac{\sum_{n=0}^{N} \sum_{n'=0}^{N} c_n c_{n'} \langle x_{t-n} x_{t-\tau-n'} \rangle}{\sum_{n=0}^{N} \sum_{n'=0}^{N} c_n c_{n'} \langle x_{t-n} x_{t-n'} \rangle} \quad . \quad (16.6)$$

If the input $x$ is an uncorrelated stochastic process ($\langle x_i x_j \rangle = 0$ for $i \neq j$) then the MA coefficients are related to the autocorrelation function by

$$\kappa_\tau = \begin{cases} \frac{\sum_{n=\tau}^{N} c_n c_{n-\tau}}{\sum_{n=0}^{N} c_n^2} & (\tau \leq N) \\ 0 & (\tau > N) \end{cases} \quad . \quad (16.7)$$

Given the $c$'s we can calculate the autocorrelation function, or this relationship can be inverted to find a set of $c$'s to match a given autocorrelation function.

Similarly, multiplying both sides of an AR model by $y_{t-\tau}$ and averaging gives

$$\langle y_t y_{t-\tau} \rangle = \sum_{m=1}^{M} b_m \langle y_{t-m} y_{t-\tau} \rangle \quad , \qquad (16.8)$$

and then after normalizing by the variance

$$\kappa_\tau = \sum_{m=1}^{M} b_m \kappa_{\tau-m} \quad . \qquad (16.9)$$

Unlike the MA case the autocorrelation function need not vanish after $M$ steps. This linear set of equations, called the *Yule–Walker* equations, can be inverted to relate the AR coefficients to the autocorrelation function. *Levinson–Durbin recursion* is an efficient way to do this [Levinson, 1947, Durbin, 1960].

Unlike the simplicity of AR and MA models there is not a unique algorithm to find the best ARMA model to describe a data set (but there is a lot of heated debate about how to do it). The *Box–Jenkins* procedure is a popular recursive solution [Box et al., 1994]. It's always possible to trade off more or less of $M$ versus $N$ in selecting the order of an ARMA model; the *Akaike Information Criteria* ($AIC$) and its Bayesian cousin the $BIC$ do this by assigning an informational cost to the number of parameters, to be minimized along with the model error [Akaike, 1979].

## 16.2 THE BREAKDOWN OF LINEAR SYSTEMS THEORY

The essence of linear systems theory is expressed by the *Wold Decomposition*: any stochastic process can be separated into the sum of two processes – a deterministic one that is a linear function of its past values, and a stochastic one that is a linear function of previous values of an uncorrelated random variable [Priestley, 1981]. Once these two pieces have been found there is nothing more that can be said about the system.

If you limit yourself to linear models, that is. Even simple nonlinearities can be completely misunderstood by a linear analysis. Consider the two simple iterated maps shown in Figure 16.1. The first one,

$$x_{n+1} = 2x_n \ (\text{mod } 1) \qquad (16.10)$$

is called the *mod map* and it shifts every bit of $x$ (written in a binary fractional expansion) over one place and then discards the most significant bit. This means that the trajectory of the system (for example, which branch it is on) is determined solely by the initial condition. If the initial condition is a real number with digits that appear to be random, say $\pi$, then this map will generate a broadband power spectrum. There is a simple truth (equation (16.10)), but a linear model is forced to find the best single straight line to fit what is really two straight lines. The mismatch can only be attributed to stochastic inputs.

The second map

$$x_{n+1} = \lambda x_n (1 - x_n) \qquad (16.11)$$

(the logistic map) arises in a variety of systems such as chemical reactions, electrical circuits, hydrodynamic flows, and population dynamics, because of the universal properties

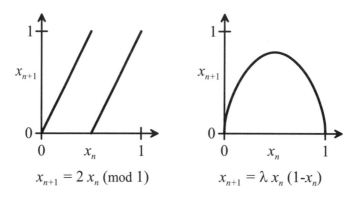

Figure 16.1. Two simple nonlinear maps.

of a smooth map with a peak (the behavior is determined by the lowest-order term in a Taylor expansion around a maximum) [Collet & Eckmann, 1980]. For $\lambda = 4$ each iteration of this simple chaotic map once again reveals one bit of information from the initial condition, and once again the output cannot be described with linear systems theory.

The problem in both cases is that the mod map and the logistic map are poorly approximated by a straight line (a linear model); the same is true in higher dimensions, where a linear model is equivalent to a hyperplane. These examples suggest two generalizations [Priestley, 1991]:

- *TAR* (Threshold Autoregressive): partition the space into two or more regions, each with a different linear model.
- *Volterra series*, *NAR* (Nonlinear Autoregressive): include bilinear and higher-order terms in the model.

The recognition of these alternatives is almost as old as Yule's original analysis, but the problem with them is that some extra kind of insight is needed to decide how to partition the space or chose the higher-order terms. Our first step in addressing these questions will be to bring in a deeper notion of the state of a nonlinear system.

## 16.3 STATE-SPACE RECONSTRUCTION

*Embedding* or *state-space reconstruction* is best introduced with a simple example. Consider the *Lorenz set* of three coupled nonlinear first-order differential equations

$$\dot{x} = \sigma(y - x)$$
$$\dot{y} = -xz + bx - y$$
$$\dot{z} = xy - rz$$

$$\sigma = 10, \ b = 8/3, \ r = 28 \quad , \tag{16.12}$$

studied by Ed Lorenz at MIT as the first terms in a Galerkin approximation of atmospheric convection [Lorenz, 1963]. Although they retain little connection to the original atmospheric problem, they do exactly describe convection in a thin annular ring heated

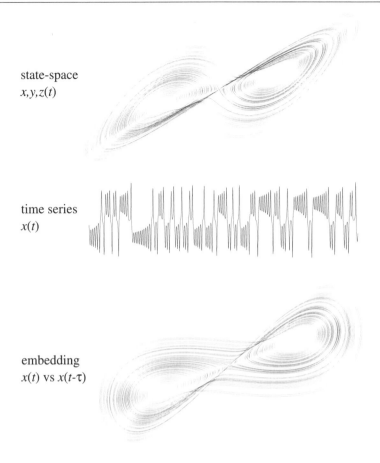

Figure 16.2. Embedding the Lorenz set.

asymmetrically, where the variables are the temperature and pressure gradient and the angular velocity. The top part of Figure 16.2 shows the trajectory in the $x, y, z$ space, looking down on the $z$ axis. The time series of $x(t)$ is shown in the middle, discarding any knowledge of $y$, $z$, or the governing equations. This looks rather random, and indeed has a broadband power spectrum. The bottom part of the figure takes the time series and plots $x(t)$ versus $x(t - \tau)$ (where $\tau$ is a fixed arbitrary delay). It looks very similar to the upper plot of $x$, $y$, and $z$: even though it is stretched, all of the details are the same. The theory of embedding explains why this is not just a remarkable coincidence but is a deep property of time-lag spaces.

It is important to be clear about the spaces relevant to embedding, shown in Figure 16.3:

(a) *The physical degrees of freedom*: For example, a convecting fluid is described by the continuous (infinite-dimensional) fields of the Navier–Stokes PDEs.
(b) *The configuration state-space*: In this space a continuous time ($\dot{\vec{x}} = \vec{f}(\vec{x})$) or discrete time ($\vec{x}_{n+1} = \vec{f}(\vec{x}_n)$) set of equations governs the evolution of the state vector $\vec{x}$, which describes a distinguishable macroscopic state of the system (such as the temperature, pressure, and angular variables in the Lorenz set). Dissipation and symmetry will reduce a PDE to such a set of coupled ODEs [Temam, 1988].

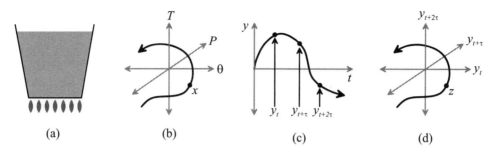

Figure 16.3. The spaces relevant to embedding.

(c) *The experimental observable*: This is an accessible continuous time $y(\vec{x}(t))$ or discrete time $y(\vec{x}_n)$ observable (or set of observables) that depends on the internal configuration variables, such as a temperature or velocity probe in a convecting fluid.

(d) *The reconstructed state-space*: This space is most simply found by time lags $\vec{z}(t) = (y_t, y_{t+\tau}, \ldots, y_{t+(D-1)\tau})$. According to the embedding theorem, the trajectory in this lag space produced from a scalar observable differs from the trajectory in the original configuration space by no more than a smooth local change of coordinates.

Unless specifically noted, all the results in this chapter apply equally to continuous time flows and discrete time maps.

Embedding is both a simple well-known prescription (use time lags) and a profound insight (an accessible variable can explicitly retrieve unseen internal degrees of freedom). Time lags have long been used in studying signals; what was new was the recognition that there is a deep connection between them and a system's inaccessible internal state. This was first proved by Floris Takens [Takens, 1981], and the idea was suggested by David Ruelle, and others [Packard et al., 1980]. In the original form it states that the configuration space vector $\vec{x}(t)$ and the one that is reconstructed from a time series $\vec{z}(t) = (y_t, y_{t+\tau}, \ldots, y_{t+(d-1)\tau})$ will *generically* differ by no more than a smooth invertible local change of coordinates (i.e., it is an *embedding*, which is locally *diffeomorphic*) for all $\tau$ as long as $d$ is large enough, $y$ depends on at least some of the components of $\vec{x}$, and the other components are coupled by the dynamics. If the system's trajectory is thought of as being printed on a rubber sheet, according to the diffeomorphic property it can be stretched but not cut. The word embedding is used in the literature both in this exact technical topological sense, and more loosely to refer to the entire procedure.

The embedding theorem is true "generically." This means that there are isolated cases for which it will fail (such as sampling a sine wave exactly at its period) but that these will be removed by a small perturbation (such as sampling a sine wave at the period plus a small $\epsilon$). In practice, the problems that embedding encounters come not from this assumption of genericity but from real-world problems such as short nonstationary data sets. Although the embedding theorem holds "for all" values of the time lag $\tau$, in practice this will not be the case. As $\tau$ increases from 0 the embedding grows away from the diagonal of the embedding space (all the lags are equal), but for small $\tau$ this can be masked by noise or a finite sampling resolution. If $\tau$ becomes too large, it can bring distant parts of the trajectory accidentally close together, which can once again be masked by noise or the sampling resolution. The last qualifier, "large enough", is explained by the *Whitney embedding theorem*: an arbitrary $D$-dimensional manifold can always be embedded in a $2D$-dimensional Euclidean space (although $D$ may be sufficient) [Guillemin & Pollack, 1974]. For example, a sheet of paper (which is two-dimensional) can be embedded into a 2D space. If the ends of the sheet of paper are joined to form a loop (or with a twist, forming a Möbius strip), then it will require a 3D embedding space. If the opposite ends are joined, forming a Klein bottle, then a 4D space is needed to avoid crossings. That's the maximum; according to the Whitney theorem, no more than 4D can be needed. We will return to these qualifiers in the Section 16.4.

The proof of the embedding theorem has two parts: one showing that local properties are preserved, such as the rate of divergence of trajectories, and the other showing that global properties are preserved, such as the linking of trajectories. The governing equations $\dot{\vec{x}} = \vec{f}(\vec{x})$ imply a solution function $\vec{x}_{t+\tau} = \vec{\varphi}_\tau(\vec{x}_t)$, which for any nontrivial system is hopelessly complicated and cannot be written down in a closed form. This (unknown) solution function in turn implies an embedding mapping function

$$\vec{z}_t = \vec{\Phi}(\vec{x}_t) = y(\vec{x}_t), y(\vec{\varphi}_\tau(\vec{x}_t)), \ldots, y(\vec{\varphi}_{t+(d-1)\tau}(\vec{x}_t)) \quad . \tag{16.13}$$

The local behavior of a point is given by the linearization of this map

$$\vec{z} + d\vec{z} = \vec{\Phi}(\vec{x}) + (D\vec{\Phi}) \cdot d\vec{x} \quad . \tag{16.14}$$

The local part of the embedding proof follows if this map is of full rank (basis vectors in the $\vec{x}$ space span the $\vec{z}$ space under the mapping). A typical term of the derivative of the map is

$$(D\vec{\Phi})_{ij} = \frac{\partial \Phi_i}{\partial x_j} = \frac{\partial y(\vec{\varphi}_{i\tau}(\vec{x}))}{\partial x_j} \quad ; \tag{16.15}$$

for the mapping not to be of full rank, one column of this matrix must be proportional to another. Since this mixes up shifts in time with shifts among variables it requires a tremendous degeneracy to occur, which will be removed by almost any small perturbation to the problem. Making this plausible argument rigorous follows from the theory of *transversality*. An example of a transversality result is the observation that the intersection between two ropes will usually remain if the ropes are moved in 2D, but not in 3D. The global part of the embedding proof follows from the uniqueness of solutions to differential equations.

The embedding result that time lags retrieve internal degrees of freedom appears similar to traditional engineering practice, which is full of the use of time lags and "state-space" models. The three key new insights are that time lags are not just convenient, they reconstruct all of the internal degrees of freedom that influence the observable; once enough time lags are used on a noise-free system (within a factor of 2 of the degrees of freedom of the system, depending on the complexity of the geometry) then there is nothing more to be learned by more lags; and even though the embedding space has been stretched by an unknown change of coordinates it is still possible to characterize, predict, and model using topologically invariant quantities.

Embedding has since been generalized in a few directions. One important result is that any linear transformation of a time-lagged vector is an embedding, with the embedding dimension being given by the rank of the (not necessarily square) transformation matrix [Sauer et al., 1991]. This means that embedding can be related to signal processing. For example, it is possible to do noise reduction along with embedding by taking an FFT, applying a filter, and then taking an inverse FFT, by using a wavelet transform and retaining some coefficients, or applying a circulant filter matrix. A related result is that if a time series consists of distinguishable events (such as pulses), the times between the events can be used for embedding.

A second generalization of embedding comes from recognizing that the probability distribution $p(\vec{z})$ in the embedding space, needed for most characterization and prediction algorithms, is defined in terms of an arbitrary test function $f(\vec{z})$ by

$$\langle f(\vec{z}_t) \rangle_t = \lim_{T \to \infty} \frac{1}{T} \int_0^T f(\vec{z}_t) dt = \int f(\vec{z}) \, p(\vec{z}) \, d\vec{z} \quad . \tag{16.16}$$

Note that this is not the same as assuming that the signal is *ergodic* (in an ergodic system all trajectories have the same long-term time average, which can be written as a space average); it is simply the definition of a probability distribution for an observed signal. If a complex exponential is used for the test function

$$\langle e^{i\vec{k}\cdot\vec{z}_t} \rangle_t = \langle e^{i\vec{k}\cdot(y_t, y_{t+\tau}, \ldots, y_{t+(d-1)\tau})} \rangle_t = \int e^{i\vec{k}\cdot\vec{z}} p(\vec{z}) \, d\vec{z} \tag{16.17}$$

we see that the time average is the Fourier transform of the embedded probability distribution, permitting embedding to be done without recording time series if the expectations can be directly measured, and providing a way to separate an unknown signal from known noise (a nonlinear generalization of a Wiener filter). This is a time-lagged *characteristic function*; its power series expansion is

$$\langle e^{i\vec{k}\cdot\vec{z}_t} \rangle_t = \int e^{i\vec{k}\cdot\vec{z}} p(\vec{z}) \, d\vec{z} \tag{16.18}$$

$$= \sum_{n_1=0}^{\infty} \sum_{n_2=0}^{\infty} \cdots \sum_{n_d=0}^{\infty} \frac{(ik_1)^{n_1}}{n_1!} \frac{(ik_2)^{n_2}}{n_2!} \cdots \frac{(ik_d)^{n_d}}{n_d!} \langle y_t^{n_1} y_{t+\tau}^{n_2} \cdots y_{t+(D-1)\tau}^{n_d} \rangle_t \quad .$$

This shows that the probability distribution depends on a infinite family of powers of multiple-time correlation functions; the measurements used in linear systems theory are based on just the first few terms of this series (single-time correlations for the mean and variance, two-time correlations for the spectrum, and three-time correlations for the *bispectrum*). It also provides a connection between embedding and deterministic expectation values of functions of the random variables of a stochastic process.

So far we have been discussing embedding *autonomous* systems that have no external inputs, but most of the world is not so obliging as to stay still while we watch it. A much more relevant case is

$$\frac{d\vec{x}}{dt} = \vec{f}(\vec{x}, \vec{u}) \quad , \tag{16.19}$$

where $\vec{u}$ is a vector of the inputs to a system. If the inputs are known then embedding

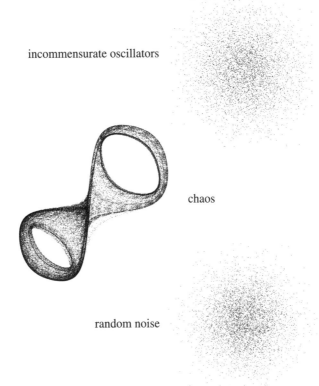

Figure 16.4. The need for characterization.

using lags of both the inputs and the outputs will still recover the internal state [Casdagli, 1992, Stark *et al.*, 1997].

## 16.4 CHARACTERIZATION

We could look at Figure 16.2 and instantly recognize the simple structure, but this will not usually be the case. Figure 16.4 shows the 2D embedding for three electrical systems [Gershenfeld, 1992]. The bottom one is Johnson noise, the thermodynamic fluctuations in an electrical resistor. This is a random Gaussian stochastic process, and the embedding is a Gaussian spot. The middle system is a chaotic nonlinear circuit, and the low-dimensional dynamics is readily apparent. However, the top example is a 2D embedding of the sum of 12 incommensurate electrical oscillators. The 2D projection of a 12D torus looks Gaussian, and in fact it approximately is (by the Central Limit Theorem). Some kind of characterization is needed to "see" in 12D to recognize that this is just a torus.

The simplest kinds of characterization are basic sanity checks, easy to forget amid advanced algorithms. To start, is the time series stationary? Are its statistical properties the same at the beginning and the end? If not, a naive analysis will mix unrelated behaviors of the system. Are there obvious features that a model must include (such as non-negativity)? It's important to look at data to recognize these basic features.

Linear correlations in a data set can mask nonlinear structure. For a 2D embedding, the moments of the probability distribution are

$$M_{ij} = \int\int x_i x_j p(x_i, x_j)\, dx_i dx_j$$
$$= \lim_{T \to \infty} \frac{1}{T} \int_0^T x_i(t) x_j(t)\, dt \quad . \tag{16.20}$$

Therefore, $M_{11} = M_{22}$ = the variance $\sigma^2$, and $M_{12} = M_{21} = \sigma^2 \kappa(\tau)$ (where $\kappa(\tau)$ is the autocorrelation function). The ratio of the eigenvalues of this correlation matrix gives the width-to-length ratio of the embedded probability distribution

$$\frac{\lambda_-}{\lambda_+} = \frac{\kappa(0) - \kappa(\tau)}{\kappa(0) + \kappa(\tau)} \quad . \tag{16.21}$$

For a power-law power spectrum $S(\omega) = |\omega|^{-\alpha}$ ($\alpha = 1 = 1/f$ noise, $\alpha = 2$ = *diffusion noise*, ...), with a bandwidth of $10^{-3}$–$10^3$ Hz and a delay $\tau = 1$ s, $\alpha = 1 \Rightarrow \lambda_-/\lambda_+ = 0.51$, $\alpha = 2 \Rightarrow \lambda_-/\lambda_+ = 0.005$, $\alpha = 3 \Rightarrow \lambda_-/\lambda_+ = 0.0001$. As $\alpha$ is increased the distribution becomes so skinny that given finite sampling resolution it will erroneously appear to be one-dimensional at a fixed $\tau$ [Gershenfeld, 1992].

An important sanity check for linear structure in a presumed nonlinear signal is to use *surrogate data* [Theiler et al., 1992]. In the simplest form, these are generated by taking the Fourier transform of the real-valued time series $y(t)$ to find $Y(\omega) = A(\omega)e^{i\theta(\omega)}$, then setting the phases $\theta(\omega)$ to random values symmetrically ($Y(-\omega) = Y(\omega)$), and transforming back. The resulting series will be real-valued, and because $A(\omega)$ is unchanged it will have the same power spectrum (and autocorrelation function), but any nonlinear relationship among the points will have been randomized. If the result of a test is the same on the original and the surrogate data, the test can only be sensitive to the linear structure in the data. More sophisticated versions fit an ARMA model and generate new realizations of the stochastic process.

### 16.4.1 Dimensions

Dimension measurement was one of the first widespread characterization techniques used in embedding spaces [Grassberger & Procaccia, 1983a]. Although it is now used less commonly in favor of more powerful and reliable techniques, it is an important concept for describing a set of points. Define a point correlation function for a data set of $N$ points by the number of neighbors in a hypersphere of radius $r$ around a reference point

$$C_i(r) \equiv \frac{1}{N} \text{ (the number of } \vec{z}_j \text{ within } r \text{ of a reference point } \vec{z}_i) \quad , \tag{16.22}$$

and average this to get the *radial correlation function*

$$C(r) \equiv \frac{1}{N} \sum_{i=1}^{N} C_i(r) \quad . \tag{16.23}$$

In general, in the limit of small $r$ this will scale as a power of the distance (the first power for points distributed along a line, the second power for points on a surface, ...), and so this exponent defines the *correlation dimension* $\nu$

$$C(r) \approx A r^{\nu} \quad . \tag{16.24}$$

Taking the logarithm,

$$\log C(r) = \log A + \nu \log r$$

$$\frac{\log C(r)}{\log r} = \frac{\log A}{\log r} + \nu \quad . \tag{16.25}$$

In the limit $r \to 0$ the first term on the right hand side will vanish, and so

$$\nu \equiv \lim_{r \to 0} \frac{\log C(r)}{\log r} \quad . \tag{16.26}$$

If $\nu$ is not an integer, the data set is a *fractal*.

The correlation dimension is invariant under a smooth change of coordinates (the embedding). In practice, the correlation dimension is measured for successively larger embedding dimensions. If the data are random they will fill the embedding space and the measured dimensions will equal the embedding dimension, but if the data are deterministic then the measured dimension will stop growing once the embedding dimension is reached. At this point the measured dimension (or the smallest integer greater than the measured dimension) gives the number of local degrees of freedom needed to describe a state of the system.

The number of data points needed to estimate a dimension can roughly be estimated by recognizing that $C(r) \sim r^{\nu}$, therefore if a reliable estimate of the slope on a log–log plot requires a decade of scaling ($r_0 \to 10\, r_0$), then the number of points required is $10^{\nu}$ (with some controversy over the base of the exponent). This simple argument is not really correct, because the real problem is that in a high-dimensional space all of the points are near the surface of the distribution, and so it becomes impossible to record a "typical" point in the interior (Section 12.5). This results in a data requirement for small dimensions that is weaker than that predicted by a simple exponential estimate, but for large dimensions it is much worse. For sane amounts of data (say, $10^7$ points) from an ideal stationary system, this limits measured dimensions to be less than roughly 10–20 degrees of freedom.

Correlation dimensions can be generalized to an infinite family of *generalized dimensions* based on the scaling of moments of the probability distribution, estimated by covering with boxes of side length $l$:

$$D_q \equiv \lim_{l \to 0} \frac{1}{q-1} \frac{\log_2 \sum_i p_d(\vec{x}_i)^q}{\log_2 l}$$

$$D_0 \equiv \lim_{l \to 0} -\frac{\log_2 \sum_i p_d(\vec{x}_i)}{\log_2 l} \quad \text{(Hausdorff dimension)}$$

$$D_1 \equiv \lim_{l \to 0} \frac{\sum_i p_d(\vec{x}_i) \log_2 p_d(\vec{x}_i)}{\log_2 l} \quad \text{(Information dimension)}$$

$$D_2 \equiv \lim_{l \to 0} \frac{\log_2 \sum_i p_d(\vec{x}_i)^2}{\log_2 l} \quad \text{(Correlation dimension)} \quad . \tag{16.27}$$

The information dimension follows from taking the limit, and the second moment is equal to the correlation dimension because the square of the number of points in a box is equal to the number of pairs of points in the box [Gershenfeld, 1989]. The $D_q$ are equal for *homogeneous fractals*, and their spread measures how singular the distribution is. Their values are closely related [Beck, 1990]

$$D_2 \geq D_\infty \geq D_2/2 \geq 0 \quad , \tag{16.28}$$

therefore any of these can be used to test for degrees of freedom.

### 16.4.2 Lyapunov Exponents

Dimension measurement provides a way to estimate the number of degrees of freedom of a system without requiring an explicit model of the system, but it provides no insight into the time evolution of the system. *Lyapunov exponents* are one common way to characterize the time dependence.

If

$$\frac{d\vec{x}}{dt} = \vec{f}(\vec{x}) \quad , \tag{16.29}$$

then a small displacement around a point

$$\frac{d\vec{x} + \vec{\delta}}{dt} = \vec{f}(\vec{x} + \vec{\delta}) \approx \vec{f}(\vec{x}) + (D\vec{f}) \cdot \vec{\delta} \tag{16.30}$$

will evolve according to the linear differential equation

$$\frac{d\vec{\delta}}{dt} = (D\vec{f}) \cdot \vec{\delta} \quad . \tag{16.31}$$

The Lyapunov exponents $\lambda_i$ are the time average of the eigenvalues of the local linearization $(D\vec{f})$.

Lyapunov exponents are preserved under a locally linear change of coordinates and thus can be measured on embedded data. Locally, positive exponents correspond to directions that expand exponentially, and negative exponents to directions that contract exponentially. If an ensemble of trajectories are started in a small hypersphere, the exponents give the growth or contraction rates $\exp(\lambda_i t)$ of the principal axes of the resulting ellipsoid. The sum of all the exponents is the volume expansion rate (zero for conservative systems, negative for dissipative systems), and the sum of the positive exponents is the rate at which new information enters the system (we will revisit this in the next section). A chaotic system has one or more positive exponents but a negative total sum, and a Hamiltonian stochastic system has one or more positive exponents and a total sum of zero. The exponents are related to the dimension by the Kaplan–Yorke conjecture [Frederickson *et al.*, 1983]

$$D_1 = j + \frac{\sum_{i=1}^{j} \lambda_i}{|\lambda_{j+1}|} \quad \left( \sum_{i=1}^{j} \lambda_i > 0 \,, \; \sum_{i=1}^{j+1} \lambda_i < 0 \right) \tag{16.32}$$

(ordering the exponents such that $\lambda_1 > \lambda_2 > \ldots$). This is an explicit statement of the connection between dissipation and dimensional reduction, underlying the applicability of embedding to systems governed by PDEs.

Dissipation in chaotic systems requires state-space volume to contract, positive exponents lead to continuous divergence of trajectories, but trajectories cannot cross because of uniqueness of solutions. The only way to satisfy these conflicting requirements is for the trajectories to lie on an infinitely interleaved *strange attractor*.

### 16.4.3 Entropies

Correlation functions, the essential tool for measuring dependencies in a linear system, are useless for nonlinear systems. Signals from even simple nonlinear systems can have broadband power spectra and hence featureless correlation structure. Information-theoretic quantities provide an elegant alternative that captures the essential features of a correlation function, and more. Entropy has a long history in dynamics starting with Boltzmann and kinetic theory, and evolving through Szilard's one-atom analysis of Maxwell's demon, to Shannon's information theory, and more recently ergodic theory due to Kolmogorov and others [Leff & Rex, 1990]. More recently, Shaw and Fraser have helped reintroduce entropy back to its roots in dynamics [Shaw, 1981, Fraser, 1989, Gershenfeld, 1993].

Let's start by assuming that we have an observable quantized to one of $N$ integer values

$$y(t) \in \{1, \ldots, N\} \tag{16.33}$$

(we will soon examine the dependence on $N$). From measurements of $y$ we can estimate the probability distribution; naively this can be done by binning, taking the ratio of the number of times a value was seen to the total number of points

$$p_1(y) = n_y/n_T \quad . \tag{16.34}$$

Chapter 14 discusses the limitations of, and alternatives to, this simple estimator.

The *entropy* of this distribution is given by

$$H_1(N) = -\sum_{y=1}^{N} p_1(y) \log_2 p_1(y) \quad . \tag{16.35}$$

It is the average number of bits required to describe a sample taken from the distribution, i.e., the expected value of the *information* in a sample. The entropy is a maximum if the distribution is flat (we don't know anything about the next point), and a minimum if the distribution is sharp (we know everything about the next point). Similarly, for a point $\vec{z}$ in the lag space we can ask for the joint probability to see the corresponding sequence in the time series

$$p_d(y_t, y_{t-\tau}, \ldots, y_{t-(d-1)\tau}) \approx n_{\vec{z}}/n_T \quad , \tag{16.36}$$

and measure the *block entropy*

$$H_d(\tau, N) = -\sum_{y_t=1}^{N} \cdots \sum_{y_{t-(d-1)\tau}=1}^{N} p_d \log_2 p_d \tag{16.37}$$

which gives the average number of bits needed to describe the sequence. In the limit of lag time going to zero, all of the points become the same and so the block entropy becomes equal to the scalar entropy

$$\tau \to 0 \Rightarrow p_d(y_t, y_{t-\tau}, \ldots, y_{t-(d-1)\tau}) = p_1(y)$$
$$\Rightarrow H_d(0, N) = H_1(N) \quad . \tag{16.38}$$

On the other hand, in the limit of long time lags, if the points become independent (the probability distribution factors), then the block entropy becomes $d$ times the scalar entropy

$$\lim_{\tau \to \infty} p_d(y_t, \ldots, y_{t-(d-1)\tau}) = p_1(y_t)p_1(y_{t-\tau}), \ldots, p_1(y_{t-(d-1)\tau}) \tag{16.39}$$

$$\Rightarrow H_d(\tau, N) = dH_1(N)$$

The connection between entropy and dimensions comes from recognizing that [Gershenfeld, 1989]

$$\lim_{q \to 1} D_q = \lim_{N \to \infty} \frac{\sum_{\vec{z}_i} p_d(\vec{z}_i) \log_2 p_d(\vec{z}_i)}{\log_2 N} = \lim_{N \to \infty} \frac{H_d(\tau, N)}{\log_2 N} \quad . \tag{16.40}$$

The scaling of the block entropy with resolution is just the (information) dimension, giving the number of local degrees of freedom of the system.

The *mutual information* is defined to be the difference in the information between two samples taken independently and taken together

$$I_2(\tau, N) = -\sum_{y_t=1}^{N} p_1(y_t) \log_2 p_1(y_t)$$
$$- \sum_{y_{t-\tau}=1}^{N} p_1(y_{t-\tau}) \log_2 p_1(y_{t-\tau})$$
$$+ \sum_{y_t=1}^{N} \sum_{y_{t-\tau}=1}^{N} p_2(y_t, y_{t-\tau}) \log_2 p_2(y_t, y_{t-\tau})$$
$$= 2H_1(\tau, N) - H_2(\tau, N) \quad . \tag{16.41}$$

If the points don't depend on each other then the mutual information is zero:

$$p_2(y_t, y_{t-\tau}) = p_1(y_t)p_1(y_{t-\tau}) \Rightarrow I_2(\tau, N) = 0 \quad , \tag{16.42}$$

and if they are completely dependent then the mutual information is equal to all of the bits in one sample:

$$p_2(y_t, y_{t-\tau}) = p_1(y_t) \Rightarrow I_2(\tau, N) = H_1 \quad . \tag{16.43}$$

Here then is an alternative to correlation functions, measuring the connection between two variables without assuming any functional form other than what is needed to estimate a probability distribution.

The *redundancy* extends mutual information to higher-dimensional spaces [Fraser, 1989]. It is equal to the information in one sample, plus the previous $d-1$ samples, minus the information in $d$ samples:

$$R_d(\tau, N) = H_1(\tau, N) + H_{d-1}(\tau, N) - H_d(\tau, N) \quad . \tag{16.44}$$

If the new point doesn't depend on the previous points,

$$p_d(y_t, \ldots, y_{t-(d-1)\tau}) = p_1(y_t) p_{d-1}(y_{t-\tau}, \ldots, y_{t-(d-1)\tau})$$

$$\Rightarrow H_d = H_1 + H_{d-1} \Rightarrow R_d = 0 \quad, \tag{16.45}$$

then the redundancy is zero, and if the point is completely determined by the previous points,

$$p_d(y_t, \ldots, y_{t-(d-1)\tau}) = p_{d-1}(y_{t-\tau}, \ldots, y_{t-(d-1)\tau})$$

$$\Rightarrow H_d = H_{d-1} \Rightarrow R_d = H_1 \quad, \tag{16.46}$$

then the redundancy is equal to the scalar entropy (all of the bits in the point). The redundancy measures how necessary the bits of the new point were given the past history.

The asymptotic growth rate of the block entropy is called the *source entropy*

$$h(\tau, N) = \lim_{N \to \infty} \lim_{d \to \infty} H_d(\tau, N) - H_{d-1}(\tau, N) \quad. \tag{16.47}$$

The limits don't need to go to infinity. The limit in $d$ is reached at the embedding dimension (if there is one), and the limit in $N$ is reached if there is a *generating partition* (which means that the intersection of all future and backwards iterates of the grid of points at the maximum resolution can be used to specify any neighborhood to an arbitrary precision). Since

$$\tau = 0 \Rightarrow H_{d-1}(0, N) = H_d(0, N) \Rightarrow R_d(0, N) = H_1(N) \quad, \tag{16.48}$$

we see that for small time lags the decay of the redundancy gives the source entropy

$$R_d(\tau, N) = H_1(N) - \tau h(1) \quad. \tag{16.49}$$

It should not be surprising that the source entropy, which is the rate at which information enters the system, is proportional to the volume divergence rate (the sum of positive exponents)

$$h(\tau) = \tau h(1) = \tau \sum_i \lambda_i^+ \quad. \tag{16.50}$$

This is called *Pesin's identity* [Liu & Qian, 1995].

Summarizing: for small time lags, the growth rate of the redundancy with resolution gives the effective sample resolution. As $\tau$ is increased, if $d$ is below the embedding dimension then the redundancy will drop quickly to zero. Once $d$ exceeds the embedding dimension (if there is one), the redundancy at small lags will jump up to the scalar entropy. The slope as $\tau$ is increased then gives the source entropy (the sum of positive exponents), and the growth rate of the entropy with resolution gives the information dimension (local degrees of freedom). The difference between the embedding dimension and the information dimension will depend on how complex the geometry is. Finally, the time lag at which the redundancy falls to zero gives the predictability horizon at that resolution, beyond which the system's state depends on information that you don't have access to. If a time series is analyzed in reverse order, positive and negative exponents switch, allowing the sum of the negative exponents to be estimated. Since flows (continuous

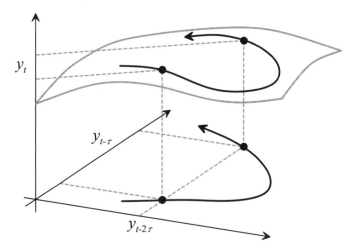

Figure 16.5. Forecasting in a lag space.

time dynamics) have a unique inverse but maps may not, the difference in forward and backward redundancy is a test of the invertibility of the dynamics.

Whew – that's a lot of information to squeeze out of a single statistic. Entropy measurements provides insight into many essential aspects of a system without imposing any specific model in advance; the results of this analysis have significant implications for how a system should best be described and understood. The only catch is that it's based on estimating entropy, which as we saw in Section 14.1 is a tricky thing to do reliably. The problem is that there's no notion of generalization built into a simple entropy estimate. Because of this it can require an enormous amount of data, and be difficult to falsify by making predictions. This is why characterization is intimately connected with forecasting, to which we turn next.

## 16.5 FORECASTING

One of the most important properties that is preserved under a smooth change of coordinates is trajectory crossing (or lack thereof): if a system's internal state is a deterministic function of its past, then the same will hold in an embedding space. As shown in Figure 16.5, this means that for a deterministic system once enough lags are used the dynamics will lie on a single-valued surface. Forecasting in this case reduces to modeling the shape of the surface. This is exactly the kind of fitting problem covered previously, and presents the same kinds of choices [Weigend & Gershenfeld, 1993].

One approach is to seek a global representation for the prediction function $y_{t+1} = f(\vec{z}_t) = f(y_t, y_{t-\tau}, ..., y_{t-(d-1)\tau})$. A convenient way to do this is to use the functional orthogonalization covered in Section 12.2 [Giona et al., 1991]. Let $p(\vec{z})$ be the probability distribution in the lag space, and $\{f_i(\vec{z})\}$ be a family of functions such as polynomials that have been constructed by Gram–Schmidt orthogonalization to be orthonormal with respect to it

$$\langle f_i(\vec{z})f_j(\vec{z})\rangle_t = \int f_i(\vec{z})f_j(\vec{z})p(\vec{z})\,d\vec{z} = \delta_{ij} \quad . \qquad (16.51)$$

The expansion coefficients that we want are

$$f(\vec{z}) = \sum_i a_i f_i(\vec{z}) \quad . \tag{16.52}$$

By construction, these can be found from the orthonormality condition by summing over the data:

$$a_i = \langle f(\vec{z}_t) f_i(\vec{z}_t) \rangle = \langle y_{t+1} f_i(\vec{z}_t) \rangle = \frac{1}{N} \sum_{t=1}^{N} y_{t+1} f_i(\vec{z}_t) \quad . \tag{16.53}$$

This is certainly easy to implement, and can work well if the surface is not too complicated, but like all global functions it has trouble capturing local features. An alternative is to replace the global predictor by a family of local models [Farmer & Sidorowich, 1987]; this requires some strategy for deciding where to place the models and how to size their neighborhoods (more about that in a moment).

So far we've been discussing *point predictions* where we try to determine future values. This makes sense if the system is nearly deterministic but is pointless if it is effectively stochastic. Then, instead of trying to forecast a random variable, it's necessary to model the expected value of observables such as the power spectral density, or the variance of a process (which is all that's needed to get rich on Wall Street [Hull, 1993]). It's possible to forecast a variance by training a second model to predict the errors of a point prediction model; better still is to do some kind of density estimation to be able to answer other questions as well. And given a model of the noise it's possible to do more than just describe it; self-consistently separating it can reduce the noise in a signal, providing a nonlinear analog to the Wold decomposition [Abarbanel *et al.*, 1993, Weigend *et al.*, 1996].

These issues of modeling stochasticity and introducing locality can be addressed by cluster-weighted modeling, using kernel density estimation with local models (Section 14.4). Figure 16.6 shows part of an example time series, a laser fluctuating near the gain threshold (data set A from [Weigend & Gershenfeld, 1993]), and Figure 16.7 shows the resulting model using linear covariance clusters.

Cluster parameters in a lag space can be related to the time series techniques described earlier. For example, in terms of the eigenvalues of the cluster-weighted covariance matrix $\mathbf{C}_m = \{\sigma_{1,m}, \ldots, \sigma_{D,m}\}$, the radial correlation integral of the probability distribution is

$$C_m(r) = \int_{-r}^{r} \int_{-r}^{r} p(x_1, \ldots, x_D | c_M) \, dx_1 \ldots dx_D$$

$$= \text{erf}\left(\frac{r}{\sqrt{2\sigma_{1,m}^2}}\right) \cdots \text{erf}\left(\frac{r}{\sqrt{2\sigma_{D,m}^2}}\right) \quad . \tag{16.54}$$

A simple calculation then shows that the cluster's correlation dimension $\nu = D_2$ is

$$\nu_m = \frac{\partial \log C_m(r)}{\partial \log r}$$

$$= \sum_{d=1}^{D} \frac{1}{\text{erf}\left(r / \sqrt{2\sigma_{d,m}^2}\right)} \sqrt{\frac{2}{\pi \sigma_{d,m}^2}} \, e^{-r^2 / 2\sigma_{d,m}^2} \, r \quad , \tag{16.55}$$

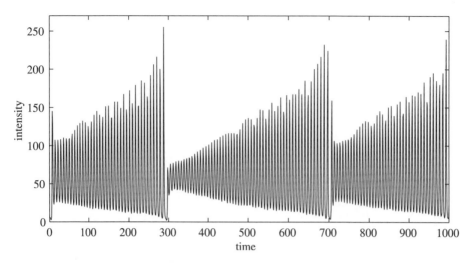

Figure 16.6. Time series of fluctuations in a laser.

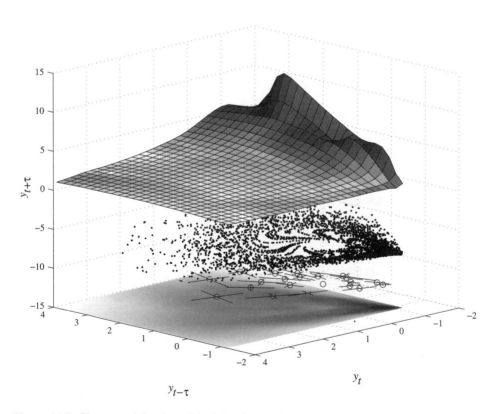

Figure 16.7. Cluster-weighted model of the time series in Figure 16.6. From bottom to top, the predicted input density estimate, the cluster means and covariances, the training data, and the conditional prediction surface with the shading showing the conditional uncertainty.

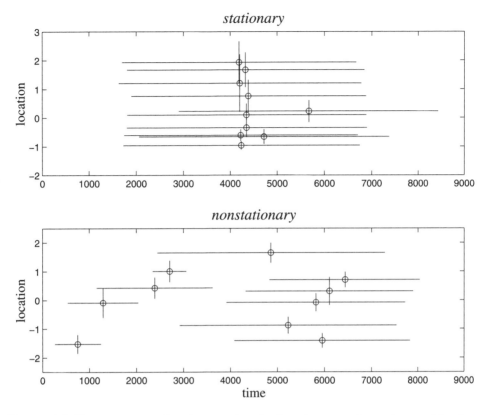

Figure 16.8. Cluster means and variances using absolute time as an input degree of freedom, shown for the time axis and first lag axis for a stationary and nonstationary time series.

and the expected dimension of the whole data set is then given by the expectation

$$\langle \nu \rangle = \sum_{m=1}^{M} \nu_m \, p(c_m) \quad . \tag{16.56}$$

This measures the average number of degrees of freedom that the system is using. Likewise, the log of the conditional output uncertainty provides an estimate of the source entropy, and hence the sum of the positive Lyapunov exponents. It's also possible to build a model recursively for on-line applications by combining new measurements with clusters that summarize old data [Gershenfeld et al., 1998].

Figure 16.8 shows another example, this time with one stationary and one nonstationary series (sets A and D from [Weigend & Gershenfeld, 1993]). The absolute time has been included as one of the input degrees of freedom. For the stationary case the clusters maximize their likelihood by expanding to cover the whole data set; for the nonstationary case they shrink down to an appropriate time scale for locally building a model.

This completes the final chapter, which fittingly has drawn on lessons from many earlier parts of the book to provide useful answers to the challenging questions asked at the beginning of the chapter about coping with the breakdown of linear systems theory. We've seen techniques for allocating local models, combining them into global nonlinear functions, and describing the essential properties of a system without making restrictive assumptions about it.

## 16.6 SELECTED REFERENCES

[Weigend & Gershenfeld, 1993] Weigend, Andreas S., & Gershenfeld, Neil A. (eds) (1993). *Time Series Prediction: Forecasting the Future and Understanding the Past*. Santa Fe Institute Studies in the Sciences of Complexity. Reading, MA: Addison–Wesley.

The results of a comparative study in which many new and old algorithms were applied to common data sets.

[Hamilton, 1994] Hamilton, J. D. (1994). *Time Series Analysis*. Princeton, NJ: Princeton University Press.

A definitive reference for time series analysis.

## 16.7 PROBLEMS

(16.1) Consider the Henon map

$$x_{n+1} = y_n + 1 - ax_n^2 \qquad (16.57)$$
$$y_{n+1} = bx_n$$

for $a = 1.4$ and $b = 0.3$.

(a) Explore embedding by plotting

1. $y_n$ versus $x_n$.
2. $x_n$ versus $n$.
3. The power spectrum of $x_n$ versus $n$.
4. $x_{n+1}$ versus $x_n$.
5. $x_{n+2}$ versus $x_n$.
6. $x_{n+1} + y_{n+1}$ versus $x_n + y_n$.

(b) Estimate as many of the nonlinear system properties as you can, such as the embedding dimension, attractor dimension, Lyapunov exponents, and source entropy.

# *Appendix 1* Graphical and Mathematical Software

Successful numerical model building depends on your ability to express algorithms in a form that a computer can understand, and on how well you can see (and perhaps hear or even feel) the model's behavior (respectively, *visualization*, *sonification*, and *haptic* interfaces). Good visualization is essential: many discoveries (such as solitons, [Miles, 1981, Zabusky, 1981]) have come first in simulations and then later been understood analytically. Fortunately, there are a number of good popular tools that help with these tasks. The most important distinction is between compiled languages with subroutine libraries, and interactive environments. The great virtue of the latter is the ease with which problems can be specified and solved, but the penalty can be a slowdown by 1–2 orders of magnitude over the intrinsic hardware speed. Obtaining performance at the limits of processors and output devices usually requires lower-level control. The interactive environments also make it easy to create impressive graphics, but once again for demanding tasks more direct control can be needed. This appendix starts by reviewing important programming languages and libraries, covers some of the interactive environments, and then also briefly introduces four useful and ubiquitous standards that can be used for graphics: Postscript, X Windows, OpenGL, and Java. Sample programs are given to draw and animate lines, images, and surfaces. These short programs are intended to show the minimal code needed to produce output (they leave out some of the steps that are not needed for such simple tasks, but that are part of the full specification for well-behaved programs), to serve as templates that can be modified for other applications, and to provide some guidance for navigating through the (large) language reference manuals.

Don't forget to check your algorithm efficiency before you check your computer efficiency. On a 10 Mflop computer, performing a $10^6$ point Fourier transform based on its definition requires $10^5$ seconds ($\sim$ a day), but it takes only $\sim 2$ seconds using a Fast Fourier Transform that eliminates hidden redundancies in the calculation. And don't forget to check your common sense as you program. Most modeling is based on software and hardware approximations that bear frequent inspection; setting out without test-problems with known solutions is a good recipe for carefully producing nonsense.

## A1.1 MATH PACKAGES

### A1.1.1 Programming Environments

#### A1.1.1.1 Languages

The most important distinction among computer languages for modeling is how far they remove you from the hardware. The fastest code possible comes from learning the instruction set of a processor and hand-coding assembly language. However, this is very slow to write and difficult to understand, and so is almost never done for general-purpose computing. Compiled languages let you express your ideas more naturally and then turn them into reasonably efficient instructions. Interpreted languages can be designed to match particular domains (such as mathematics), and so can allow algorithms to be expressed still more rapidly and conveniently, but usually pay a penalty in producing the least efficient code. The overall goal is to reduce the time to develop and then execute a program. If it takes months to write a program that executes in minutes, it's worth looking for a better language; conversely, if you're going to wait weeks for the output from your program then you should make sure that it's as efficient as possible. It's important to remember that abstraction alone is not a desirable goal unless it helps to make the entire development process more efficient; many graceful programs have ground to a halt from excessive parameter passing.

The two primary compiled languages used for mathematical computing are *C* and *Fortran*. Fortran is much older; it was developed at IBM in 1954 as the first commerical high-level language, and for many years was equivalent to scientific computing. That is what everyone used, and what every computer supported. C was developed later at Bell Labs (the language reference was first published in 1978), originally as the language to write the UNIX operating system and its utilities. C was not intended for mathematics and so some important features from Fortran are missing (such as the `complex` data type). However, C has grown to replace Fortran for most computing on workstations. There are two reasons for this: it has much richer abstractions for data structures and flow control, and it continues to be the language most closely associated with operating systems and so is usually the first and best-supported language on any platform. The exception to this is supercomputers, where Fortran lives on because of the large existing base of programs and programmers.

Both Fortran and C have important extensions that address some of their limitations. The successor to C is C++, which provides much more flexibility in defining operators and objects that can be inherited, modified, and reused. The new version of Fortran is Fortran 90 (named after the year of the IEEE committee that developed it), and the closely related High Performance Fortran (HPF). These add more modern data types and flow control, and add important syntactical conveniences that let many explicit loops be replaced by operators on more complex data structures (such as saying `A=B+C`, where A, B, and C are arrays). This helps the programmer, because the code is simpler to read and write, and the compiler, because it makes the parallelism explicit.

C is used on everything from micro-controllers to supercomputers and so there is an enormous range of commerical and freely available compilers. The best known is GCC from the GNU project; this is perhaps the most thoroughly tested and debugged compiler of any kind because so many people from around the world have contributed to

its development and testing. GCC is available at ftp://prep.ai.mit.edu/pub/gnu/.
Commercial C compilers can rarely beat GCC for robustness and reliability, but many
do offer much more capable interactive visual environments for coding, debugging, and
project management, such as Microsoft's Visual C++ (http://www.microsoft.com).
Because Fortran is more closely associated with large computers there has been less
diversity in its implementations.

The common interpreted languages are rarely used for mathematical computing. Most
*BASIC* implementations are too slow to be useful for anything but the simplest problems, and floating point math can be truly painful in *LISP*. This is beginning to change
with rapid prototyping environments such as Microsoft's Visual BASIC, which permit
easy integration of sophisticated interfaces and operating system features within a mature coding environment. These can be particularly convenient for building complex
user interfaces linked to compiled code. But one interesting long-standing exception is
APL, which has a curious place in the history of computer languages. It was originally
developed to be a more coherent replacement for conventional mathematical notation,
and happened to be useful as a programming language. Therefore, it is ideally suited
for expressing algorithms. It is possible to write one (cryptic) line of APL that expresses
the same thing as many lines of conventional C. Because, unlike C, it is better matched
to the minds of algorithm writers rather than compiler writers, it has languished with
relatively poor support. Although it is still available, its greatest impact may be via its
many fans on the evolution of languages such as Fortran 90 and Mathematica that have
incorporated many of its best features.

One final language with great promise (and equally great hype) that merges many of
the strengths of these other languages is *Java*, covered in Section A1.2.4.

### A1.1.1.2 Subroutine Libraries

There are many algorithms, such as matrix inversion, that are difficult to write well
but that are used routinely. For this reason there are a number of collections of useful
mathematical subroutines.

Perhaps the best-known and most widely used are those found in *Numerical Recipes*
(http://nr.harvard.edu/numerical-recipes). This marvelous collection of about
400 routines covers everything from addint to zroots, and is available in C, Fortran
(77 and 90), Pascal, and BASIC, with more languages coming. The routines are carefully
documented and explained in the accompanying text [Press *et al.*, 1992], and because the
source code is available they are easily modified and incorporated into other programs.
Their best feature is that they reflect the working experience of a number of leading
scientists rather than the theoretical opinions of a few numerical analysts and so replace
rigor with relevance.

Among the major commerical subroutine libraries available are *NAG* (http://www.nag.co.uk) and *IMSL* (http://www.vni.com). These are much more expensive than Numerical Recipes, and do not release the source code and hence are closer to
mysterious black boxes, and so they are less useful for anything other than production
applications which might need the extra support and algorithm sophistication.

Finally, *Netlib* (http://www.netlib.org/) is a broad collection of more specialized
routines that are freely available, as well as related information such as benchmarks on
many machines.

### A1.1.2 Interactive Environments

Interactive environments all provide a programming language, useful built-in primitives, and some kind of graphical capabilities. The best can compete with the performance of much more laboriously written compiled programs, and add capabilities such as animation, sonification, and symbolic math.

Symbolic math is the ability to do math as you would by manipulating symbols instead of numbers wherever possible. $\pi$ remains $\pi$ rather than 3.14159, $\int \sin = \cos$, and so forth. The first major symbolic math package was *Macsyma*, written at MIT in 1968. It languished when the government reduced funding for symbolic math (among other reasons), but is now being actively developed again and has grown into a full featured environment (http://www.macsyma.com).

When first encountered, symbolic math is liberating, because it promises to replace forever your integral tables and pages of calculus. The reality is a bit more limited. The number of math problems that cannot easily be done by hand and can be done by a symbolic package is surprisingly small. Most problems that look hard are hard, and cannot be solved analytically. This means that the greatest strength of symbolic math environments is book-keeping for calcuations that are large but straightforward (such as analytical perturbation theory).

*Maple* is another symbolic math environment. It was originally developed at the University of Waterloo in Canada, and continuing work is also being done by the ETH in Zürich. Because of this ongoing academic grounding, Maple has perhaps the strongest of the symbolic math engines (although people fight religiously about this). It is available in a stand-alone environment (http://www.maplesoft.com), and is commonly used as a module in Matlab or Mathcad (described below).

*Mathematica* (http://www.wri.com) is the creation of Stephen Wolfram, intended to meet an unfilled need that he saw for software that would reflect how people (including himself) do math. It is distinguished by allowing almost any style of programming. Symbolic and variable precision numerical calculations can be freely intermixed, and the language is equally adept at APL-like array processing, C-like coding, and LISP-like data structure manipulation. For this reason Mathematica is very common in physics research. Figure A1.1 shows a simple example, plotting $\sin(x)/x$. Note that there is no need to specify where to plot the points; Mathematica is able to examine the function and decide how to sample it (including the division of 0/0 at the origin that has a limiting value of 1).

*Matlab* (http://www.mathworks.com) has a very different lineage. Cleve Moler was one of the authors of *Linpack* and *Eispack*, now-standard Fortran libraries for matrix problems (respectively, linear systems of equations and eigenvalue problems, available from Netlib). Matlab was originally written as an instructional interactive front end to the routines, and has grown into the most commonly used engineering math environment. The underlying language reflects its lineage from the early days of Fortran linear algebra, but its programming constructs and data structures are catching up with more modern programming style. Many toolboxes are available for application domains such as optimization, neural networks, signal processing, and Maple for symbolic math. To produce a $\sin(x)/x$ plot in Matlab, the input is

```
>> x = -10.5:.2:10.5;
>> plot(x,sin(x)./x)
```

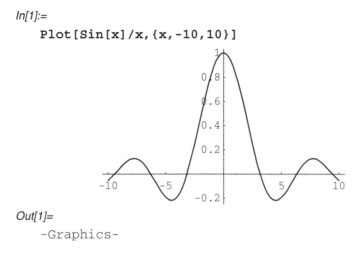

Figure A1.1. Output from Mathematica.

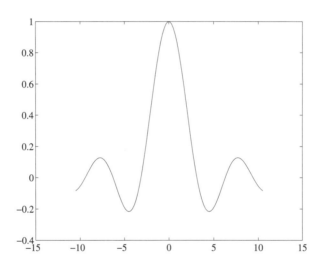

Figure A1.2. Output from Matlab.

and the output is shown in Figure A1.2. It is necessary to explicitly construct an array of points to be plotted, and to avoid the tricky point at the origin, but it is easy to express the construction of a vector and the parallel element-by-element operations on that vector. When it is possible to express an algorithm as such a vectorized construct in Matlab the performance is within an order of magnitude of optimized compiled code, but if looping is needed then the interpretation overhead increases the performance penalty to 1–2 orders of magnitude.

*Excel* (http://www.microsoft.com) started out as a *spreadsheet* aimed at manipulating tables of numbers for financial analysis, and has grown into a widely-used environment for working with all kinds of information. Because of this lineage it has particularly good features for managing and annotating data, and because of its open architecture and popularity there are a large number of extensions available for it.

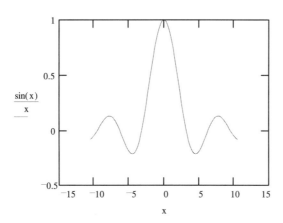

Figure A1.3. Output from Mathcad.

The last math environment to be mentioned is *Mathcad* (`http://www.mathsoft.com`). This has completely rethought the interface to do math on a computer, dispensing with any notion of a command line. Instead, the metaphor is an active piece of paper that can be written on anywhere. When valid expressions are written they are evaluated, and if they reference expressions anywhere else on the sheet they become linked and will update each other if any one of them is changed. Our $\sin(x)/x$ example in Mathcad is shown in Figure A1.3. The expression above the graph defines a vector, and then filling in expressions on axes drawn on the page evaluates them based on the definition of the vector. Changing the values in the vector would automatically update the plot.

## A1.2 GRAPHICS TOOLS

### A1.2.1 Postscript

*PostScript* was developed by John Warnock and Chuck Geschke (at Xerox's Palo Alto Research Center and then at Adobe Systems) as a page description language, motivated by the difficulty of using pixel-based approaches to communicate with increasingly high-resolution printers. It is a true programming language, so that the processing needed to describe a page can be shared between the host and the printer, and it is *device-independent*, so that the same program can be used on any PostScript printer and will take advantage of the available capabilities (most importantly, the resolution). With the advent of Display PostScript, and, more modestly, good software interpreters such as Ghostscript and previewers such as Ghostview (available from `http://www.cs.wisc.edu/~ghost`), it can be used interactively as well as for producing hardcopy. PostScript programs are usually ASCII text. One of the reasons for the success of Postscript is the careful balance it strikes between the generality of recognizing that describing a page is really a programming problem, and the specificity of making it easy to describe a page. Postscript is documented in [Taft & Walden, 1990].

PostScript comes in two flavors: ordinary PostScript (the files often end in .ps), and *Encapsulated PostScript* (these files often end in .eps). Regular PostScript is intended to be sent to a printer; Encapsulated PostScript (sometimes called *EPS*) is a set of conventions for PostScript figures to be included in other documents. EPS does not have a showpage command to eject a page, and it must have *DSC* (Document Structuring Convention) information such as the size of a bounding box that encloses the image (although it is good form for all PostScript programs to include this). Both types of files start with the characters %! to indicate that it is PostScript. An offshoot from PostScript is Adobe's *Portable Document Format*, *PDF*, which eliminates the programmability from PostScript to make a much less flexible language that is better suited for online reading of electronic documents.

An amusing question that is often asked is how to write a filter to examine a PostScript program and determine the number of pages that it will produce. If the program follows the DSC rules then it will include a comment that gives the number of pages it contains and so this is easy, but if it doesn't then this problem is insoluble: determining the number of pages produced would be equivalent to a solution to the *halting problem*, which Turing showed is impossible (page 115). This is the cost of using a universal language in a printer. In return, it's possible to do significant computing in PostScript directly. At one time there were more RISC processors in PostScript printers than in workstations; it's even possible to do 3D rendering directly in a PostScript program!

Most people use PostScript without thinking about it, as something that programs use to talk to printers. However, there are many reasons to learn to write PostScript directly. It is easy to do, it allows you to bypass intervening programs and specify exactly what you want, and knowledge of PostScript is frequently useful for patching up anomalies in the output from other programs.

Our first example C program, psline, shows how to write a PostScript file that sets up a coordinate system, sets the linewidth and darkness, and then draws a line (once again, the function $\sin(x)/x$). The default coordinate system for PostScript has the origin in the lower-left corner of the page, and one unit is equal to 1/72 inch (a *printer's point*). This program is certainly longer and less transparent than Plot[Sin[x]/x], but once you understand this program you will be able to write directly to any PostScript device.

```
/*
 * psline.c
 * (c) Neil Gershenfeld 9/1/97
 * demonstrates drawing lines in PostScript by drawing
 *    sin(k*x)/k*x
 */

#include <math.h>
#include <stdio.h>

int i;
float x,y;
```

```
FILE *outfile;

#define NPTS 1000
#define PAGE_WIDTH 8
#define PAGE_HEIGHT 10

main() {
    outfile = fopen("psline.eps","w");
    fprintf(outfile,"%%! psline output\n");
    fprintf(outfile,"%%%%BoundingBox: 0 0 %f %f\n",
        72.0*PAGE_WIDTH,72.0*PAGE_HEIGHT);
    fprintf(outfile,"gsave\n");
    fprintf(outfile,"/l {lineto} def\n");
    fprintf(outfile,"%f %f scale\n",72.0*PAGE_WIDTH*0.5,
        72.0*PAGE_HEIGHT*0.5);
    fprintf(outfile,"90 rotate\n");
    fprintf(outfile,"1 -1 translate\n");
    fprintf(outfile,"0.5 setgray 0.02 setlinewidth\n");
    x = (1.0 - NPTS)/NPTS;
    y = sin(50.0*x)/(50.0*x);
    fprintf(outfile,"%.3f %.3f moveto\n",x,y);
    for (i = 1; i < NPTS; ++i) {
        x = (2.0*i + 1.0 - NPTS)/NPTS;
        y = sin(50.0*x)/(50.0*x);
        fprintf(outfile,"%.3f %.3f l\n",x,y);
    }
    fprintf(outfile,"stroke\n");
    fprintf(outfile,"grestore\n");
    fclose(outfile);
}
```

On most UNIX systems it can be compiled by `cc psline.c -o psline -lm`. The output that it writes to the file `psline.eps` looks like

```
%! psline output
%%BoundingBox: 0 0 576.000000 720.000000
gsave
/l {lineto} def
288.000000 360.000000 scale
90 rotate
1 -1 translate
0.5 setgray 0.02 setlinewidth
-0.999 -0.006 moveto
-0.997 -0.008 l
-0.995 -0.010 l
...
```

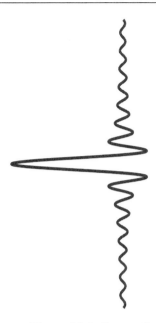

Figure A1.4. Output from psline.

```
0.997 -0.008 l
0.999 -0.006 l
stroke
grestore
```

and sending this to a PostScript interpreter produces the output shown in Figure A1.4. Note that it needs a showpage command added at the end to produce a page if it is sent to a printer.

The second program, psimage, shows the use of the image command to produce a grayscale image of $\sin(r)/r$.

```
/*
 * psimage.c
 * (c) Neil Gershenfeld   9/1/97
 * demonstrates PostScript images by drawing sin(r)/r
 */

#include <math.h>
#include <stdio.h>

int i,j;
float x,y,r,z;
unsigned char grey;
FILE *outfile;

#define NPTS 100
```

```
#define PAGE_WIDTH 8.0

main() {
   outfile = fopen("psimage.eps","w");
   fprintf(outfile,"%%! psimage output\n");
   fprintf(outfile,"%%%%BoundingBox: 0 0 %f %f\n",
      72.0*PAGE_WIDTH,72.0*PAGE_WIDTH);
   fprintf(outfile,"gsave\n");
   fprintf(outfile,"%f %f scale\n",72.0*PAGE_WIDTH,
      72.0*PAGE_WIDTH);
   fprintf(outfile,"%d %d 8 [%d 0 0 %d 0 0] {<\n",
      NPTS,NPTS,NPTS,NPTS);
   for (i = 0; i < NPTS; ++i)
      for (j = 0; j < NPTS; ++j) {
         x = (2.0*i + 1.0 - NPTS)/NPTS;
         y = (2.0*j + 1.0 - NPTS)/NPTS;
         r = 20.0*sqrt(x*x + y*y);
         z = sin(r)/r;
         grey = (unsigned char) (255.0 * (z + 0.3)/1.3);
         fprintf(outfile,"%.2x",grey);
      }
   fprintf(outfile,">} image\n");
   fprintf(outfile,"grestore\n");
   fclose(outfile);
}
```

Its output

```
%! psimage output
%%BoundingBox: 0 0 576.000000 576.000000
gsave
576.000000 576.000000 scale
100 100 8 [100 0 0 100 0 0] {<
3c3e4041
...
44444443
>} image
grestore
```

is shown in Figure A1.5.

### A1.2.2  X Windows

The great contribution of *X Windows* (originally developed at MIT in the mid 1980s) is that it decouples a computer's display from the programs that use it, so that any processor

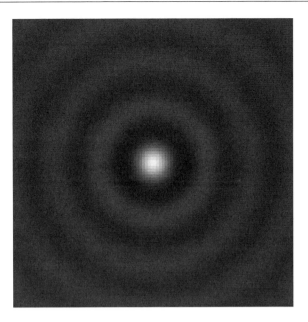

Figure A1.5. Output from `psimage`.

on a network can produce output on any display (as long as it has permission). For this reason it has emerged as the low-level graphics standard for workstations. Its equally great liabilities are that it is pixel-based rather than device-independent (like PostScript), so that a program needs to know about the display that it is using and the output will appear different on different displays, and since it is aimed at low-level graphics communications simple high-level tasks (such as drawing a line) require either many initial function calls or the use of a software toolkit. Nevertheless, X has become a lowest common denominator for network graphics and is available on most platforms, and so here again a modest investment in learning some basic routines will allow you to directly write very portable graphics programs.

The X program that manages a computer's display is called the *server*, and the programs associated with individual windows (such as a terminal emulator or a numerical model that produces graphics) are called *clients*. If I am using an X server running on a machine that has an Internet address of mach.univ.edu, then I can a run any program anywhere else on Internet and see the output if I tell the remote machine to setenv DISPLAY mach.univ.edu:0 (or an equivalent; this sets the environment for the UNIX shell csh). The :0 identifies which screen to use (in case the server has more than one screen). Depending on how permissions are set on the server it may be necessary to give it the command xhost +, which permits clients from any other system to access its screen.

The first program, xline, is similar to psline, but shows how it is possible to animate curves by repeatedly drawing and then erasing them (now showing the variation of $\sin(kx)/(kx)$ with $k$). It is compiled by cc xline.c -o xline -lm -lX11; Figure A1.6 shows the output. The default coordinate system for X has the origin at the upper-left corner, with $y$ increasing down and $x$ increasing to the right measured in pixel units (this is traditional in display graphics because of the raster pattern in CRTs).

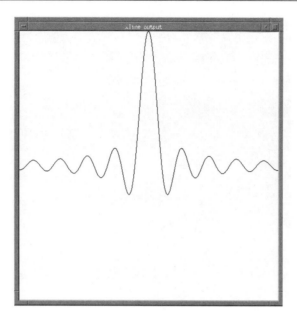

Figure A1.6. Snapshot of the output from xline.

```
/*
 * xline.c
 * (c) Neil Gershenfeld  9/1/97
 * demonstrates drawing X lines by animating sin(k*x)/k*x
 */

#include <X11/Xlib.h>
#include <math.h>

Display *D;
int     S,Loop,Point;
Window  W;
GC      Gc,GcRev;
XEvent  Event;
XPoint  PointBuf[1000],OldPointBuf[1000];
float   r;

#define WIDTH 500
#define HEIGHT 500
#define NPTS 500

main() {
   D = XOpenDisplay("");
   S = DefaultScreen (D);
   W = XCreateSimpleWindow (D, DefaultRootWindow (D), 0, 0,
      WIDTH, HEIGHT, 1, WhitePixel(D,S), WhitePixel(D,S));
```

```
        XStoreName(D, W, "xline output");
        XMapRaised(D, W);
        Gc = XCreateGC (D, W, 0L, (XGCValues *) 0);
        XSetForeground(D, Gc, BlackPixel(D, S));
        XSetBackground(D, Gc, WhitePixel(D, S));
        XSetLineAttributes(D, Gc, 0, LineSolid,
           CapButt, JoinMiter);
        GcRev = XCreateGC (D, W, 0L, (XGCValues *) 0);
        XSetForeground(D, GcRev, WhitePixel(D, S));
        XSetBackground(D, GcRev, BlackPixel(D, S));
        XSetLineAttributes(D, GcRev, 0, LineSolid,
           CapButt, JoinMiter);
        for (Loop = 1; Loop <= NPTS; ++Loop) {
           for (Point = 0; Point < NPTS; ++Point) {
              r = 0.1 * (2.0*Point + 1.0 - NPTS)/NPTS;
              OldPointBuf[Point].x = PointBuf[Point].x;
              OldPointBuf[Point].y = PointBuf[Point].y;
              PointBuf[Point].x = Point;
              PointBuf[Point].y = (int) (NPTS * (1.0 -
                 sin(Loop*r)/(Loop*r)) / 2.0);
           }
           XDrawLines(D,W,Gc,PointBuf,NPTS,CoordModeOrigin);
           XDrawLines(D,W,GcRev,OldPointBuf,NPTS,CoordModeOrigin);
           XFlush(D);
        }
}
```

The connection between an X client and server is *asynchronous*. For efficiency, drawing requests are buffered and sent in groups. The XFlush command instructs the client to send all pending requests to the server. A fast client can overwhelm a slow server; the XSync command instead makes sure that the server and client states are identical. It requires more communication to check this and so can slow down a program, but may be necessary for slow networks or servers.

The second program, ximage, shows the use of X images for 2D grayscale animation (Figure A1.7). Because most X displays are capable of specifying many more colors than they can show at once, X applications that need many colors can define *colormaps* that map pixel values into colors according to their needs. ximage defines a colormap for the grayscale mapping. The client program's colormap will usually only be active when its window has the focus; it will otherwise use the system's existing colortable and appear rather random. When the pointer is moved into the window and the colormap is switched, all of the other windows will appear odd because they are now using the new colortable (with much more programming effort, it is possible to be more careful about sharing colortables).

```
/*
 * ximage.c
```

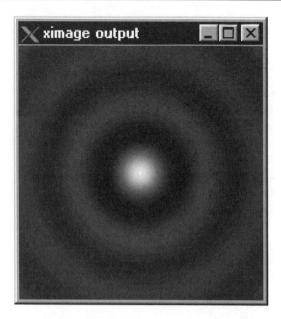

Figure A1.7. Snapshot of the output from ximage.

```
* (c) Neil Gershenfeld  9/1/97
* demonstrates drawing X images by animating sin(k*r)/k*r
*/

#include <X11/Xlib.h>
#include <X11/Xutil.h>
#include <math.h>

#define WIDTH 200
#define HEIGHT 200
#define NPTS 200
#define NCOLORS 256
#define DEPTH 8

Display   *D;
Visual    *V;
Colormap  C;
int       S,Loop,Point;
Window    W;
GC        Gc,GcRev;
XImage    *I;
XColor    Colors[NCOLORS];
XSetWindowAttributes Attr;
XVisualInfo Info;
char      Data[NPTS][NPTS];
float     x,y,r,z;
int       i,j;
```

```
main() {
  D = XOpenDisplay("");
  S = DefaultScreen(D);
  if (XMatchVisualInfo(D,S,DEPTH,PseudoColor,&Info) == 0) {
     printf("Display can not handle %d bit pseudo color\n",
        DEPTH);
     return;
  }
  V = Info.visual;
  for (i = 0; i < NCOLORS; ++i) {
     Colors[i].pixel = i;
     Colors[i].red = (int) ((65535*i)/NCOLORS);
     Colors[i].green = (int) ((65535*i)/NCOLORS);
     Colors[i].blue = (int) ((65535*i)/NCOLORS);
     Colors[i].flags = DoRed | DoGreen | DoBlue;
  }
  C = XCreateColormap(D,RootWindow(D,S),V,AllocAll);
  XStoreColors(D,C,Colors,NCOLORS);
  Attr.colormap = C;
  Attr.background_pixel = WhitePixel(D,S);
  Attr.border_pixel = BlackPixel(D,S);
  W = XCreateWindow(D,DefaultRootWindow (D),0,0,WIDTH,
     HEIGHT,100,DEPTH,InputOutput,V,
     CWColormap|CWBackPixel|CWBorderPixel,&Attr);
  XStoreName(D, W, "ximage output");
  XMapRaised(D, W);
  Gc = XCreateGC (D, W, 0L, (XGCValues *) 0);
  I = XCreateImage(D,V,DEPTH,ZPixmap,0,Data,WIDTH,HEIGHT,
     DEPTH,0);
  for (Loop = 1; Loop <= NPTS; ++Loop) {
     for (i = 0; i < NPTS; ++i)
        for (j = 0; j < NPTS; ++j) {
           x = (2.0*i + 1.0 - NPTS)/NPTS;
           y = (2.0*j + 1.0 - NPTS)/NPTS;
           r = Loop*20.0*sqrt(x*x + y*y)/NPTS;
           z = sin(r)/r;
           Data[i][j] = (unsigned char) (NCOLORS *
              (z + 0.3)/1.3);
        }
     XPutImage(D,W,Gc,I,0,0,0,0,NPTS,NPTS);
  }
}
```

In addition to manufacturers' often cryptic documentation, O'Reilly and Associates publish a good comprehensive series of X reference manuals [Nye, 1992].

### A1.2.3 OpenGL

X Windows is a windowing system, and it is oriented towards 2D graphics. *OpenGL* is the converse: it is not a windowing system, and it is designed for 3D graphics. It has descended from Silicon Graphics' proprietary IRIS GL standard which ran on their graphics computers. IRIS GL was difficult to port to other platforms; OpenGL introduced a number of incompatibilities with IRIS GL in order to make it much more portable and usable. It runs within whatever windowing system is native to the operating system; versions are available for many platforms. It is now a mature and widely used 3D graphics programming standard. OpenGL is documented in [Woo *et al.*, 1997, Kempf & Frazier, 1997], and is maintained by the OpenGL Architecture Review Board (information is available from http://www.sgi.com).

Writing an ordinary OpenGL program requires mastering not only the language, but also the native windowing system on each platform you want to use to manage putting up a window and getting user input. This chore is significantly simplified by the *GLUT* toolkit, which hides these details and hence makes it straightforward to write GL programs that are platform-independent. GLUT has been ported to all the platforms that GL is available on; information is available at http://www.sgi.com/Technology/OpenGL/glut.html.

Here's a first example of a GLUT program:

```
/*
 * glexample.c
 * (c) Neil Gershenfeld  8/30/97
 * draw a sphere and a cube with GLUT
 */

#include <GL/glut.h>

void display(void) {
  glClear(GL_COLOR_BUFFER_BIT | GL_DEPTH_BUFFER_BIT);
  glMatrixMode(GL_MODELVIEW);
  glPushMatrix();
  glTranslatef(-.3,0,0);
  glutSolidSphere(0.5,50,50);
  glTranslatef(.5,0,0);
  glutSolidCube(1.0);
  glPopMatrix();
  glFlush();
  }

void mouse(int button, int state, int x, int y) {
   exit(0);
  }

void main(int argc, char **argv) {
   glutInit(&argc,argv);
```

Figure A1.8. Snapshot of the output from glexample.

```
    glutInitDisplayMode(GLUT_RGB | GLUT_DEPTH);
    glutInitWindowSize(500,500);
    glutCreateWindow("GLUT example");
    glutDisplayFunc(display);
    glutMouseFunc(mouse);
    glEnable(GL_LIGHTING);
    glEnable(GL_LIGHT0);
    glEnable(GL_DEPTH_TEST);
    glClearColor(1.0,1.0,1.0,1.0);
    glMatrixMode(GL_PROJECTION);
    glRotatef(-140.0,1.0,1.0,0.0);
    glutMainLoop();
}
```

GL is based on a state machine model: successive function calls update the state of a virtual machine that renders the scene. The main routine puts up a window, registers a routine named display to be called whenever the display needs updating (for example, when the window is raised) and a routine mouse to be called when there is mouse input, turns on lighting, sets the viewpoint, and then starts running a loop to process events. The display routine clears the window (to the white background set in main), draws a sphere and a cube displaced relative to the current coordinate system, and flushes any pending display requests. The mouse routine simply exits when a mouse is clicked in the window. The same program will compile unchanged on any platform that runs GLUT; Figure A1.8 shows the output it produces on a Windows computer.

Now here's a more complex example:

```c
/*
 * glsurf.c
 * (c) Neil Gershenfeld   9/1/97
 * example of GL and GLUT, drawing sin(kr)/kr surface
 */

#include <GL/glut.h>
#include <math.h>

void normal(GLfloat,GLfloat,GLfloat,GLfloat,GLfloat,GLfloat,
            GLfloat,GLfloat,GLfloat,GLfloat*,GLfloat*,GLfloat*);

#define NGRID 50
#define KMIN 0.0
#define KMAX 20.0
float k=0.0, dk=0.2;

#define r(x,y) (k*sqrt(x*x+y*y))
#define height(x,y) (sin(r(x,y))/r(x,y))

GLfloat x[NGRID][NGRID],y[NGRID][NGRID],z[NGRID][NGRID];

void display(void) {
   int i,j;
   GLfloat nx,ny,nz;
   glClear(GL_COLOR_BUFFER_BIT | GL_DEPTH_BUFFER_BIT);
   glBegin(GL_QUADS);
   for (i = 0; i < (NGRID-1); ++i) {
      for (j = 0; j < (NGRID-1); ++j) {
         normal(x[i][j],y[i][j],z[i][j],
                x[i+1][j],y[i+1][j],z[i+1][j],
                x[i][j+1],y[i][j+1],z[i][j+1],
                &nx,&ny,&nz);
         glNormal3f(nx,ny,nz);
         glVertex3f(x[i][j],y[i][j],z[i][j]);
         glVertex3f(x[i+1][j],y[i+1][j],z[i+1][j]);
         glVertex3f(x[i+1][j+1],y[i+1][j+1],z[i+1][j+1]);
         glVertex3f(x[i][j+1],y[i][j+1],z[i][j+1]);
         }
      }
   glEnd();
   glFlush();
   }

void idle(void) {
   GLfloat rx,ry,rz;
```

```
      int i,j;
      if ((k > KMAX) | (k < KMIN))
         dk = -dk;
      k += dk;
      for (i = 0; i < NGRID; ++i)
         for (j = 0; j < NGRID; ++j) {
            x[i][j] = 2.0*((float) j + 0.5)/NGRID - 1.0;
            y[i][j] = 2.0*((float) i + 0.5)/NGRID - 1.0;
            z[i][j] = height(x[i][j],y[i][j]);
   }
   glMatrixMode(GL_MODELVIEW);
   rx = rand();
   ry = rand();
   rz = rand();
   glRotatef(1.0,rz,ry,rz);
   glutSwapBuffers();
   glutPostRedisplay();
   }

void mouse(int button, int state, int x, int y) {
   exit(0);
   }

void normal(GLfloat x1, GLfloat y1, GLfloat z1,
            GLfloat x2, GLfloat y2, GLfloat z2,
            GLfloat x3, GLfloat y3, GLfloat z3,
            GLfloat *xn, GLfloat *yn, GLfloat *zn) {
   *xn = (y2-y1)*(z3-z1) - (z2-z1)*(y3-y1);
   *yn = (z2-z1)*(x3-x1) - (x2-x1)*(z3-z1);
   *zn = (x2-x1)*(y3-y1) - (y2-y1)*(x3-x1);
   }

void main(int argc, char **argv) {
   GLfloat matl_ambient[] = {.25, .22, .06, 1.0};
   GLfloat matl_diffuse[] = {.35, .31, .09, 1.0};
   GLfloat matl_specular[] = {.80, .72, .21, 1.0};
   GLfloat light_ambient[] = {0.9, 0.9, 0.9, 1.0};
   GLfloat light_diffuse[] = {0.8, 0.8, 0.8, 1.0};
   GLfloat light_specular[] = {1.0, 1.0, 1.0, 1.0};
   GLfloat light_position[] = {0,0,1, 1.0};

   glutInit(&argc,argv);
   glutInitDisplayMode(GLUT_DOUBLE | GLUT_RGB | GLUT_DEPTH);
   glutInitWindowSize(500, 500);
   glutCreateWindow("GLUT sin(kr)/kr example");
   glutDisplayFunc(display);
   glutMouseFunc(mouse);
```

```
    glutIdleFunc(idle);
    glMaterialfv(GL_FRONT_AND_BACK,GL_AMBIENT,matl_ambient);
    glMaterialfv(GL_FRONT_AND_BACK,GL_DIFFUSE,matl_diffuse);
    glMaterialfv(GL_FRONT_AND_BACK,GL_SPECULAR,matl_specular);
    glMaterialf(GL_FRONT_AND_BACK,GL_SHININESS, 95.0);
    glEnable(GL_LIGHTING);
    glLightModelf(GL_LIGHT_MODEL_TWO_SIDE, 1.0);
    glLightModelfv(GL_LIGHT_MODEL_AMBIENT,light_ambient);
    glLightfv(GL_LIGHT0,GL_AMBIENT,light_ambient);
    glLightfv(GL_LIGHT0,GL_DIFFUSE,light_diffuse);
    glLightfv(GL_LIGHT0,GL_SPECULAR,light_specular);
    glLightfv(GL_LIGHT0,GL_POSITION,light_position);
    glEnable(GL_LIGHT0);
    glEnable(GL_DEPTH_TEST);
    glEnable(GL_NORMALIZE);
    glClearColor(1.0,1.0,1.0,1.0);
    glMatrixMode(GL_PROJECTION);
    glLoadIdentity();
    glOrtho(-1.5,1.5,-1.5,1.5,-1.5,1.5);
    glutMainLoop();
    }
```

This program draws $\sin(kr)/kr$ as a 3D surface, animates it as $k$ is changed, and slowly rotates the orientation from which it is viewed. The main routine now does a bit more work, defining the properties of the materials and lights in the scene (glsphere.c used the defaults for these). It also registers a new routine, idle, that is called whenever there are CPU cycles free. This is what does the work of calculating the animation. It updates the surface, swaps the buffer being displayed with the one being calculated, and asks for a redisplay. The display routine draws the surface with quadrilaterals, and calls another routine to calculate the normals of the quadrilaterals (which GL needs to determine the shading). Figure A1.9 shows a frame of the output.

GL is a low-level procedural language for describing 3D graphics. *Open Inventor* is built on top of GL; it is a high-level language for describing scenes. There is a file format associated with Open Inventor, and the *VRML* 3D extension (pronounced "vermul") to the World Wide Web is based on this. A good reference for Inventor is [Wernecke, 1994].

### A1.2.4 Java

Even though code written with GLUT is portable, it still needs to be recompiled for each platform on which it needs to be run. *Java* takes portability one big step further so that there's no need for recompilation. For this reason it is rapidly emerging as the standard for programming associated with the Internet. It started as a project at Sun to develop a programming language for appliances, and grew from there to encompass the whole Net. It is now being developed by a company spun-off from Sun, JavaSoft (a base development

Figure A1.9. Snapshot of the output from `glsurf`.

environment is available from their home page at `http://www.javasoft.com`), and is being supported by many other companies.

The language is completely *object oriented*, so that instead of writing conventional procedural programs, a Java program consists of the definition of a set of *classes* that contain data structures and *methods* associated with them (a method is the object oriented programming name for a function). This style of programming makes it much easier to create new programs by specifying differences from old ones, and to handle many kinds of abstractions. For example, many routines with the same name could be defined that differ in the expected data type. The object orientation of Java is similar to that of C++, but it dispenses with many aspects of C++ that reflect the C legacy or that have proved themselves to be rich sources of bugs. Unfortunately the single aspect of C++ most useful for mathematical programming was dropped because it was deemed to fit into the latter category. In C++ dyadic operators can be overloaded, so that for example $A + B$ can work for matrices, complex numbers, members of a group, and so forth. In Java this is not possible; overloading is restricted to functions.

The portability of Java comes because it gets compiled into a byte code representation that is machine-independent, which can then be run by a Java interpreter native to a particular machine. This representation, which is between traditional compiled and interpreted languages, is also in-between in performance (roughly an order of magnitude slower than compiled code, although note that Java can run much slower still in some browsers than it does as a stand-alone application). There is ongoing work on just-in-time compilers to improve the performance of Java. An *applet* is a small Java program designed to be downloaded from a Web server and executed in a local Web browser. X Windows sends a description of a window over the net; Java sends the whole program.

Java defines a basic set of classes including simple graphics. There are many class libraries becoming available to extend Java's graphical capabilities, including variants of OpenGL and VRML. The first sample program, JavaLine, defines a class that uses Java's basic graphics to animate $\sin(kx)/(kx)$ as an applet. Using Sun's Java Development Kit (JDK), the command javac JavaLine.java compiles it and produces a binary byte code file JavaLine.class. An *HTML* (HyperText Markup Language) file containing the following instructions:

```
<html>
<body>
<applet code="JavaLine.class" width=500 height=500>
</applet>
</body>
</html>
```

will load this class and execute it (in a Web browser, or in a stand-alone interpreter such as Sun's appletviewer).

Since Java programs almost always are run with many other processes, time-sharing is very important. This example uses *threads* to periodically execute the program to update the display and then go back to sleep. The Java interpreter calls the start method to set the applet up, the run method to execute it, the stop method to halt it, and the paint method to redraw the window.

```
//
// JavaLine.java
// (c) Neil Gershenfeld  9/1/97
// demonstrates Java by animating sin(k*x)/k*x
//

import java.awt.Graphics;

public class JavaLine extends java.applet.Applet
   implements Runnable {

   Thread T;
   final int NPTS = 500;
   final int NSTEPS = 100;
   int point,step;
   int x[] = new int[NPTS];
   int y[] = new int[NPTS];

   public void start() {
      if (T == null) {
         T = new Thread(this);
         T.start();
         }
      }
```

```
   public void stop() {
      if (T != null) {
         T.stop();
         T = null;
      }
   }
   public void run() {
      double r;
      while (true) {
         for (step = 1; step < NSTEPS; ++step) {
            for (point = 0; point < (NPTS-1); ++point) {
               r = 100 * (step*(point+0.5-NPTS/2))
                  / (NPTS*NSTEPS);
               x[point] = point;
               y[point] = (int) ((NPTS/2) -
                  (NPTS/2)*Math.sin(r)/r);
            }
            repaint();
            try {Thread.sleep(10);}
            catch (InterruptedException e) { }
         }
      }
   }
   public void paint(Graphics g) {
      for (point = 1; point < (NPTS-1); ++point) {
         g.drawLine(x[point-1],y[point-1],x[point],y[point]);
      }
   }
}
```

The next example, JavaImage, shows the use of images to animage the $\sin(kr)/kr$ example. This time it is defined as a stand-alone Java program by including a main method, so that in addition to being runnable as an applet it can also be executed as an application from a command line (in Sun's JDK, with the java command). There are severe security restrictions on what applets can and cannot do (such as read or write files); these do not apply to applications.

In JavaLine the image flickered each time that it was updated. JavaImage uses an extra step, called *double buffering*, to avoid that: a complete image is prepared off-screen, and then that image is sent to the screen in a single call. In this example a simple and slow command is used to write into the image; Java has much more complex but faster methods for updating images.

```
//
// JavaImage.java
// (c) Neil Gershenfeld   9/1/97
// demonstrates Java by animating sin(k*r)/k*r
```

```java
//

import java.awt.Graphics;
import java.awt.Color;
import java.awt.Image;
import java.awt.Frame;

public class JavaImage extends java.applet.Applet
   implements Runnable {

   static final int NPTS = 100;
   static final int NSTEPS = 100;
   static final int NGRAYS = 100;
   Thread T;
   Image I;
   Graphics G;
   Color ColorTable[] = new Color[NGRAYS];

  public static void main(String args[]) {
     Frame F = new Frame("JavaImage");
     F.setSize(2*NPTS,2*NPTS);
     F.show();
     JavaImage J = new JavaImage();
     F.add("Center",J);
     J.setSize(2*NPTS,2*NPTS);
     J.init();
     J.start();
     }
  public void init() {
     int i,gray;
     I = createImage(2*NPTS,2*NPTS);
     G = I.getGraphics();
     for (i = 0; i < NGRAYS; ++i) {
        gray = (int) (255.0 * ((double) i)/NGRAYS);
        ColorTable[i] = new Color(gray,gray,gray);
        }
     }
  public void start() {
     if (T == null) {
        T = new Thread(this);
        T.start();
        }
     }
  public void stop() {
     if (T != null) {
        T.stop();
        T = null;
```

```
      }
    }
    public void run() {
      double r, xr, yr;
      int gray, i, j, step;
      while (true) {
        for (step = 1; step < NSTEPS; ++step) {
          for (i = 0; i < NPTS; ++i) {
            for (j = 0; j < NPTS; ++j) {
              xr = (2.0*i + 1.0 - NPTS)/NPTS;
              yr = (2.0*j + 1.0 - NPTS)/NPTS;
              r = step * 20.0 * Math.sqrt(xr*xr + yr*yr)
                  / NSTEPS;
              gray = (int) (NGRAYS*(0.5+Math.sin(r)/r)
                  / 1.5);
              G.setColor(ColorTable[gray]);
              G.fillRect(2*i,2*j,2,2);
            }
          }
          repaint();
          try {Thread.sleep(10);}
          catch (InterruptedException e) { }
        }
      }
    }
    public synchronized void paint(Graphics g) {
      g.drawImage(I,0,0,this);
    }
    public void update(Graphics g) {
      paint(g);
    }
}
```

## A1.3 PROBLEMS

(A1.1) Write a program to directly produce PostScript output for the trajectory of a ball bouncing on a floor, assuming that at each reflection the ball's kinetic energy decreases by a constant fraction.

(A1.2) Write an X program to animate the motion of a bouncing ball.

(A1.3) Write a OpenGL program to animate the motion of a bouncing ball.

(A1.4) Write a Java program to animate the motion of a bouncing ball.

(A1.5) Repeat these problems in as many different interactive environments as you can.

# *Appendix 2*  Network Programming

It is much easier to find problems that exceed available computational resources than the converse. The obvious solution, using a faster computer, is far from universally applicable, if for no other reason than physical limits will keep scalar processors from running much faster than 1 GHz. Parallel computing is then needed to run problems faster (unless of course you have a quantum computer [Gershenfeld, 1999a]). One approach is to use a massively parallel computer, but as network bandwidths begin to approach Gbit/s speeds they become comparable to the bandwidth of computer backplanes. Therefore, for problems that don't require too much interprocessor communication, a building (or planet) full of frequently idle workstations can become a very effective supercomputer. As workstation access is much more common than supercomputer access, this appendix introduces the techniques needed to distribute programs over computers on networks.

## A2.1 OSI, TCP/IP, AND ALL THAT

The *ISO* (International Standards Organization) develops international standards in many areas; the US member organization is *ANSI*, the American National Standards Institute. ISO adopted *OSI*, the Open Systems Interconnection model, in 1984 as a general framework for communications that recognizes seven layers (Table A2.1). This was originally intended to result in explicit standards for protocols at each layer. That grand vision ended up being settled more in the marketplace than in the standards committees, but this still remains a useful guide for understanding networks.

Layers 1 and 2 refer to the physical transmission medium, and the logical encoding of signals in the physical medium. Layer 3 is responsible for routing a message from its origin to its destination, possibly across many physical networks (*internetworking*). Layer 4 takes care of managing an end-to-end link to make sure that both ends agree to communicate and are satisfied with the result. The higher layers provide the interface to users and their programs.

Perhaps the most important network protocol is *IP* (Internet Protocol), which was developed as part of *ARPANET* and has since become the foundation of the *Internet*. An IP *packet* or *datagram* consists of data, an address, and some extra administrative information. The maximum size of a packet is hardware-dependent, but is guaranteed to be at least 576 bytes. A machine that generates an IP packet, which can range from a laptop to a supercomputer, need not know how to reach the final address – it need only know how to put the packet onto the local network. From there, it gets passed between machines

Table A2.1. *OSI communications layers.*

| layer | name | example |
|---|---|---|
| 7 | application | Telnet |
|   |   | FTP |
| 6 | presentation |   |
| 5 | session |   |
| 4 | transport | TCP |
|   |   | UDP |
| 3 | network | IP |
|   |   | PPP |
| 2 | data link | ethernet |
|   |   | V.32/42 |
| 1 | physical | coaxial cable |
|   |   | twisted pair |

and across routers and gateways until it reaches its destination. Internet addresses are of the form 198.49.45.10, and can have a name registered to go along with them (in this case, www.internic.net, a directory of Internet machines). In addition to an Internet address, a packet also needs to have specified a *port*. This is an integer address that the sending and receiving program agree in advance to use. Many port numbers have special meanings and must be avoided; in particular, most of the ports below $\sim 1000$ are reserved for system functions (such as telnet and ftp; on a UNIX system the known ports are listed in /etc/services).

*TCP* (Transmission Control Protocol), and *UDP* (User Datagram Protocol), are two options for managing communications with IP packets. Both wrap extra information around the packets. TCP takes a large message and splits it into pieces that are small enough to fit in packets, and it makes sure that all of the packets are correctly received and reassembled by the recipient. UDP is *connectionless*, so any receiver can accept a packet from any sender without an advance agreement to communicate, and it is *unreliable*, so no guarantee is made that a packet is received. This apparent liability is not as bad as it sounds because it means that there is much less communications overhead and so the bandwidth and latency are better. If a lower network layer is taking care of error control and the network is not overloaded, then UDP can actually be quite reliable. Almost all communications of any kind on the Internet are transported in TCP/IP or UDP/IP packets. The next section shows how to use UDP packets. For more information, see [Tanenbaum, 1988] for a very readable introduction to computer networks.

## A2.2 SOCKET I/O

The most common programming approach for sending and receiving IP packets is through *sockets*, developed at Berkeley in the early 1980s. Setting up a socket requires defining the protocol (such as TCP or UDP), address, and port. Once that is done, the program simply writes to, or reads from, the socket to exchange data with the remote machine. Socket programming started in UNIX (see [Stevens, 1990] for more information about UNIX network programming), and has since expanded to most platforms.

Here is a simple program that opens a socket and writes a string from its command line to a remote machine:

```c
/*
 * sendudp.c
 * (c) Neil Gershenfeld 9/1/97
 *
 * synatx: sendudp hostname string
 *
 * send this string to hostname in a udp packet on port 6543
 *
 */

#include <stdio.h>
#include <sys/types.h>
#include <sys/socket.h>
#include <netdb.h>
#include <netinet/in.h>
#include <arpa/inet.h>
#include <string.h>

main(argc, argv)
   int argc;
   char *argv[]; {

   struct hostent *IPAddress;
   struct sockaddr_in Remote, Local;
   int Port = 6543, UDPtr, IPSize;
   char Address[100];

   IPAddress = gethostbyname(argv[1]);
   if (IPAddress == NULL) {
      printf("%s isn't in the name server\n",argv[1]);
      return; }
   strcpy(Address, inet_ntoa(*((struct in_addr *)
      *(IPAddress->h_addr_list))));
   bzero((char *) &Local, sizeof(Local));
   bzero((char *) &Remote, sizeof(Remote));
   Remote.sin_family = AF_INET;
   Remote.sin_addr.s_addr = inet_addr(Address);
   Remote.sin_port = htons(Port);
   Local.sin_family = AF_INET;
   Local.sin_addr.s_addr = htonl(INADDR_ANY);
   Local.sin_port = htons(Port);
   UDPtr = socket(AF_INET,SOCK_DGRAM,0);
```

```
    if (UDPtr < 0) {
       puts("Can't create the local socket");
       return; }
    if (bind(UDPtr, (struct sockaddr *)
       &Local, sizeof(Local)) < 0) {
       puts("Can't bind local address");
       return; }
    IPSize = sendto(UDPtr, argv[2], strlen(argv[2]), 0,
       &Remote, sizeof(Remote));
    if (IPSize < 0) {
       puts("sendto error");
       return; }
    close(UDPtr);
    printf("Sent %d bytes to %s\n",IPSize, Address);
    }
```

and this program listens for the packet:

```
/*
 * recvudp.c
 * (c) Neil Gershenfeld 9/11/94
 *
 * waits for an incoming udp packet on port 6543
 * and print it out
 *
 */

#include <stdio.h>
#include <sys/types.h>
#include <sys/socket.h>
#include <netdb.h>
#include <netinet/in.h>
#include <arpa/inet.h>
#include <string.h>

#define BufferSize 100

main(argc, argv)
   int argc;
   char *argv[]; {

   struct hostent *IPAddress;
   struct sockaddr_in Remote, Local;
   int Port = 6543, UDPtr, IPSize, RemoteSize;
   char Buffer[BufferSize];
```

```
    UDPtr = socket(AF_INET,SOCK_DGRAM,0);
    if (UDPtr < 0) {
      puts("Can't create the local socket");
      return; }
    bzero((char *) &Local, sizeof(Local));
    Local.sin_family = AF_INET;
    Local.sin_addr.s_addr = htonl(INADDR_ANY);
    Local.sin_port = htons(Port);
    if (bind(UDPtr, (struct sockaddr *)
       &Local, sizeof(Local)) < 0) {
      puts("UDPRecv: Can't bind local address");
      return; }
    IPSize = recvfrom(UDPtr, Buffer, BufferSize, 0,
                     &Remote, &RemoteSize);
    if (IPSize < 0) {
      puts("UDPRecv: recvfrom error");
      return; }
    close(UDPtr);
    printf("Received %d bytes: %s\n",IPSize, Buffer);
    }
```

Both programs compile on a standard UNIX system by `cc prog.c -o prog`. Here is what their output looks like:

```
rho.media.mit.edu> sendudp media.mit.edu "test message"
Sent 12 bytes to 18.85.13.107

media-lab.media.mit.edu> recvudp
Received 12 bytes: test message
```

## A2.3 PARALLEL PROGRAMMING

Using sockets it is possible to manually split a program into multiple pieces that can execute on available machines and communicate their results to each other. This has been done with great success for particular important problems, such as applying many machines around the Internet to the problem of finding prime factors of a cryptographic key (which is a very parallelizable algorithm). This kind of programming is something like writing assembly language: you must decide exactly when and how messages get passed. An important level of abstraction above that is *MPI*, the Message Passing Interface (http://www.netlib.org/mpi). This standard is widely supported on many platforms: the same MPI program can run whether the processors are housed in a supercomputer or in workstations in a building. It still requires explicitly identifying when messages need to be passed in a parallel algorithm, but it hides the details of socket programming and includes routines to handle more advanced functions such as multiway communications and implementing effective connectivity topologies.

In Appendix 3 we will look at the series representation

$$\pi \approx \sum_{i=1}^{N} \frac{0.5}{(i - 0.75)(i - 0.25)} \qquad (A2.1)$$

as a simple benchmark task. The following program shows how to use MPI to parallelize it:

```
/*
 * mpipi.c
 * (c) Neil Gershenfeld 9/1/97
 * use MPI to evaluate pi by summation
 */

#include <stdio.h>
#include <mpi.h>

void main(int argc, char** argv) {
   int rank,nproc,tag,i,istart,iend,N;
   double sum,pi;

   tag = 0;
   sum = 0.0;
   N = 1000000;
   MPI_Init(&argc, &argv);
   MPI_Comm_rank(MPI_COMM_WORLD, &rank);
   MPI_Comm_size(MPI_COMM_WORLD, &nproc);
   if (rank == 0) {
      MPI_Reduce(&sum,&pi,1,
         MPI_DOUBLE,MPI_SUM,0,MPI_COMM_WORLD);
      printf("Using %d processes, pi = %f\n",nproc,pi);
   } else {
      istart = 1 + (rank-1)*N/(nproc-1);
      iend = rank*N/(nproc-1);
      for (i = istart; i <= iend; ++i)
         sum += 0.5/((i-0.75)*(i-0.25));
      MPI_Reduce(&sum,&pi,1,
         MPI_DOUBLE,MPI_SUM,0,MPI_COMM_WORLD);
   }
   MPI_Finalize();
}
```

The same program is run on all the processors being used. After initializing MPI the first two calls determine how many processes are being run, and the process number of each program. Then all of the copies with a process number not equal to 0 add up part of the sum based on their process number, and make an MPI call that asks to send the results to process 0 and add them up. Process 0 takes a different branch, waiting to receive the sum and then printing it out. Asking this to be run with six processors gives the output

```
% mpirun -np 6 mpitest.exe

Using 6 processes, pi = 3.141592
Process [0] exited with status [0]
Process [1] exited with status [0]
Process [2] exited with status [0]
Process [3] exited with status [0]
Process [4] exited with status [0]
Process [5] exited with status [0]
MPIRUN exited with status [0]
```

MPI has taken care of all of the message passing, hiding it behind a few simple calls. Because this program is so easy to parallelize, the time to run it will decrease roughly linearly with the number of processors used.

The most convenient parallel programming of all would be to have a compiler rather than the programmer do the work of dividing a program among multiple processors. The style of parallel programming used in MPI is *coarse-grained*: a relatively small number of powerful processors execute complex programs. Although there have been attempts to automate this kind of programming they have not been very successful, which is not surprising given that it is a difficult task even for a skilled programmer. A *fine-grained* parallel computer is one that has a large number of processors each doing a simple job. Loops that do the same operation on many elements, such as adding corresponding components of two vectors, are easy to recognize and automatically parallelize on such a machine. This is particularly true in a language like Fortran 90 that can express such operations without loops. However, such machines are currently less popular, in part because of the remarkable improvements in scalar processing speed, and in part because many complex problems don't naturally map onto such an architecture. These are active research questions; in the meantime MPI provides a convenient way to make good use of available resources to speed up a program.

# Appendix 3  Benchmarking

Benchmarking is a subject that receives both too little and too much attention. Too little, because knowing the relative speeds of machines, languages, and algorithms can have an enormous impact on your ability to obtain timely results. Too much, because tests that may have little bearing on practical problems can dominate manufacturers' advertising and ultimately users' purchase decisions.

Amid all of the hype, a simple recurring truth is that the best benchmark is a problem that you are interested in. An early standard was the LINPACK set of subroutines, which have been run on an enormous range of machines (see the listing at http://www.netlib.org). Because there is a great deal of specialized structure in these routines, some aggressive compilers started to have switches that recognized them and used carefully hand-tuned assembly code to appear faster on this benchmark. To prevent that, as well as to cover a much broader range of applications, an industry-wide group has defined a suite of test problems called the *SPEC* benchmark (http://www.specbench.org/). This is a comprehensive set of programs covering many types of numerical algorithms. Where it's available, it's a reliable guide to machine speed. However, it may not be available for a particular machine that you're interested in, and it is not freely accessible so that you can use it to evaluate alternative compilers (for example). For this reason it's useful to have a simple test program that can provide a rough order-of-magnitude estimate of speed (which can often be as good as much more complex benchmarks).

I've found it convenient to use a series expansion of $\pi$,

$$\pi = 4\tan^{-1}(1)$$
$$\approx \sum_{i=1}^{N} \frac{0.5}{(i-0.75)(i-0.25)} .$$

Summing this series requires five floating point operations per step, making it easy to estimate the computational speed (usually measured in millions of floating point operations per second, megaflops or Mflops) by measuring the time to sum it. The total time should be linear in the number of terms used once $N$ is large enough (there is always some overhead associated with starting and finishing execution). And since the correct answer is known, it is easy to check the validity and precision of the result.

Further, it's instructive to write the series two ways, one as a sum of a scalar term

$$\pi(N) = \sum_{i=1}^{N} \frac{0.5}{(i-0.75)(i-0.25)} , \qquad (A3.1)$$

Table A3.1. *Selected execution speeds to sum a series expansion of $\pi$.*

| computer | speed (Mflops) | program version |
|---|---|---|
| Connection Machine CM-2 | 851 | scalar |
| Cray Y-MP4/464 | 118 | scalar |
|  | 10.0 | array |
| DEC AlphaStation 500/500 | 70.0 | scalar |
|  | 69.4 | array |
| HP 735 | 18.2 | scalar |
|  | 14.1 | array |
| SUN Sparc 10/40 | 9.86 | scalar |
|  | 7.97 | array |
| 200MHz Pentium Pro | 13.3 | scalar |
| 90 MHz Pentium | 6.33 | scalar |
| Sun SPARCStation 2 | 4.50 | scalar |
|  | 0.34 | array |
| Sun SPARCStation 1 | 1.21 | scalar |
|  | 1.01 | array |
| DEC VAX 8650 | 1.72 | scalar |
| 25 MHz 486 | 0.70 | scalar |
| 33 MHz 386/387 | 0.35 | scalar |
| 25MHz 486/87 | 0.23 | scalar |
| Sun 3/60 (68020) | 0.036 | scalar |
|  | 0.036 | array |
| 12.5 MHz 286 | 0.034 | scalar |
| 10 MHz 8088 | 0.0011 | scalar |

and the other recursively in terms of an array

$$\pi(N) = \pi(N-1) + \frac{0.5}{(N-0.75)(N-0.25)} \quad . \tag{A3.2}$$

While these are formally equivalent, the terms of the former can be calculated in parallel and require little storage, while the latter must be done in series to fill the array and requires much more storage. These distinctions can help point to strengths and weaknesses of serial and parallel architectures.

Table A3.1 shows some sample times (in all cases using the machine's native floating-point precision). It is NOT in any way a thorough characterization of these machines, but it is an easily generated estimate that is typically surprisingly close to much more careful benchmarks.

The single most remarkable feature of this table is that it spans roughly six orders of magnitude. That's the difference between an algorithm taking one day and 3000 years, about the duration of recorded history. For some big problems, literally the fastest way to solve them can be to wait for a faster computer to be built. Notice also how the array version of the program inhibits the Cray's vector processing, slowing the speed down by an order of magnitude, but has little effect on the scalar processing machines. For some problems, what might naively appear to be a slower computer can be preferable.

# Appendix 4  Problem Solutions

## A4.1 INTRODUCTION

There are no problems in Chapter 1.

## A4.2 ORDINARY DIFFERENTIAL AND DIFFERENCE EQUATIONS

(2.1) *Consider the motion of a damped, driven harmonic oscillator (such as a mass on a spring, a ball in a well, or a pendulum making small motions):*

$$m\ddot{x} + \gamma\dot{x} + kx = e^{i\omega t} \quad . \tag{A4.1}$$

(a) *Under what conditions will the governing equations for small displacements of a particle around an arbitrary 1D potential minimum be simple undamped harmonic motion?*

Let the potential be $V(x)$, and the (possibly local) minimum be at $x = 0$. For small displacements the potential can be expanded in a Taylor series, and because this is a minimum the linear term must vanish:

$$V(x) = V|_{x=0} + \frac{1}{2}\frac{d^2V}{dx^2}\bigg|_{x=0} x^2 + \mathcal{O}(x^3)$$

$$\equiv V_0 + \frac{1}{2}V_2 x^2 + \mathcal{O}(x^3) \quad . \tag{A4.2}$$

Ignoring the higher order terms ($\lim_{x \to 0} x^2/x^3 = \infty$) and taking the gradient of the potential to find the force, $F = ma$ becomes

$$m\ddot{x} = -\frac{dV}{dx} = -V_2 x + \mathcal{O}(x^3)$$

$$m\ddot{x} + V_2 x = 0 \quad . \tag{A4.3}$$

Therefore, as long as the quadratic term in the expansion of the potential is nonzero, for small displacements the governing equation reduces to simple harmonic motion.

(b) *Find the solution to the homogeneous equation, and comment on the possible cases. How does the amplitude depend on the frequency?*

Plugging $\exp(rt)$ into the homogeneous equation gives

$$mr^2 + \gamma r + k = 0$$

$$r^2 + \underbrace{\frac{\gamma}{m}}_{\equiv 2\lambda} r + \underbrace{\frac{k}{m}}_{\equiv \omega_0^2} = 0 \qquad (A4.4)$$

$$r = \frac{1}{2}\left[-2\lambda \pm \sqrt{4\lambda^2 - 4\omega_0^2}\right] = -\lambda \pm \underbrace{\sqrt{\lambda^2 - \omega_0^2}}_{\equiv \eta} \quad .$$

There are three cases:

- $\lambda^2 > \omega_0^2$ (*overdamped*) In this case the radical is real, so the general solution is a sum of two decaying exponentials

$$x = A_1 e^{(-\lambda+\eta)t} + A_2 e^{(-\lambda-\eta)t} \quad . \qquad (A4.5)$$

- $\lambda^2 < \omega_0^2$ (*underdamped*) Now the radical is imaginary, so the general solution consists of a periodic oscillation with a decaying amplitude

$$x = A_1 e^{-\lambda t} e^{i|\eta|t} + A_2 e^{-\lambda t} e^{-i|\eta|t} \quad . \qquad (A4.6)$$

- $\lambda^2 = \omega_0^2$ (*critically damped*) The two roots merge into one real root (exponential decay with a single time constant). Generating two linearly independent solutions according to equation (2.10) gives

$$x = (A_1 + A_2 t)e^{-\lambda t} \quad . \qquad (A4.7)$$

For all of these cases the amplitude is independent of the frequency (this is not true of a nonlinear oscillator).

(c) *Find a particular solution to the inhomogeneous problem by assuming a response at the driving frequency, and plot its magnitude and phase as a function of the driving frequency for $m = k = 1, \gamma = 0.1$.*

Substituting $A\exp(i\omega t)$ into (2.57) and cancelling the exponentials gives

$$-m\omega^2 A + i\gamma\omega A + kA = 1$$

$$A = \frac{1}{k + i\gamma\omega - m\omega^2} \quad . \qquad (A4.8)$$

At low frequencies this is real (the response is in phase with the driving). When $\omega = \sqrt{k/m} = \omega_0$ the real parts of the denominator cancel and there is a resonant peak with a magnitude determined by the dissipation, and a phase shift of $i = \pi/2$ relative to the drive. At high frequencies the denominator is dominated by the $\omega^2$ term, and so the amplitude returns to zero and there is a phase shift of $-1 = \pi$. This is plotted in Figure A4.1.

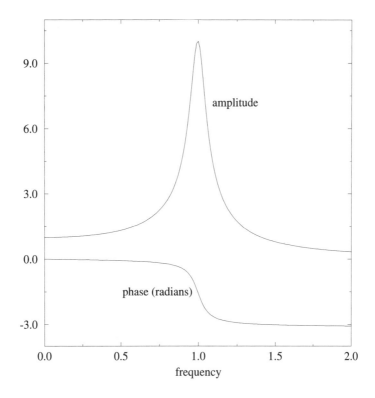

Figure A4.1. Phase and amplitude of a damped driven harmonic oscillator.

(d) *For a driven oscillator the Q or Quality factor is defined as the ratio of the center frequency to the width of the curve of the average energy (kinetic + potential) in the oscillator versus the driving frequency (the width is defined by the places where the curve falls to half its maximum value). For an undriven oscillator the Q is defined to be the ratio of the energy in the oscillator to the energy lost per radian (one cycle is $2\pi$ radians). Show that these two definitions are equal, assuming that the damping is small. How long does it take the amplitude of a 100 Hz oscillator with a Q of $10^9$ to decay by $1/e$?*

The average energy is the sum of the average potential energy $V$ and average kinetic energy $U$

$$\langle E \rangle = \langle V \rangle + \langle U \rangle = \frac{1}{2} k \langle x^2 \rangle + \frac{1}{2} m \langle \dot{x}^2 \rangle \quad . \tag{A4.9}$$

If $x = A \exp(i\omega t)$ (as found in part (c)), and recalling that we take the real part of a complex exponential to get the real position, and that $\langle \sin^2 \rangle = \langle \cos^2 \rangle = 1/2$, the average energy is

$$\langle E \rangle = \frac{1}{2}kA^2\langle\cos^2(\omega t)\rangle + \frac{1}{2}m\omega^2 A^2\langle\sin^2(\omega t)\rangle$$

$$= \frac{1}{4}(k + m\omega^2)A^2$$

$$= \frac{1}{4}\frac{k + m\omega^2}{(k - m\omega^2 + i\gamma\omega)(k - m\omega^2 - i\gamma\omega)}$$

$$= \frac{1}{4}\frac{k + m\omega^2}{(k - m\omega^2)^2 + \gamma^2\omega^2}$$

$$= \frac{1}{4m}\frac{\omega_0^2 + \omega^2}{(\omega_0^2 - \omega^2)^2 + \omega^2\gamma^2/m^2} \quad . \tag{A4.10}$$

If we put in $\omega = \omega_0 + \delta$ (weak damping implies that the resonance is narrow, and so we need only look at frequencies near the resonance), and keep only the lowest order terms in $\delta$ and $\gamma$, we find

$$\langle E \rangle \approx \frac{2}{4\delta^2 + \gamma^2/m^2} \quad . \tag{A4.11}$$

This falls to half its peak when

$$4\delta^2 = \frac{\gamma^2}{m^2} \Rightarrow \delta = \pm\frac{1}{2}\frac{\gamma}{m} \quad . \tag{A4.12}$$

Therefore the $Q$ is

$$Q = \frac{\omega}{\Delta\omega} \approx \frac{\omega_0}{\gamma/m} \quad . \tag{A4.13}$$

For the undriven oscillator, the amplitude falls off as $A = \exp(-\gamma t/2m)$, and for light damping the frequency is $\omega \approx \omega_0$ (equation 2.1(b)). In the period of $2\pi/\omega$ the oscillator goes through $2\pi$ radians, and so oscillating through 1 radian requires a time of $\Delta t = (2\pi/\omega)/(2\pi) = 1/\omega$. Since we just saw that the energy is proportional to the square of the amplitude,

$$Q = \frac{E}{\Delta E} = \frac{E}{-\frac{dE}{dt}\Delta t} = \frac{e^{-\gamma t/m}}{\frac{\gamma}{m}e^{-\gamma t/m}\frac{1}{\omega_0}} = \frac{\omega_0}{\gamma/m} \quad . \tag{A4.14}$$

Since $\exp(-\gamma t/2m) = \exp(-\omega t/2Q)$, an $e$-folding time is $t = 2Q/\omega = 2 \times 10^9/100$ Hz $= 2 \times 10^7$ seconds (1 year $\approx 3 \times 10^7$ seconds). $Q$'s this large can exist is some physical systems (such as nuclear spins, atomic excitations, and resonant cavities), requiring very precise instruments to measure the small widths or decay rates.

(e) *Now find the solution to equation (2.57) by using Laplace transforms. Take the initial condition as $x(0) = \dot{x}(0) = 0$.*
Transforming equation (2.57),

$$m[s^2 X(s) - sx(0) - \dot{x}(0)] + \gamma[sX(s) - x(0)] + kX(s)$$

$$= \frac{1}{s - i\omega}$$

$$X(s)[ms^2 + \gamma s + k] = \frac{1}{s - i\omega} + \underbrace{msx(0) + m\dot{x}(0) + \gamma x(0)}_{0}$$

$$\begin{aligned}
X(s) &= \frac{1}{s - i\omega} \frac{1}{ms^2 + \gamma s + k} \\
&= \frac{1}{s - i\omega} \frac{1}{s^2 + 2\lambda s + \omega_0^2} \frac{1}{m} \\
&= \frac{1}{s - i\omega} \frac{1}{(s + \lambda + \eta)(s + \lambda - \eta)} \frac{1}{m} \\
x(t) &= \frac{e^{i\omega t}}{(-i\omega - \lambda - \eta)(-i\omega - \lambda + \eta)m} \\
&\quad - \frac{e^{-(\lambda+\eta)t}}{(-i\omega - \lambda - \eta)(\lambda + \eta - \lambda + \eta)m} \\
&\quad - \frac{e^{-(\lambda-\eta)t}}{(-i\omega - \lambda + \eta)(\lambda - \eta - \lambda - \eta)m} \\
&= \frac{e^{i\omega t}}{k + i\omega\gamma - m\omega^2} + \frac{e^{-(\lambda+\eta)t}}{(i\omega + \lambda + \eta)2\eta m} \\
&\quad + \frac{e^{-(\lambda-\eta)t}}{(-i\omega - \lambda + \eta)2\eta m}
\end{aligned}$$
(A4.15)

(using Table 2.1). We see the oscillatory solutions and the two exponentially decaying ones (that arise from the initial transient due to starting the particle moving).

(f) *For an arbitrary potential minimum, work out the form of the lowest-order correction to simple undamped unforced harmonic motion.*

The next term in equation (A4.2) is a cubic:

$$V(x) = V_0 + \frac{1}{2}V_2 x^2 + \frac{1}{6}V_3 x^3 + \mathcal{O}(x^4) \quad . \tag{A4.16}$$

Taking this correction to be small (as it will be for small displacements), and rewriting $V_2 = k$ and $V_3/2 = \epsilon$, the equation of motion is

$$m\ddot{x} + kx + \epsilon x^2 = 0 \quad . \tag{A4.17}$$

Expanding $x = x_0 + \epsilon x_1 + \mathcal{O}(\epsilon^2)$ and collecting orders of $\epsilon$ gives the unperturbed equation

$$m\ddot{x}_0 + kx_0 = 0 \tag{A4.18}$$

and the lowest-order correction equation

$$m\ddot{x}_1 + kx_1 + x_0^2 = 0 \quad . \tag{A4.19}$$

Taking the initial conditions to give an unperturbed solution of $x_0 = A \exp(i\omega_0 t)$ (where $\omega_0 = \sqrt{k/m}$) yields a correction equation

$$m\ddot{x}_1 + kx_1 + A^2 e^{2i\omega_0 t} = 0 \quad , \tag{A4.20}$$

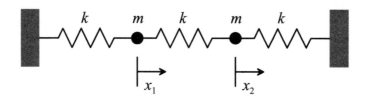

Figure A4.2. Two coupled harmonic oscillators.

which is solved by

$$x_1 = \frac{A^2}{3m\omega_0^2} e^{2i\omega_0 t} \quad . \tag{A4.21}$$

This next order term gives a harmonic at twice the natural frequency of the oscillator.

(2.2) *Explicitly solve (and try to simplify) the system of differential equations for two coupled harmonic oscillators (see Figure A4.2; don't worry about the initial transient), and then find the normal modes by matrix diagonalization.*

Adding up the forces on each mass gives

$$m\ddot{x}_1 = -kx_1 - k(x_1 - x_2)$$
$$m\ddot{x}_2 = -kx_2 - k(x_2 - x_1) \quad , \tag{A4.22}$$

and rearranging terms and substituting $\omega_0^2 = k/m$ results in two coupled ODEs:

$$\ddot{x}_1 + 2\omega_0^2 x_1 - \omega_0^2 x_2 = 0$$
$$\ddot{x}_2 + 2\omega_0^2 x_2 - \omega_0^2 x_1 = 0 \quad . \tag{A4.23}$$

Trying $x_1 = \exp(i\omega t)$ and $x_2 = A\exp(i\omega t)$ reduces this to two equations in two unknowns

$$-\omega^2 + 2\omega_0^2 - \omega_0^2 A = 0$$
$$-A\omega^2 + 2\omega_0^2 A - \omega_0^2 = 0$$

$$(\omega^2 - 2\omega_0^2) + \omega_0^2 A = 0$$
$$\omega_0^2 + (\omega^2 - 2\omega_0^2)A = 0 \quad . \tag{A4.24}$$

The first equation can be rearranged to show that $\omega^2 = \omega_0^2(2 - A)$, and plugging this into the second gives $A^2 = 1 \Rightarrow A = \pm 1$. The possible frequencies can then be found by plugging this back into the first equation: $\omega = \omega_0$ ($A = 1$), and $\omega = \sqrt{3}\omega_0$ ($A = -1$). The first solution corresponds to both masses moving in the same direction (the central spring has no effect), and the second solution corresponds to them moving in opposite directions (loading the central spring, which raises the oscillation frequency).

Motivated by this, define new variables $x_1 = y_1 + y_2$, and $x_2 = y_1 - y_2$. In terms of these variables, the governing equations become

$$\ddot{y}_1 + \ddot{y}_2 + 2\omega_0^2(y_1 + y_2) - \omega_0^2(y_1 - y_2) = 0$$
$$\ddot{y}_1 - \ddot{y}_2 + 2\omega_0^2(y_1 - y_2) - \omega_0^2(y_1 + y_2) = 0 \quad . \tag{A4.25}$$

If these two equations are added and subtracted, the resulting two decoupled equations are

$$\ddot{y}_1 + \omega_0^2 y_1 = 0$$
$$\ddot{y}_2 + 3\omega_0^2 y_2 = 0 \quad . \tag{A4.26}$$

Equations (A4.26) were found through a clever coordinate change, but they can also be found by inverting the coupling matrix. Equations (A4.23) can be written as the matrix equation

$$\frac{d^2 \vec{x}}{dt^2} + \underbrace{\begin{pmatrix} 2\omega_0^2 & -\omega_0^2 \\ -\omega_0^2 & 2\omega_0^2 \end{pmatrix}}_{\mathbf{A}} \cdot \vec{x} = 0 \quad . \tag{A4.27}$$

The eigenvalues are found by solving ([Strang, 1988])

$$\begin{aligned} 0 &= |\mathbf{A} - \lambda \mathbf{I}| \\ &= \begin{vmatrix} 2\omega_0^2 - \lambda & -\omega_0^2 \\ -\omega_0^2 & 2\omega_0^2 - \lambda \end{vmatrix} \\ &= \lambda^2 - 4\omega_0^2 \lambda + 3\omega_0^4 \\ \Rightarrow \lambda &= \omega_0^2, 3\omega_0^2 \quad . \end{aligned} \tag{A4.28}$$

Solving for the corresponding eigenvectors gives (1 1) for $\lambda = \omega_0^2$, and (1 −1) for $\lambda = 3\omega_0^2$. This means that the coordinate transformation matrix is

$$\mathbf{M} = \begin{pmatrix} 1 & 1 \\ 1 & -1 \end{pmatrix} \quad . \tag{A4.29}$$

The inverse can be found by scaling and adding rows to transform the matrix into the identity; by definition the same operations performed on the identity transform it into the inverse:

$$\overbrace{\begin{pmatrix} 1 & 1 \\ 1 & -1 \end{pmatrix}}^{\mathbf{M}} \overbrace{\begin{pmatrix} 1 & 0 \\ 0 & 1 \end{pmatrix}}^{\mathbf{I}}$$
$$\begin{pmatrix} 1 & 0 \\ 1 & -1 \end{pmatrix} \begin{pmatrix} \frac{1}{2} & \frac{1}{2} \\ 0 & 1 \end{pmatrix}$$
$$\underbrace{\begin{pmatrix} 1 & 0 \\ 0 & 1 \end{pmatrix}}_{\mathbf{I}} \underbrace{\begin{pmatrix} \frac{1}{2} & \frac{1}{2} \\ \frac{1}{2} & -\frac{1}{2} \end{pmatrix}}_{\mathbf{M}^{-1}} \tag{A4.30}$$

The eigenvalues gives the frequencies of the normal modes ($\omega_0$ and $\sqrt{3}\omega_0$), and $\mathbf{M}^{-1}$ gives the same coordinate transformation we saw above ($y_1 + y_2$ and $y_1 - y_2$).

(2.3) *A common simple digital filter used for smoothing a signal is*

$$y(k) = \alpha y(k-1) + (1-\alpha)x(k) \quad , \tag{A4.31}$$

*where $\alpha$ is a parameter that determines the response of the filter. Use z-transforms to solve for $y(k)$ as a function of $x(k)$ (assume $y(k < 0) = 0$). What is the amplitude of the frequency response?*

Rearranging the $z$-transform of equation (A4.31)

$$Y(z) = \alpha[z^{-1}Y(z) + \underbrace{y(-1)}_{0}] + (1-\alpha)X(z) \tag{A4.32}$$

gives the transfer function

$$H(z) = \frac{Y(z)}{X(z)} = (1-\alpha)\frac{z}{z-\alpha} \ . \tag{A4.33}$$

This can be inverted (using Table 2.2) to give the impulse response

$$h(k) = (1-\alpha)\alpha^k \ . \tag{A4.34}$$

Therefore, the output for an arbitrary input is found from the convolution with the impulse response

$$y(k) = (1-\alpha)\sum_{n=0}^{k}\alpha^n x(k-n) \tag{A4.35}$$

and is an exponentially decaying sum of the history of the input.

The amplitude of the frequency response is found from

$$\begin{aligned}
\left|H(e^{i\omega\delta_t})\right| &= \left|(1-\alpha)\frac{e^{i\omega\delta_t}}{e^{i\omega\delta_t}-\alpha}\right| \\
&= (1-\alpha)\left[\frac{e^{i\omega\delta_t}}{e^{i\omega\delta_t}-\alpha}\frac{e^{-i\omega\delta_t}}{e^{-i\omega\delta_t}-\alpha}\right]^{1/2} \\
&= (1-\alpha)\left[\frac{1}{1-\alpha e^{-i\omega\delta_t}-\alpha e^{i\omega\delta_t}+\alpha^2}\right]^{1/2} \\
&= \frac{1-\alpha}{\sqrt{1+\alpha^2-2\alpha\cos\omega\delta_t}} \ .
\end{aligned} \tag{A4.36}$$

This is a low-pass filter: as $\omega$ increases the magnitude decreases. It has a periodic response at higher frequencies because of *aliasing*, the wrapping around of frequencies in discretely sampled systems.

## A4.3 PARTIAL DIFFERENTIAL EQUATIONS

(3.1) *Consider a round drumhead of radius L. For small displacements its motion is described by a linear wave equation. Find the frequencies of the six lowest oscillation modes, and plot the shape of the modes.*

We need to solve a 2D wave equation with circular symmetry, which is a special case of cylindrical symmetry with no $z$ dependence. The separation equations (3.29) will not depend on $z$ if $\alpha = k$. Therefore, the general form of the solution will be a

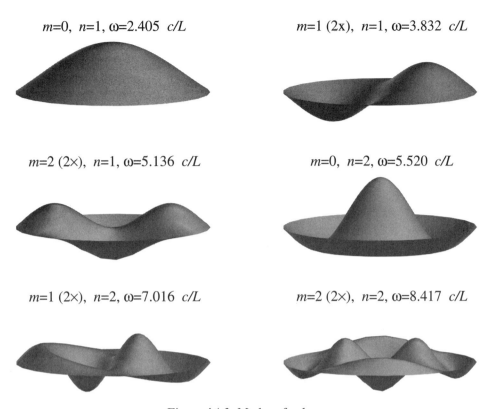

Figure A4.3. Modes of a drum.

sum of terms depending on the radial separation constant $k$ and the angular one $m$:

$$\psi(r, \theta, t) = R(r)\Theta(\theta)T(t)$$
$$= \sum_{k,m} A_m J_m(kr)[B_m \sin(m\theta) + C_m \cos(m\theta)]$$
$$\times [D_k \sin(kct) + E_k \cos(kct)] \quad . \tag{A4.37}$$

(leaving off the radial Bessel function $N_m(kr)$ that diverges at the origin). $m$ is determined by the requirement that the solution be single-valued:

$$\Theta(m\theta) = \Theta(m(\theta + 2\pi)) \Rightarrow m = 0, 1, 2, \ldots \quad . \tag{A4.38}$$

The radial solution must vanish at the boundary at $r = L$, which requires that

$$J_m(kL) = 0 \Rightarrow k = \frac{x_m(n)}{L} \quad , \tag{A4.39}$$

where $x_m(n)$ is the $n$th root of the $m$th Bessel function. Since the time dependence is of the form $\sin(kct)$, this implies that the frequencies are

$$\omega = x_m(n)\frac{c}{L} \quad . \tag{A4.40}$$

The $m = 0$ mode will have only the cos component ($\sin(0) = 0$); for $m \neq 0$ there will be two degenerate modes (the sin and cos terms will have the same frequency and shape, differing only by a rotation of $90°$ (Figure A4.3).

(3.2) *Solve a 1D diffusion equation with Fourier transforms.*

$$\frac{\partial^2 \varphi(x,t)}{\partial x^2} = \frac{1}{D}\frac{\partial \varphi(x,t)}{\partial t}$$

$$\varphi(x,t) = \int_{-\infty}^{\infty} \Phi(k,t)e^{ikx}\,dk \qquad \Phi(k,t) = \frac{1}{2\pi}\int_{-\infty}^{\infty} \varphi(x,t)e^{-ikx}\,dx$$

$$\int_{-\infty}^{\infty} e^{ikx}\left[\frac{1}{D}\frac{\partial \Phi(k,t)}{\partial t} + k^2 \Phi(k,t)\right] = 0$$

$$\frac{\partial \Phi(k,t)}{\partial t} = -Dk^2 \Phi(k,t) \Rightarrow \Phi(k,t) = A(k)e^{-Dk^2 t}$$

$$\Phi(k,0) = \frac{1}{2\pi}\int_{-\infty}^{\infty} \varphi(x,0)e^{-ikx}\,dx = A(k)e^0 = A(k)$$

$$\varphi(x,t) = \int_{-\infty}^{\infty}\left[\frac{1}{2\pi}\int_{-\infty}^{\infty}\varphi(x',0)e^{-ikx'}\,dx'\right]e^{-Dk^2 t}e^{ikx}\,dk$$

$$= \frac{1}{2\pi}\int_{-\infty}^{\infty}\varphi(x',0)\int_{-\infty}^{\infty} e^{-Dk^2 t - ikx' + ikx}\,dk\,dx'$$

$$= \frac{1}{\sqrt{4\pi Dt}}\int_{-\infty}^{\infty} e^{-\frac{(x-x')^2}{4Dt}}\varphi(x',0)\,dx' \quad . \qquad (A4.41)$$

The solution is the initial condition convolved with a spreading Gaussian.

(3.3) *Assume a crowded room full of generous children who have varying amounts of candy. Let $\varphi_{n,m}(t_i)$ be the amount of candy held by the $n,m$-th child at time $t_i$. Because of the crowding, the children are approximately close-packed on a square grid. The children want to equalize the amount of candy, but it is so noisy that they can only talk to their nearest neighbors (although they are wearing watches). Find a simple strategy for them to use that results in the candy being evenly distributed, and in the continuum limit find a familiar PDE that is equivalent to this strategy.*

At each time step, each child first announces how much candy he or she has. Then, each child either gives to (or takes from) the four nearest neighbors a fraction of the difference between the amount held by the child and the average amount held by the neighbors.

$$\varphi_{n,m}(t+1) - \varphi_{n,m}(t) =$$

$$D[\varphi_{n-1,m}(t) + \varphi_{n+1,m}(t) + \varphi_{n,m-1}(t) + \varphi_{n,m+1}(t) - 4\varphi_{n,m}(t)] \quad . \qquad (A4.42)$$

The term on the left is a discrete approximation to the time derivative, and the term on the right is a the discrete form of the 2D Laplacian:

$$\nabla^2 \varphi = \frac{1}{D}\frac{\partial \varphi}{\partial t} \quad . \qquad (A4.43)$$

This is just a diffusion equation, and is a simple example of a finite-difference approximation to a PDE (to be covered in Chapter 6).

## A4.4 VARIATIONAL PRINCIPLES

**(4.1)** *Consider a chain of length L and density $\rho$ that is hanging between two posts. Find the general form for the shape that minimizes the potential energy. Remember that the potential energy of a segment ds is $\rho\, ds\, g\, y$.*

The total potential energy in the chain is found by integrating the energy in a segment:

$$E = \int \rho g y \, ds = \int \rho g y \sqrt{1+\dot{y}^2}\, dx \quad . \tag{A4.44}$$

The length of the chain is fixed at

$$L = \int \sqrt{1+\dot{y}^2}\, dx \quad . \tag{A4.45}$$

Adding the constraint equation times a Lagrange multiplier to the energy equation gives the function to be minimized:

$$f = \rho g y \sqrt{1+\dot{y}^2} + \lambda \sqrt{1+\dot{y}^2} \quad . \tag{A4.46}$$

Because this does not depend on $x$, equation (4.14) can be used to find the first integral:

$$f - \dot{y}\frac{\partial f}{\partial \dot{y}} = A$$

$$\rho g y \sqrt{1+\dot{y}^2} + \lambda \sqrt{1+\dot{y}^2} - \frac{\rho g y \dot{y}^2}{\sqrt{1+\dot{y}^2}} - \frac{\lambda \dot{y}^2}{\sqrt{1+\dot{y}^2}} = A$$

$$\frac{\rho g y + \lambda}{\sqrt{1+\dot{y}^2}} = A$$

$$\frac{dy}{dx} = \frac{1}{A}\sqrt{(\rho g y + \lambda)^2 - A^2}$$

$$\frac{dy}{\sqrt{(\rho g y + \lambda)^2 - A^2}} = \frac{dx}{A} \quad . \tag{A4.47}$$

Since the left hand side depends only on $y$, and the right side only on $x$, the indefinite integrals are equal:

$$\int \frac{dy}{\sqrt{(\rho g y + \lambda)^2 - A^2}} = \int \frac{dx}{A} + B \quad . \tag{A4.48}$$

Defining $z = \rho g y + \lambda$, $dz = \rho g\, dy$,

$$\frac{1}{\rho g}\int \frac{dz}{\sqrt{z^2 - A^2}} = \int \frac{dx}{A} + B$$

$$\frac{1}{\rho g} \cosh^{-1}\left(\frac{z}{A}\right) = \frac{x}{A} + B$$

$$z = A \cosh\left(\frac{\rho g x}{A} + \rho g B\right)$$

$$\Rightarrow y = \frac{1}{\rho g}\left[A \cosh\left(\frac{\rho g x}{A} + \rho g B\right) - \lambda\right] \quad . \tag{A4.49}$$

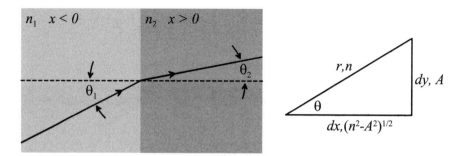

Figure A4.4. Refraction of light at an interface.

The three constants $A$, $B$, and $\lambda$ are determined by the location of the two ends of the chain and from equation (A4.45). This solution is called a *catenary*; it is also the curve that bounds a surface of revolution with the least area.

(4.2) *Consider a light ray travelling in a medium that has index of refraction $n_1$ for $x < 0$ and $n_2$ for $x > 0$. As it crosses the line $x = 0$ its angle with respect to the $x$ axis will change. Solve Euler's equation to find the simple relationship between the two angles.*

The starting point is equation (4.23), which can easily be integrated in this problem because the index of refraction does not depend on $y$:

$$\underbrace{\frac{\partial n}{\partial y}}_{0}\sqrt{1+\dot{y}^2} - \frac{d}{dx}\frac{n\dot{y}}{\sqrt{1+\dot{y}^2}} = 0 \quad .$$

$$\frac{d}{dx}\frac{n\dot{y}}{\sqrt{1+\dot{y}^2}} = 0 \Rightarrow \frac{n\dot{y}}{\sqrt{1+\dot{y}^2}} = A \Rightarrow \dot{y} = \frac{A}{\sqrt{n^2 - A^2}} \quad , \tag{A4.50}$$

where the integration constant $A$ is determined by the boundary conditions. This relates the slope to the index of refraction (Figure A4.4):

$$\frac{dy}{dx} = \tan\theta = \frac{A}{\sqrt{n^2 - A^2}} \Rightarrow \sin\theta = \frac{A}{n} \quad . \tag{A4.51}$$

The ratio of this equation between the regions is independent of the constant $A$; it is called *Snell's Law*

$$\frac{\sin\theta_1}{\sin\theta_2} = \frac{n_2}{n_1} \quad . \tag{A4.52}$$

(4.3) *Find the Lagrangian for a mass on a spinning hoop (Figure A4.5), and use it to find the equation of motion. What is its form for small oscillations?*

The Lagrangian is

$$\mathcal{L} = U - V = \left(\frac{1}{2}mr^2\dot{\theta}^2 + \frac{1}{2}mr^2\sin^2\theta\omega^2\right) - (-mgr\cos\theta) \tag{A4.53}$$

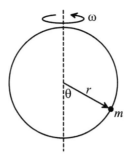

Figure A4.5. Mass on a hoop.

Applying Lagrange's equation gives the equation of motion:

$$mr^2 \sin\theta \cos\theta \omega^2 - mgr \sin\theta - mr^2\ddot\theta = 0$$

$$mr^2\ddot\theta + mgr\sin\theta - mr^2 \sin\theta \cos\theta \omega^2 = 0 \quad . \tag{A4.54}$$

For small motions $\sin\theta \approx \theta$, $\cos\theta \approx 1$, and so the approximate governing equation is

$$r\ddot\theta + (g - r\omega^2)\theta = 0 \quad . \tag{A4.55}$$

This is a simple harmonic oscillator, and the effect of the rotation is to decrease the effective gravitational constant.

## A4.5 RANDOM SYSTEMS

(5.1) (a) *Work out the first three cumulants $C_1, C_2,$ and $C_3$.*
Plugging in equations (5.25) and (5.26), we have

$$\begin{aligned}
&1 + \left(ikC_1 - \frac{k^2}{2}C_2 - \frac{ik^3}{6}C_3 + \cdots\right) \\
&+ \frac{1}{2}\left(ikC_1 - \frac{k^2}{2}C_2 - \frac{ik^3}{6}C_3 + \cdots\right)^2 \\
&+ \frac{1}{6}\left(ikC_1 - \frac{k^2}{2}C_2 - \frac{ik^3}{6}C_3 + \cdots\right)^3 + \cdots \\
&= 1 + ik\langle x\rangle - \frac{k^2}{2}\langle x^2\rangle - \frac{ik^3}{6}\langle x^3\rangle + \cdots \quad .
\end{aligned} \tag{A4.56}$$

Collecting terms with the same power of $k$, this gives

$$C_1 = \langle x \rangle \tag{A4.57}$$

for the first cumulant,

$$-\frac{k^2}{2}C_1^2 - \frac{k^2}{2}C_2 = -\frac{k^2}{2}\langle x^2\rangle \tag{A4.58}$$

or

$$C_2 = \langle x^2\rangle - \langle x\rangle^2 \tag{A4.59}$$

(the variance) for the second, and

$$-\frac{ik^3}{6}C_1^3 - \frac{ik^3}{2}C_1C_2 - \frac{ik^3}{6}C_3 = \frac{ik^3}{6}\langle x^3\rangle \quad (A4.60)$$

or

$$C_3 = \langle x^3\rangle - 3\langle x\rangle\langle x^2\rangle + 2\langle x\rangle^3 \quad (A4.61)$$

for the third.

(b) *Evaluate the first three cumulants for a Gaussian distribution*

$$p(x) = \frac{1}{\sqrt{2\pi\sigma^2}}e^{-(x-\bar{x})^2/2\sigma^2} \quad . \quad (A4.62)$$

$$\int_{-\infty}^{\infty} x \frac{1}{\sqrt{2\pi\sigma^2}}e^{-(x-\bar{x})/2\sigma^2}\, dx = \bar{x}$$

$$\int_{-\infty}^{\infty} x^2 \frac{1}{\sqrt{2\pi\sigma^2}}e^{-(x-\bar{x})/2\sigma^2}\, dx = \bar{x}^2 + \sigma^2$$

$$\int_{-\infty}^{\infty} x^3 \frac{1}{\sqrt{2\pi\sigma^2}}e^{-(x-\bar{x})/2\sigma^2}\, dx = \bar{x}^3 + 3\bar{x}\sigma^2. \quad (A4.63)$$

This gives for the first three cumulants

$$C_1 = \bar{x}$$
$$C_2 = \bar{x}^2 + \sigma^2 - \bar{x}^2$$
$$= \sigma^2$$
$$C_3 = \bar{x}^3 + 3\bar{x}\sigma - 3\bar{x}(\bar{x}^2 + \sigma^2) + 2\bar{x}^3$$
$$= 0 \quad . \quad (A4.64)$$

In fact, all of the higher cumulants vanish for a Gaussian (this can be shown by going back to the definition of cumulants, taking the log of the Fourier transform of a Gaussian). It can be shown for an arbitrary distribution that either all of the cumulants above 2 vanish (in which case it is a Gaussian), or there is an infinite number of nonzero cumulants [Marcienkiewicz, 1939]. The cumulants can be used as a test for non-Gaussianity, and to define an expansion for the deviation from a Gaussian distribution.

(5.2) (a) *For an order 4 maximal LFSR write down the bit sequence.*

The taps are at 1 and 4, therefore the recursion is

$$x_n = x_{n-1} + x_{n-4} \quad .$$

This produces the output shown in Table A4.1. Notice that runs of 0's and 1's of all possible lengths are present.

(b) *If an LFSR has a clock rate of 1 GHZ, how long must the register be for the time between repeats to be the age of the universe ($\sim 10^{10}$ years)?*

$$(2^N - 1)\, 10^{-9}\, \text{(seconds)} = 10^{10}\, \text{(years)} \cdot 10^7 \left(\frac{\text{seconds}}{\text{year}}\right)$$

$$\Rightarrow N \approx 86 \quad .$$

Table A4.1. *Sequence for an order 4 maximal LFSR.*

| $x_n$ | $x_{n-1}$ | $x_{n-2}$ | $x_{n-3}$ | $x_{n-4}$ |
|---|---|---|---|---|
| 0 | 1 | 1 | 1 | 1 |
| 1 | 0 | 1 | 1 | 1 |
| 0 | 1 | 0 | 1 | 1 |
| 1 | 0 | 1 | 0 | 1 |
| 1 | 1 | 0 | 1 | 0 |
| 0 | 1 | 1 | 0 | 1 |
| 0 | 0 | 1 | 1 | 0 |
| 1 | 0 | 0 | 1 | 1 |
| 0 | 1 | 0 | 0 | 1 |
| 0 | 0 | 1 | 0 | 0 |
| 0 | 0 | 0 | 1 | 0 |
| 1 | 0 | 0 | 0 | 1 |
| 1 | 1 | 0 | 0 | 0 |
| 1 | 1 | 1 | 0 | 0 |
| 1 | 1 | 1 | 1 | 0 |
| 0 | 1 | 1 | 1 | 1 |

(5.3) (a) *Use a Fourier transform to solve the diffusion equation (5.46) (assume that the initial condition is a normalized delta function at the origin).*

The spatial Fourier transform pair is

$$F(k,t) = \int_{-\infty}^{\infty} e^{ikx} f(x,t)\, dx$$

$$f(x,t) = \frac{1}{2\pi} \int_{-\infty}^{\infty} e^{-ikx} F(k,t)\, dk \quad . \tag{A4.65}$$

Substituting the definition of the transform into the diffusion equation gives

$$\frac{1}{2\pi} \int_{-\infty}^{\infty} e^{-ikx} \left[\frac{\partial F}{\partial t} + k^2 D F\right] dk = 0 \quad . \tag{A4.66}$$

Either $F = 0$ or the integrand must vanish, resulting in an ordinary differential equation for $F$

$$\frac{dF}{dt} + k^2 D F = 0 \quad . \tag{A4.67}$$

This can be easily solved to find

$$F = A e^{-k^2 D t} \quad , \tag{A4.68}$$

where $A$ is an arbitrary constant. This is a Gaussian in $k$; the Fourier transform of a Gaussian is still a Gaussian, but with the variance inverted. Therefore, the inverse transform of $F$ is

$$f = A' e^{-x^2/2\sigma^2} = A' e^{-x^2/4Dt} \tag{A4.69}$$

(where $A'$ is another constant). Remembering that a normalized Gaussian is

$$\frac{1}{\sqrt{2\pi\sigma^2}} e^{-x^2/2\sigma^2} \quad , \tag{A4.70}$$

our solution is

$$f = \frac{1}{\sqrt{4\pi Dt}} e^{-x^2/4Dt} \quad . \tag{A4.71}$$

(b) *What is the variance as a function of time?*
From part (a), $\sigma^2 = 2Dt$.

(c) *How is the diffusion coefficient for Brownian motion related to the viscosity of a fluid?*
Equating the variance found from the Langevin equation and the result from part (b),

$$2Dt = \frac{kTt}{3\pi\mu d} \tag{A4.72}$$

and so

$$D = \frac{kT}{6\pi\mu d} \quad . \tag{A4.73}$$

(d) *Write a program (including the random number generator) to plot the position as a function of time of a random walker in 1D that at each time step has an equal probability of making a step of $\pm 1$. Plot an ensemble of 10 trajectories, each 1000 points long, and overlay error bars of width $3\sigma(t)$ on the plot.*
In deriving the equation (5.46) we found that

$$D = \frac{\langle \delta^2 \rangle}{2\tau} \quad , \tag{A4.74}$$

where $\delta$ is the step in a time $\tau$. For this problem

$$p(\delta = 1) = p(\delta = -1) = 1/2 \quad , \tag{A4.75}$$

therefore $\langle \delta^2 \rangle = 1$. Since $\tau = 1$, we have

$$\sigma^2 = 2Dt$$
$$= 2\frac{\langle \delta^2 \rangle}{2\tau}t$$
$$= t \quad . \tag{A4.76}$$

Therefore, $3\sigma$ error bars go from $\pm 1.5\sqrt{t}$. Figure A4.6 plots these on the ensemble of trajectories, generated with a linear congruential generator by the following Matlab file:

```
%
% stochwalk.m
% (c) Neil Gershenfeld  9/1/97
% generate an ensemble of random walks
%
npts = 1000;
```

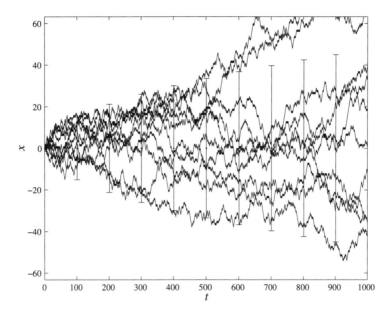

Figure A4.6. Trials of a 1D random walker.

```
nplot = 10;
seed = 1;
for i = 1:nplot
   xplot = [];
   x = 0;
   for j=1:npts
      seed = rem(8121 * seed + 28411, 134456);
      if (seed > (134456/2))
         x = x + 1;
      else
         x = x - 1;
         end
      xplot = [xplot x];
      end
   plot(xplot,'g')
   hold on
   end
t = 1:(npts/10):npts;
error = sqrt(t)*1.5;
errorbar(t,zeros(size(t)),error,error,'.')
xlabel('t')
ylabel('x')
axis([0 npts -2*sqrt(npts) 2*sqrt(npts)])
hold off
print -deps plot.eps
```

Table A4.2. *Errors with the Euler method.*

| step size | average error | final error | final slope |
|---|---|---|---|
| 1 | $1.55^{45}$ | $1.82 \times 10^{47}$ | $-1.82 \times 10^{47}$ |
| 0.1 | 247040 | $3.32 \times 10^6$ | $5.17 \times 10^6$ |
| 0.01 | 0.91 | 3.81 | 0.046 |
| 0.001 | 0.053 | 0.17 | $-0.00074$ |
| $10^{-4}$ | 0.0050 | 0.016 | $-3.41 \times 10^{-5}$ |
| $10^{-5}$ | 0.00050 | 0.0016 | $-4.64 \times 10^{-6}$ |

(e) *What fraction of the trajectories should be contained in the error bars?*

$$\int_{-1.5\sigma}^{1.5\sigma} \frac{1}{\sqrt{2\pi\sigma^2}} e^{-x^2/2\sigma^2}\, dx = \operatorname{erf}\left(\frac{3}{2\sqrt{2}}\right) = 0.866 \qquad (A4.77)$$

(erf is the error function).

## A4.6 FINITE DIFFERENCES: ORDINARY DIFFERENTIAL EQUATIONS

(6.1) *What is the second-order approximation error of the* Heun *method, which averages the slope at the beginning and the end of the interval?*

$$y(x+h) = y(x) + \frac{h}{2}\{f(x,y(x)) + f[x+h, y(x) + hf(x,y(x))]\} \qquad (A4.78)$$

Expanding the right hand side,

$$y(x+h) = y(x) + \frac{h}{2}\bigg[f(x,y(x)) + f(x,y(x))$$

$$+ h\frac{\partial f}{\partial x} + hf\frac{\partial f}{\partial y}\bigg] \qquad (A4.79)$$

$$= y(x) + hf(x,y(x)) + \frac{h^2}{2}\left[\frac{\partial f}{\partial x} + f\frac{\partial f}{\partial y}\right] + \mathcal{O}(h^3)$$

Like the midpoint method, this matches the expansion of the solution (equation (6.5)) up to the second order term, therefore there is no error at this order.

(6.2) *For a simple harmonic oscillator $\ddot{y}+y = 0$, with initial conditions $y(0) = 1$, $\dot{y}(0) = 0$, find $y(t)$ from $t = 0$ to $100\pi$. Use an Euler method and a fixed-step fourth-order Runge–Kutta method. For each method, what step size is needed for the average error over the interval to be less than 0.001? Check also the error in the value and slope at the last point.*

The solution to this problem is $y = \cos t$. Using double precision arithmetic, Table A4.2 shows the results for the Euler method (boom!), and Table A4.3 for the Runge–Kutta method. The extra effort is clearly justified.

Table A4.3. *Errors with the Runge–Kutta method.*

| step size | average error | final error | final slope |
|---|---|---|---|
| 1 | 0.46 | 0.58 | 0.12 |
| 0.1 | $8.33 \times 10^{-5}$ | $1.12 \times 10^{-5}$ | -0.040 |
| 0.01 | $8.31 \times 10^{-9}$ | $2.00 \times 10^{-10}$ | -0.00073 |

Here is the Euler program:

```c
/*
 * euler.c
 * (c) Neil Gershenfeld 9/1/97
 * integrate simple harmonic motion with an Euler algorithm
 */

#include <math.h>

double error,final_err,t,h,tmax,pi,y[2],ynew[2];
int nsteps;

main() {
   printf("Step size?   "); scanf("%lf",&h);
   pi = 4.0 * atan(1.0);
   tmax = 100*pi;
   t = 0;
   nsteps = ((int) (tmax / h));
   y[0] = 1;
   y[1] = 0;
   error = 0;
   while (t < tmax) {
      ynew[0] = y[0] + h * y[1];
      ynew[1] = y[1] - h * y[0];
      y[0] = ynew[0];
      y[1] = ynew[1];
      t += h;
      error = error + fabs(cos(t) - y[0]);
      }
   error = error / nsteps;
   final_err = fabs(cos(t) - y[0]);
   printf("error: %g   final error %g   slope: %g\n",
      error, final_err, y[1]);
}
```

And here is the Runge–Kutta program:

```
/*
 * rk.c
 * (c) Neil Gershenfeld  9/1/97
 * integrate simple harmonic motion
 *    with 4th order Runge-Kutta
 */

#include <math.h>

double error,final_err,t,h,tmax,pi,y[2],ynew[2],
    A[2],B[2],C[2],D[2];
int nsteps;

main() {
    printf("Step size? "); scanf("%lf",&h);
    pi = 4.0 * atan(1.0);
    tmax = 100*pi;
    t = 0;
    nsteps = ((int) (tmax / h));
    y[0] = 1;
    y[1] = 0;
    error = 0;
    while (t < tmax) {
       A[0] = h * y[1];
       A[1] = -h * y[0];
       B[0] = h * (y[1] + A[1]/2.0);
       B[1] = -h * (y[0] + A[0]/2.0);
       C[0] = h * (y[1] + B[1]/2.0);
       C[1] = -h * (y[0] + B[0]/2.0);
       D[0] = h * (y[1] + C[1]);
       D[1] = -h * (y[0] + C[0]);
       y[0] = y[0] + (A[0]+2.0*B[0]+2.0*C[0]+D[0])/6.0;
       y[1] = y[1] + (A[1]+2.0*B[1]+2.0*C[1]+D[1])/6.0;
       t += h;
       error = error + fabs(cos(t) - y[0]);
    }
    error = error / nsteps;
    final_err = fabs(cos(t) - y[0]);
    printf("error: %g  final error: %g  slope: %g\n",
       error, final_err, y[1]);
}
```

(6.3) *If the step size for a fourth-order Runge–Kutta algorithm is decreased by a factor of 10, by approximately how much will the local error decrease? Verify*

Table A4.4. *Errors with an adaptive stepper.*

| target error | average step size | average error | final error | average local error |
|---|---|---|---|---|
| 1 | 3.02 | 0.41 | 0.99 | 0.20 |
| 0.1 | 1.27 | 0.18 | 0.029 | 0.026 |
| 0.01 | 0.93 | 0.043 | 0.022 | 0.0045 |
| 0.001 | 0.57 | 0.0071 | 0.0028 | 0.00042 |
| 0.0001 | 0.35 | 0.0011 | 0.00067 | $3.97 \times 10^{-5}$ |
| $10^{-5}$ | 0.22 | 0.00018 | $1.66 \times 10^{-5}$ | $3.74 \times 10^{-6}$ |
| $10^{-6}$ | 0.14 | $3.34 \times 10^{-5}$ | $4.09 \times 10^{-7}$ | $4.20 \times 10^{-7}$ |
| $10^{-7}$ | 0.091 | $5.44 \times 10^{-6}$ | $6.13 \times 10^{-7}$ | $3.89 \times 10^{-8}$ |
| $10^{-8}$ | 0.059 | $9.44 \times 10^{-7}$ | $6.64 \times 10^{-8}$ | $3.93 \times 10^{-9}$ |
| $10^{-9}$ | 0.038 | $1.62 \times 10^{-7}$ | $7.12 \times 10^{-9}$ | $4.35 \times 10^{-10}$ |
| $10^{-10}$ | 0.025 | $2.94 \times 10^{-8}$ | $1.61 \times 10^{-9}$ | $5.21 \times 10^{-11}$ |

*this functional relationship by writing an adaptive variable stepper to find the average step size necessary for a given local error over the interval in the preceding problem. Check also the dependence of the average actual error, and the final error, on the step size.*

Since the local error is proportional to $h^5$, decreasing the step by a factor of 10 changes the local error by $h^5/(h/10)^5 = 10^5$. As Table A4.4 shows, this is the case. The actual error is worse than that because of the accumulation of errors.

Here is the adaptive step-size program:

```
/*
 * rkstep.c
 * (c) Neil Gershenfeld  9/1/97
 * integrate simple harmonic motion with
 *    an adaptive stepper 4th order Runge-Kutta
 */

#include <math.h>
void step();

void step(ystart,yend,step)
   double ystart[2], yend[2], step; {
   double A[2],B[2],C[2],D[2];
   A[0] = step * ystart[1];
   A[1] = -step * ystart[0];
   B[0] = step * (ystart[1] + A[1]/2.0);
   B[1] = -step * (ystart[0] + A[0]/2.0);
   C[0] = step * (ystart[1] + B[1]/2.0);
   C[1] = -step * (ystart[0] + B[0]/2.0);
   D[0] = step * (ystart[1] + C[1]);
```

```
      D[1] = -step * (ystart[0] + C[0]);
      yend[0] = ystart[0] +
         (A[0]+2.0*B[0]+2.0*C[0]+D[0])/6.0;
      yend[1] = ystart[1] +
         (A[1]+2.0*B[1]+2.0*C[1]+D[1])/6.0;
   }

   main() {
      double error,final_err,local_err,err_ave,t,h,
         h_ave,tmax,pi,target,y1[2],y2[2],y3[2],y[2];
      int nsteps;
      printf("Desired error? ");
      scanf("%lf",&target);
      pi = 4.0 * atan(1.0);
      tmax = 100*pi;
      t = 0;
      nsteps = 0;
      y[0] = 1;
      y[1] = 0;
      error = 0;
      err_ave = 0;
      h = 0.1;
      h_ave = 0;
      while (t < tmax) {
         nsteps += 1;
         step(y,y1,(h/2.0));
         step(y1,y2,(h/2.0));
         step(y,y3,h);
         local_err = y3[0] - y2[0];
         if (fabs(local_err) > target) {
            h = h / 1.3;
            continue;
         }
         y[0] = y2[0];
         y[1] = y2[1];
         t += h;
         err_ave = err_ave + fabs(local_err);
         error = error + fabs(cos(t) - y[0]);
         h_ave = h_ave + h;
         if (fabs(local_err) < (target/10.0))
            h = 1.2 * h;
      }
      error = error / nsteps;
      h_ave = h_ave / nsteps;
      err_ave = err_ave / nsteps;
      final_err = fabs(cos(t) - y[0]);
      printf(
```

```
                    "step: %g   error: %g   end error %g   loc error: %g\n",
                    h_ave, error, final_err, err_ave);
}
```

## A4.7 FINITE DIFFERENCES: PARTIAL DIFFERENTIAL EQUATIONS

(7.1) *Consider the 1D wave equation*

$$\frac{\partial^2 u}{\partial t^2} = v^2 \frac{\partial^2 u}{\partial x^2} \quad . \tag{A4.80}$$

(a) *Write down the straightforward finite-difference approximation.*

$$\frac{u_j^{n+1} - 2u_j^n + u_j^{n-1}}{(\Delta t)^2} = v^2 \frac{u_{j+1}^n - 2u_j^n + u_{j-1}^n}{(\Delta x)^2} \tag{A4.81}$$

$$u_j^{n+1} = 2u_j^n - u_j^{n-1} + \frac{v^2(\Delta t)^2}{(\Delta x)^2}[u_{j+1}^n - 2u_j^n + u_{j-1}^n] \quad .$$

(b) *What order approximation is this in time and in space?*
    The method is second order in $x$ and in $t$.

(c) *Use the von Neumann stability criterion to find the mode amplitudes.*

$$A = 2 - \frac{1}{A} + \frac{v^2(\Delta t)^2}{(\Delta x)^2}[e^{ikx} - 2 + e^{-ikx}]$$

$$= 2 - \frac{1}{A} + \frac{v^2(\Delta t)^2}{(\Delta x)^2} 2[\cos(kx) - 1]$$

$$= 2 - \frac{1}{A} - 4\frac{v^2(\Delta t)^2}{(\Delta x)^2} \sin^2 \frac{kx}{2}$$

$$A^2 + \underbrace{\left[\frac{4v^2(\Delta t)^2}{(\Delta x)^2} \sin^2 \frac{kx}{2} - 2\right]}_{\equiv 2b} A + 1 = 0$$

$$A = -b \pm \sqrt{b^2 - 1} \quad . \tag{A4.82}$$

(d) *Use this to find a condition on the velocity, time step, and space step for stability (hint: consider the product of the two amplitude solutions).*

$$A_+ A_- = b^2 - (b^2 - 1) = 1 \Rightarrow |A| = 1 \quad . \tag{A4.83}$$

The product of the solutions equals 1; unless both solutions have a magnitude of 1 this implies that one of the solutions will be larger than 1 (and hence unstable). For them both to have a magnitude of 1 they must be complex conjugates, therefore the square root must be imaginary:

$$A = -b \pm i\sqrt{1-b^2}$$
$$|A| = b^2 + (1-b^2) = 1 \quad . \tag{A4.84}$$

As long as $1 - b^2$ is positive, $|A| = 1$. This requires that

$$1 - \left(\frac{2v^2(\Delta t)^2}{(\Delta x)^2} \sin^2 \frac{kx}{2} - 1\right)^2 \geq 0 \tag{A4.85}$$

and so

$$\frac{v\Delta t}{\Delta x} \leq 1 \quad . \tag{A4.86}$$

This is just the Courant condition again.

(e) *Do different modes decay at different rates for the stable case?*

$|A| = 1$ independent of $k$ (as long as the Courant condition is satisfied), and so there is no damping of the magnitude of the amplitudes.

(f) *Numerically solve the wave equation for the evolution from an initial condition with $u = 0$ except for one nonzero node, and verify the stability criterion.*

The following program solves the problem for this and the last part. It uses a simple library that hides the details of X Windows, available at http://www.cup.cam.ac.uk/online/nmm. Initial conditions are input by clicking the mouse.

```
/*
 * str.c
 * (c) Neil Gershenfeld 9/1/97
 * integrate the equations of motion for a damped string
 */

#include <math.h>
#include <stdio.h>
#include "xng.h"

#define MAX_NODES 1000
#define WIN_SIZE 500
#define WORLD_MIN -3.0
#define WORLD_MAX 3.0

int i,n,nnodes,node,npts;
float d2_1,d2_0,Spring,Dissipation,
      X[MAX_NODES],Y2[MAX_NODES],
      Y1[MAX_NODES],Y0[MAX_NODES],Yold[MAX_NODES],Force;

main() {
   printf("Number of nodes?   ");
   scanf("%d",&nnodes);
```

```
        printf("Spring constant?  ");
        scanf("%f",&Spring);
        printf("Bending dissipation?  ");
        scanf("%f",&Dissipation);
        InitWindow("str",WIN_SIZE,WIN_SIZE);
        Scale(0.0,WORLD_MIN,((float) nnodes),WORLD_MAX);
        for (i = 0; i <= nnodes; ++i) {
           Yold[i] = Y2[i] = Y1[i] = Y0[i] = 0;
           X[i] = (float) i;
           }
        WLines(X,Y2,(nnodes+1));
        while (1) {
           for (i = 0; i < nnodes; ++i) {
              Y0[i] = Y1[i];
              Y1[i] = Y2[i];
              }
           for (i = 2; i < (nnodes-1); ++i) {
              d2_1 = Y1[i+1] + Y1[i-1] - 2 * Y1[i];
              d2_0 = Y0[i+1] + Y0[i-1] - 2 * Y0[i];
              Y2[i] = Spring*d2_1 + Dissipation*(d2_1-d2_0) +
                      2 * Y1[i] - Y0[i];
              }
           if (CheckButton() == TRUE) {
              node = (int) (0.5 + Xworld(XButton));
              if ((node > 1) && (node < (nnodes-1)))
                 Y2[node] = Yworld(YButton);
              }
           WLines(X,Y2,(nnodes+1));
           WLinesRev(X,Yold,(nnodes+1));
           for (i = 0; i <= nnodes; ++i)
              Yold[i] = Y2[i];
           }
        }
```

(g) *If the equation is replaced by*

$$\frac{\partial^2 u}{\partial t^2} = v^2 \frac{\partial^2 u}{\partial x^2} + \gamma \frac{\partial}{\partial t} \frac{\partial^2 u}{\partial x^2} \quad , \tag{A4.87}$$

*assume that*

$$u(x,t) = Ae^{i(kx-\omega t)} \tag{A4.88}$$

*and find a relationship between $k$ and $\omega$, and simplify it for small $\gamma$. Comment on the relationship to the preceeding question.*
Plugging in,

$$-\omega^2 = -v^2 k^2 + i\gamma k^2 \omega$$
$$0 = \omega^2 + i\gamma k^2 \omega - v^2 k^2$$

$$\omega = \frac{1}{2}[-i\gamma k^2 \pm \sqrt{-\gamma^2 k^4 + 4v^2 k^2}]$$

$$= -\frac{1}{2}i\gamma k^2 \pm vk\sqrt{1 - \frac{\gamma^2 k^2}{4v^2}}$$

$$\approx -\frac{1}{2}i\gamma k^2 \pm vk \quad . \tag{A4.89}$$

The two solutions correspond to travelling waves in either direction that are exponentially damped in time, with a time constant proportional to the square of 1/wavelength (short wavelengths are damped much faster). This wavelength-dependent damping is absent in the undamped exact problem and in the finite difference approximation, but is helpful in rounding off high-frequency artifacts in a numerical solution.

(h) See solution to part (f).

(7.2) *Write a program to solve a 1D diffusion problem on a lattice of 500 sites, with an initial condition of zero at all the sites, except the central site which starts at the value 1.0. Take $D = \Delta x = 1$, and use fixed boundary conditions set equal to zero.*

(a) *Use the explicit finite difference scheme, and look at the behavior for $\Delta t = 1, 0.5,$ and $0.1$. What step size is required by the Courant condition?*

Plugging in,

$$\frac{2D\Delta t}{\Delta x^2} = \frac{2 \times 1 \times \Delta t}{1^2} \leq 1 \Rightarrow \Delta t \leq \frac{1}{2} \quad . \tag{A4.90}$$

The following program shows the solution.

```
/*
 * diffexp.c
 * (c) Neil Gershenfeld 9/1/97
 * 1D diffusion explicit differences
 */

#include <math.h>
#include "xng.h"

float d,dx,dt,alpha,u[1000],unew[1000],x[1000];
int i,n,nsteps,step;

main() {
   printf("Diffusion constant? ");
   scanf("%f",&d);
   printf("Space step? ");
   scanf("%f",&dx);
   printf("Time step? ");
   scanf("%f",&dt);
   printf("Number of time steps? ");
   scanf("%d",&nsteps);
```

```
    printf("Lattice size?  ");
    scanf("%d",&n);

    alpha = d * dt / (dx * dx);

    InitWindow("Explicit Differences",n,n);
    Scale(0.0,-0.1,((float) n-1),1.0);

    for (i = 0; i < n; ++i) {
        u[i] = 0;
        x[i] = i;}
    u[n/2] = 1.0;
    WLines(x,u,n);

    for (step = 1; step <= nsteps; ++step) {
        for (i = 1; i < (n-1); ++i)
            unew[i] = u[i] + alpha *
                (u[i-1]+u[i+1]-2.0*u[i]);
        for (i = 1; i < (n-1); ++i)
            u[i] = unew[i];
        PClear();
        PLines(x,u,n);
        PShow();
    }
}
```

(b) *Now repeat this using implicit finite differences and compare the stability.*

```
/*
 * diffimp.c
 * (c) Neil Gershenfeld 9/1/97
 * 1D diffusion, implicit differences
 */

#include <math.h>
#include "xng.h"

float d,dx,dt,alpha,u[1000],x[1000],a[1000],b[1000],c[1000],
        cprime[1000],uprime[1000];
int i,n,nsteps,step;

main() {
    printf("Diffusion constant?  "); scanf("%f",&d);
    printf("Space step?  "); scanf("%f",&dx);
    printf("Time step?  "); scanf("%f",&dt);
    printf("Number of time steps?  "); scanf("%d",&nsteps);
```

```
            printf("Lattice size?   "); scanf("%d",&n);

            alpha = d * dt / (dx * dx);

            InitWindow("Implicit Differences",n,n);
            Scale(0.0,-0.1,((float) n-1),1.0);

            for (i = 0; i < n; ++i) {
               u[i] = 0;
               x[i] = i;
               a[i] = -alpha;
               b[i] = 1+2.0*alpha;
               c[i] = -alpha;}
            u[n/2] = 1.0;
            WLines(x,u,n);

            for (step = 1; step <= nsteps; ++step) {
               cprime[0] = c[0] / b[0];
               uprime[0] = u[0] / b[0];
               for (i = 1; i < n; ++i) {
                  cprime[i] = c[i] / (b[i] - a[i] * cprime[i-1]);
                  uprime[i] = (u[i] - a[i] * uprime[i-1]) /
                     (b[i] - a[i] * cprime[i-1]);
               }
               u[n-1] = uprime[n-1];
               for (i = (n-2); i >= 0; --i)
                  u[i] = uprime[i] - cprime[i] * u[i+1];
               PClear();
               PLines(x,u,n);
               PShow();
            }
         }
```

(7.3) *Use ADI to solve a 2D diffusion problem on a lattice, starting with randomly seeded values.*

```
/*
 * diffadi.c
 * (c) Neil Gershenfeld  9/1/97
 * 2D diffusion with ADI
 */

#include <math.h>
#include <stdlib.h>
#include "xng.h"
```

```
float d,dx,dt,alpha,u[1000][1000],a[1000],b[1000],c[1000],
    cprime[1000],uprime[1000];
int j,k,i,n,nsteps,step;

main() {
  printf("Diffusion constant?  "); scanf("%f",&d);
  printf("Space step?  "); scanf("%f",&dx);
  printf("Time step?  "); scanf("%f",&dt);
  printf("Number of time steps?  "); scanf("%d",&nsteps);
  printf("Lattice size?  "); scanf("%d",&n);

  alpha = d * dt / (2.0 * dx * dx);

  InitRainbow("ADI",n,0.0,1.0,255);

  for (i = 1; i < (n-1); ++i) {
    a[i] = -alpha;
    b[i] = 1+2.0*alpha;
    c[i] = -alpha;
  }
  a[0] = c[0] = a[n-1] = c[n-1] = 0;
  b[0] = b[n-1] = 1;

  for (j = 0; j < n; ++j)
    for (k = 0; k < n; ++k)
      u[j][k] = rand()/((float) RAND_MAX);

  for (step = 1; step <= nsteps; ++step) {
    for (k = 1; k < (n-1); ++k) {
      cprime[0] = c[0] / b[0];
      uprime[0] = ( (1.0 - 2.0*alpha) * u[j][k] +
          alpha * (u[j][k+1] + u[j][k-1]) ) / b[0];
      for (j = 1; j < n; ++j) {
        cprime[j] = c[j] / (b[j] - a[j] * cprime[j-1]);
        uprime[j] = ( ( (1.0 - 2.0*alpha) * u[j][k] +
          alpha * (u[j][k+1] + u[j][k-1]) ) - a[j] *
          uprime[j-1]) / (b[j] - a[j] * cprime[j-1]);
      }
      for (j = (n-2); j > 0; --j)
        u[j][k] = uprime[j] - cprime[j] * u[j+1][k];
    } /* end of k loop */
    for (j = 1; j < (n-1); ++j) {
      cprime[0] = c[0] / b[0];
      uprime[0] = ( (1.0 - 2.0*alpha) * u[j][k] +
          alpha * (u[j+1][k] + u[j-1][k]) ) / b[0];
      for (k = 1; k < n; ++k) {
        cprime[k] = c[k] / (b[k] - a[k] * cprime[k-1]);
```

```
            uprime[k] = ( ( (1.0 - 2.0*alpha) * u[j][k] +
                alpha * (u[j+1][k] + u[j-1][k]) ) - a[k] *
                uprime[k-1]) / (b[k] - a[k] * cprime[k-1]);
          }
          for (k = (n-2); k > 0; --k) {
            u[j][k] = uprime[k] - cprime[k] * u[j][k+1];
            SetImage(j,k,u[j][k]);
          }
        } /* end of j loop */
      UpdateImage();
      } /* end of step loop */
    }
```

(7.4) *Use SOR to solve Laplace's equation in 2D, with boundary conditions* $u_{j,1} = u_{1,k} = 0, u_{N,k} = -1, u_{j,N} = 1$, *and explore how the convergence rate depends on* $\alpha$, *and how the best choice for* $\alpha$ *depends on the lattice size.*

```
/*
 * sor.c
 * (c) Neil Gershenfeld  9/1/97
 * Successive Overrelaxation of elliptic boundary value problem
 */

#include "xng.h"

float alpha,u[1000][1000];
int i,j,n,nsteps,step;

main() {
   printf("Number of time steps?  ");
   scanf("%d",&nsteps);
   printf("Lattice size?  ");
   scanf("%d",&n);
   printf("Alpha?  ");
   scanf("%f",&alpha);

   InitRainbow("SOR",n,-1.0,1.0,255);

   for (i = 0; i < n; ++i) {
      u[i][0] = 0.0;
      SetImage(i,0,0.0);
      u[i][n-1] = -1.0;
      SetImage(i,(n-1),-1.0);
   }
   for (j = 0; j < n; ++j) {
      u[0][j] = 0;
```

```
            SetImage(0,j,0.0);
            u[n-1][j] = 1;
            SetImage((n-1),j,1.0);
        }
    UpdateImage();

    for (step = 1; step <= nsteps; ++step) {
        for (i = 1; i < (n-1); ++i)
            for (j = 1; j < (n-1); ++j) {
                u[i][j] = (1.0 - alpha) * u[i][j] +
                    alpha * 0.25 * (u[i+1][j] + u[i-1][j] +
                                    u[i][j+1] + u[i][j-1]);
                SetImage(i,j,u[i][j]);
            }
        UpdateImage();
    }
}
```

## A4.8 FINITE ELEMENTS

(8.1) *Use the Galerkin method to find a system of differential equations to approximate the wave equation*

$$\frac{\partial^2 u}{\partial t^2} = v^2 \frac{\partial^2 u}{\partial x^2} + \gamma \frac{\partial}{\partial t}\frac{\partial^2 u}{\partial x^2} \quad . \tag{A4.91}$$

*Take the solution domain to be the interval* $[0, 1]$.

Substituting

$$u = \sum_i a_i(t)\varphi_i(x) \tag{A4.92}$$

into the residual weighting equation gives

$$\begin{aligned}
0 &= \int_0^1 \left( \frac{\partial^2 u}{\partial t^2} - v^2 \frac{\partial^2 u}{\partial x^2} - \gamma \frac{\partial}{\partial t}\frac{\partial^2 u}{\partial x^2} \right) \varphi_j(x)\, dx \\
&= \sum_i \frac{d^2 a_i}{dt^2} \underbrace{\int_0^1 \varphi_i(x)\varphi_j(x)\, dx}_{\equiv A_{ij}} \\
&\quad - \sum_i \left( v^2 a_i + \gamma \frac{da_i}{dt} \right) \underbrace{\int_0^1 \frac{d^2\varphi_i}{dx^2}\varphi_j(x)\, dx}_{\underbrace{\left.\frac{d\varphi_i}{dx}\varphi_j\right|_0^1 - \int_0^1 \frac{d\varphi_i}{dx}\frac{d\varphi_j}{dx}\, dx}_{\equiv -B_{ij}}} \\
&= \mathbf{A}\cdot\frac{d^2\vec{a}}{dt^2} + \gamma\mathbf{B}\cdot\frac{d\vec{a}}{dt} + v^2 \mathbf{B}\cdot\vec{a} \quad . \tag{A4.93}
\end{aligned}$$

This is a system of equations for coupled, damped, driven oscillators.

**(8.2)** *Evaluate the matrix coefficients from the previous problem for linear hat basis functions, using elements with a fixed size of h.*

$$A_{i,i} = \int_{x_i-h}^{x_i} \left[\frac{x-(x_i-h)}{h}\right]^2 dx + \int_{x_i}^{x_i+h} \left[\frac{x_i+h-x}{h}\right]^2 dx$$

$$= \frac{2h}{3}$$

$$A_{i-1,i} = A_{i+1,i} = A_{i,i-1} = A_{i,i+1}$$

$$= \int_{x_i-h}^{x_i} \frac{x_i-x}{h} \frac{x-(x_i-h)}{h} dx$$

$$= \frac{h}{6}$$

$$B_{i,i} = \int_{x_i-h}^{x_i} \frac{1}{h^2} dx + \int_{x_i}^{x_i+h} \frac{1}{h^2} dx$$

$$= \frac{2}{h}$$

$$B_{i-1,i} = B_{i+1,i} = B_{i,i-1} = B_{i,i+1}$$

$$= \int_{x_i-h}^{x_i} \frac{-1}{h} \frac{1}{h} dx$$

$$= -\frac{1}{h} \quad . \tag{A4.94}$$

The corner elements will depend on the choice of boundary conditions.

**(8.3)** *Now find the matrix coefficients for Hermite polynomial interpolation basis functions, once again using elements with a fixed size of h. A symbolic math environment is useful for this problem.*

There are two types of basis functions, one having the value of $u$ at the element boundary as its expansion coefficient, and the other having the slope of $u$ as the coefficient. Writing, for example, A_(u_(i-1), du_i) to represent the term for the overlap integral between the basis function with coefficient $u_{i-1}$ at $x_{i-1}$ and the basis function with coefficient $\dot{u}_i$ at $x_i$, the nontrivial matrix elements are

```
A_(u_i,u_i) =
            26
            -- h
            35
A_(du_i,du_i) =
                  3
            2/105 h
A_(u_i,du_i) =
            0
A_(u_(i-1),u_i) =
            9/70 h
A_(du_(i-1),du_i) =
                  3
```

$$A\_(du\_(i-1),u\_i) = \frac{-1/140\ h}{} $$

$$A\_(du\_(i-1),u\_i) = \frac{13}{420} h^2$$

$$B\_(u\_i,u\_i) = 12/5\ 1/h$$

$$B\_(du\_i,du\_i) = 4/15\ h$$

$$B\_(u\_i,du\_i) = 0$$

$$B\_(u\_(i-1),u\_i) = -6/5\ 1/h$$

$$B\_(du\_(i-1),du\_i) = -1/30\ h$$

$$B\_(du\_(i-1),u\_i) = \frac{-1}{10}$$

.............................................................

These were found by the following Matlab program, using the symbolic toolkit:

.............................................................

```
%
% femat.m
% finite element problem matrices
% (c) Neil Gershenfeld   9/1/97
%
clear all
syms x h;
M = [1 0 0 0;0 1 0 0;1 h h^2 h^3;0 1 2*h 3*h^2];
Minv = inv(M);
for i = 1:4
   psi(i) = sum(Minv(:,i) .* [1; x; x^2; x^3]);
   dpsi(i) = diff(psi(i),x);
   end
fprintf('A_(u_i,u_i) = ')
pretty(int(psi(3)*psi(3),x,0,h)+int(psi(1)*psi(1),x,0,h),30)
fprintf('A_(du_i,du_i) = ')
pretty(int(psi(4)*psi(4),x,0,h)+int(psi(2)*psi(2),x,0,h),30)
fprintf('A_(u_i,du_i) = ')
pretty(int(psi(3)*psi(4),x,0,h)+int(psi(1)*psi(2),x,0,h),30)
fprintf('A_(u_(i-1),u_i) = ')
pretty(int(psi(1)*psi(3),x,0,h),30)
fprintf('A_(du_(i-1),du_i) = ')
pretty(int(psi(2)*psi(4),x,0,h),30)
fprintf('A_(du_(i-1),u_i) = ')
```

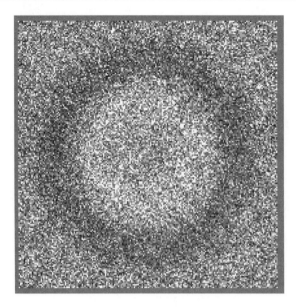

Figure A4.7. Square lattice gas, starting with a randomly filled lattice and a filled central square.

```
pretty(int(psi(2)*psi(3),x,0,h),30)
fprintf('B_(u_i,u_i) = ')
pretty(int(dpsi(3)*dpsi(3),x,0,h)+int(dpsi(1)*dpsi(1),x,0,h),30)
fprintf('B_(du_i,du_i) = ')
pretty(int(dpsi(4)*dpsi(4),x,0,h)+int(dpsi(2)*dpsi(2),x,0,h),30)
fprintf('B_(u_i,du_i) = ')
pretty(int(dpsi(3)*dpsi(4),x,0,h)+int(dpsi(1)*dpsi(2),x,0,h),30)
fprintf('B_(u_(i-1),u_i) = ')
pretty(int(dpsi(1)*dpsi(3),x,0,h),30)
fprintf('B_(du_(i-1),du_i) = ')
pretty(int(dpsi(2)*dpsi(4),x,0,h),30)
fprintf('B_(du_(i-1),u_i) = ')
pretty(int(dpsi(2)*dpsi(3),x,0,h),30)
```

..................................................................................................

## A4.9 CELLULAR AUTOMATA AND LATTICE GASES

(9.1) *Simulate the HPP lattice gas model. Take as a starting condition a randomly seeded lattice, but completely fill a square region in the center and observe how it evolves. Use periodic boundary conditions. Now repeat the calculation, but leave the lattice empty outside of the centeral filled square.*

This problem is solved by the following program, with output shown in Figures A4.7 and A4.8. The first part of the question shows the remarkable result that a spherical sound wave forms, even though spherical symmetry isn't apparent in the square lattice or governing rules. In the second part of the question, because the

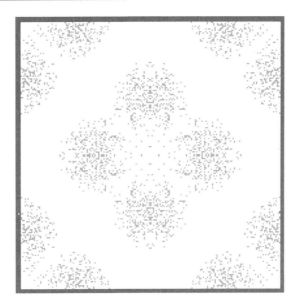

Figure A4.8. Square lattice gas, starting with an empty lattice except for a filled central square.

initial condition reflects the symmetry of the square lattice, the anisotropic viscocity leads a nonphysical solution of perpendicularly traveling packets of particles.

```
/*
 * lgssq.c
 * (c) Neil Gershenfeld  9/1/97
 * simple square lattice gas
 */

#include <stdio.h>
#include <math.h>
#include <X11/Xlib.h>
#include <X11/Xutil.h>
#include <X11/Xos.h>

#define SIZE 200
#define MAX_BITS 16
#define DISPLAY_DEPTH 8

#define NumBits(i) ((i&1)+((i&2)>>1)+((i&4)>>2)+((i&8)>>3))

/*
 * X variables *
 */
Display *D;
Window  W;
```

```
        XVisualInfo Info;
        Visual   *V;
        Colormap Cmap;
        XSetWindowAttributes Attr;
        XColor   Colors[MAX_BITS];
        GC       Gc;
        XImage *I1,*I2;
        unsigned char L1[SIZE*SIZE],L2[SIZE*SIZE];
        int      S;

        /*
         * math variables
         */
        int i,j,m,a,c,ran;
        unsigned char Table[MAX_BITS] = {
            0, 1, 2, 3, 4,      /*  0, 1, 2, 3, 4, */
           10, 6, 7, 8, 9,      /*  5, 6, 7, 8, 9, */
            5,11,12,13,14,      /* 10,11,12,13,14, */
           15};                 /* 15              */

        main() {
           /***************
            * set up for X *
            ***************/
           D = XOpenDisplay("");
           S = DefaultScreen(D);
           if (XMatchVisualInfo(D,S,DISPLAY_DEPTH,PseudoColor,&Info)
              == 0) {
              printf("Display can not handle %d bit pseudo color\n",
                     DISPLAY_DEPTH);
              return;
           }
           V = Info.visual;
           Colors[0].pixel = 0;
           Colors[0].red = 0;
           Colors[0].green = 0;
           Colors[0].blue = 0;
           Colors[0].flags = DoRed | DoGreen | DoBlue;
           for (i = 1; i < MAX_BITS; ++i) {
                   Colors[i].pixel = i;
                   Colors[i].red = (int) (65535*NumBits(i))/4;
                   Colors[i].green = (int) (65535*NumBits(i))/4;
                   Colors[i].blue = (int) (65535*NumBits(i))/4;
                   Colors[i].flags = DoRed | DoGreen | DoBlue;
              }
             Cmap = XCreateColormap(D,RootWindow(D,S),V,AllocAll);
             XStoreColors(D,Cmap,Colors,MAX_BITS);
```

```
      Attr.colormap = Cmap;
      Attr.background_pixel = WhitePixel(D,S);
      Attr.border_pixel = BlackPixel(D,S);
      W = XCreateWindow(D,DefaultRootWindow (D),0,0,SIZE,
            SIZE,1,DISPLAY_DEPTH,InputOutput,V,
      CWColormap|CWBackPixel|CWBorderPixel,&Attr);
      XStoreName(D,W,"Square Lattice");
      XMapRaised(D,W);
      Gc = XCreateGC (D, W, 0L, (XGCValues *) 0);
      I1 = XCreateImage(D,V,DISPLAY_DEPTH,ZPixmap,0,L1,SIZE,
      SIZE,DISPLAY_DEPTH,0);
      I2 = XCreateImage(D,V,DISPLAY_DEPTH,ZPixmap,0,L2,SIZE,
      SIZE,DISPLAY_DEPTH,0);
/********************
 * initialize lattice *
 ********************/
m = 134456;
a = 8121;
c = 28411;
ran = 1;
for (i = 0; i < SIZE; ++i)
   for (j = 0; j < SIZE; ++j) {
      ran = (ran*a+c) % m;
      L1[i+SIZE*j] = 2 << ((unsigned char) (6 *
         ran/((double) m)));
   }
L1[0] = L1[SIZE-1] = 0;
L1[SIZE*(SIZE-1)] = L1[SIZE+SIZE*(SIZE-1)] = 0;
for (i = 4*SIZE/10; i < 6*SIZE/10; ++i)
   for (j = 4*SIZE/10; j < 6*SIZE/10; ++j)
      L1[i+SIZE*j] = 15;
while (1) {
/****************
 * collision step *
 ****************/
for (i = 0; i < SIZE; ++i)
   for (j = 0; j < SIZE; ++j) {
      L2[i+SIZE*j] = Table[L1[i+SIZE*j]];
   }
/******************
 * propagation step *
 ******************/
for (i = 1; i < (SIZE-1); ++i)
   for (j = 1; j < (SIZE-1); ++j) {
      L1[i+SIZE*j] =
         (L2[i+SIZE*(j+1)] & 1) +
         (L2[(i-1)+SIZE*j] & 2) +
```

```
            (L2[i+SIZE*(j-1)] & 4) +
            (L2[(i+1)+SIZE*j] & 8);
      }
      /*******************************
       * periodic boundary conditions *
       *******************************/
      for (i = 1; i < (SIZE-1); ++i) {
         L1[i+SIZE*0] =
            (L2[i+SIZE*(0+1)] & 1) +
            (L2[(i-1)+SIZE*0] & 2) +
            (L2[i+SIZE*(SIZE-1)] & 4) +
            (L2[(i+1)+SIZE*0] & 8);
         L1[i+SIZE*(SIZE-1)] =
            (L2[i+SIZE*0] & 1) +
            (L2[(i-1)+SIZE*(SIZE-1)] & 2) +
            (L2[i+SIZE*(SIZE-1-1)] & 4) +
            (L2[(i+1)+SIZE*(SIZE-1)] & 8);
      }
      for (j = 1; j < (SIZE-1); ++j) {
         L1[0+SIZE*j] =
            (L2[0+SIZE*(j+1)] & 1) +
            (L2[(SIZE-1)+SIZE*j] & 2) +
            (L2[0+SIZE*(j-1)] & 4) +
            (L2[1+SIZE*j] & 8);
         L1[SIZE-1+SIZE*j] =
            (L2[SIZE-1+SIZE*(j+1)] & 1) +
            (L2[(SIZE-1-1)+SIZE*j] & 2) +
            (L2[SIZE-1+SIZE*(j-1)] & 4) +
            (L2[0+SIZE*j] & 8);
      }
      XPutImage(D,W,Gc,I1,0,0,0,0,SIZE,SIZE);
      }
}
```

(9.2) *Simulate the FHP model, with the same conditions as the previous model, and describe the difference. Alternate between the two possible two-body rules on alternate site updates.*

The next program implements the triangular lattice, using two look-up tables for the two possible pair-collision rules. Since the viscocity is isotropic, the solution to the second part no longer reflects the lattice symmetry (Figures A4.9 and A4.10).

```
/*
 * lgshex.c
 * (c) Neil Gershenfeld  9/1/97
 * simple hexagonal lattice gas
```

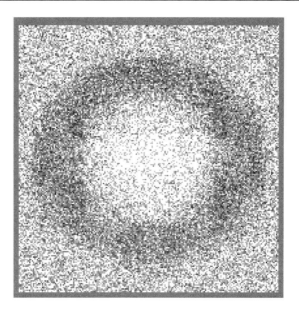

Figure A4.9. Hexagonal lattice gas, starting with a randomly filled lattice and a filled central square.

Figure A4.10. Hexagonal lattice gas, starting with an empty lattice except for a filled central square.

```
*/

#include <stdio.h>
#include <math.h>
#include <X11/Xlib.h>
#include <X11/Xutil.h>
#include <X11/Xos.h>

#define SIZE 200
#define MAX_BITS 64
#define DISPLAY_DEPTH 8

#define NumBits(i) ((i&1)+((i&2)>>1)+((i&4)>>2)+((i&8)>>3)+\
                   ((i&16)>>4)+((i&32)>>5))

/*
 * X variables *
 */
Display *D;
Window  W;
XVisualInfo Info;
Visual  *V;
Colormap Cmap;
XSetWindowAttributes Attr;
XColor  Colors[MAX_BITS];
GC      Gc;
XImage  *I1,*I2;
unsigned char L1[SIZE*SIZE],L2[SIZE*SIZE];
int     S;

/*
 * math variables
 */
int i,j,m,a,c,ran;
unsigned char Table[MAX_BITS] = {
    0, 1, 2, 3, 4,      /*  0, 1, 2, 3, 4, */
    5, 6, 7, 8,36,      /*  5, 6, 7, 8, 9, */
   10,11,12,13,14,      /* 10,11,12,13,14, */
   15,16,17, 9,19,      /* 15,16,17,18,19, */
   20,42,22,23,24,      /* 20,21,22,23,24, */
   25,26,27,28,29,      /* 25,26,27,28,29, */
   30,31,32,33,34,      /* 30,31,32,33,34, */
   35,18,37,38,39,      /* 35,36,37,38,39, */
   40,41,21,43,44,      /* 40,41,42,43,44, */
   45,46,47,48,49,      /* 45,46,47,48,49, */
   50,51,52,53,54,      /* 50,51,52,53,54, */
   55,56,57,58,59,      /* 55,56,57,58,59, */
```

```
      60,61,62,63};       /* 60,61,62,63    */
  unsigned char Table1[MAX_BITS] = {
       0, 1, 2, 3, 4,     /*  0, 1, 2, 3, 4, */
       5, 6, 7, 8,18,     /*  5, 6, 7, 8, 9, */
      10,11,12,13,14,     /* 10,11,12,13,14, */
      15,16,17,36,19,     /* 15,16,17,18,19, */
      20,42,22,23,24,     /* 20,21,22,23,24, */
      25,26,27,28,29,     /* 25,26,27,28,29, */
      30,31,32,33,34,     /* 30,31,32,33,34, */
      35, 9,37,38,39,     /* 35,36,37,38,39, */
      40,41,21,43,44,     /* 40,41,42,43,44, */
      45,46,47,48,49,     /* 45,46,47,48,49, */
      50,51,52,53,54,     /* 50,51,52,53,54, */
      55,56,57,58,59,     /* 55,56,57,58,59, */
      60,61,62,63};       /* 60,61,62,63    */

  main() {
    /***************
     * set up for X *
     ***************/
    D = XOpenDisplay("");
    S = DefaultScreen(D);
    if (XMatchVisualInfo(D,S,DISPLAY_DEPTH,PseudoColor,&Info)
       == 0) {
       printf("Display can not handle %d bit pseudo color\n",
              DISPLAY_DEPTH);
       return;
    }
    V = Info.visual;
    Colors[0].pixel = 0;
    Colors[0].red = 0;
    Colors[0].green = 0;
    Colors[0].blue = 0;
    Colors[0].flags = DoRed | DoGreen | DoBlue;
    for (i = 1; i < MAX_BITS; ++i) {
            Colors[i].pixel = i;
            Colors[i].red = (int) (65535*NumBits(i))/6;
            Colors[i].green = (int) (65535*NumBits(i))/6;
            Colors[i].blue = (int) (65535*NumBits(i))/6;
            Colors[i].flags = DoRed | DoGreen | DoBlue;
         }
      Cmap = XCreateColormap(D,RootWindow(D,S),V,AllocAll);
      XStoreColors(D,Cmap,Colors,MAX_BITS);
      Attr.colormap = Cmap;
      Attr.background_pixel = WhitePixel(D,S);
      Attr.border_pixel = BlackPixel(D,S);
      W = XCreateWindow(D,DefaultRootWindow (D),0,0,SIZE,
```

```
          SIZE,1,DISPLAY_DEPTH,InputOutput,V,
    CWColormap|CWBackPixel|CWBorderPixel,&Attr);
    XStoreName(D,W,"Hex Lattice");
    XMapRaised(D,W);
    Gc = XCreateGC (D, W, 0L, (XGCValues *) 0);
    I1 = XCreateImage(D,V,DISPLAY_DEPTH,ZPixmap,0,L1,SIZE,
    SIZE,DISPLAY_DEPTH,0);
    I2 = XCreateImage(D,V,DISPLAY_DEPTH,ZPixmap,0,L2,SIZE,
    SIZE,DISPLAY_DEPTH,0);
/*********************
* initialize lattice *
*********************/
m = 134456;
a = 8121;
c = 28411;
ran = 1;
for (i = 0; i < SIZE; ++i)
   for (j = 0; j < SIZE; ++j) {
      ran = (ran*a+c) % m;
      L1[i+SIZE*j] = 2 << ((unsigned char) (6 *
         ran/((double) m)));
   }
L1[0] = L1[SIZE-1] = 0;
L1[SIZE*(SIZE-1)] = L1[SIZE+SIZE*(SIZE-1)] = 0;
for (i = 4*SIZE/10; i < 6*SIZE/10; ++i)
   for (j = 4*SIZE/10; j < 6*SIZE/10; ++j)
      L1[i+SIZE*j] = 63;
while (1) {
/*****************
* collision step *
*****************/
for (i = 0; i < SIZE; ++i)
   for (j = 0; j < SIZE; j += 2) {
      L2[i+SIZE*j] = Table[L1[i+SIZE*j]];
      L2[i+SIZE*(j+1)] = Table1[L1[i+SIZE*(j+1)]];
   }
/*******************
* propagation step *
*******************/
for (i = 1; i < (SIZE-1); i += 2)
   for (j = 1; j < (SIZE-1); ++j) {
      L1[i+SIZE*j] =
         (L2[i+SIZE*(j+1)] & 1) +
         (L2[(i-1)+SIZE*(j+1)] & 2) +
         (L2[(i-1)+SIZE*j] & 4) +
         (L2[i+SIZE*(j-1)] & 8) +
         (L2[(i+1)+SIZE*j] & 16) +
```

```
              (L2[(i+1)+SIZE*(j+1)] & 32);
        L1[(i+1)+SIZE*j] =
              (L2[(i+1)+SIZE*(j+1)] & 1) +
              (L2[i+SIZE*j] & 2) +
              (L2[i+SIZE*(j-1)] & 4) +
              (L2[(i+1)+SIZE*(j-1)] & 8) +
              (L2[(i+2)+SIZE*(j-1)] & 16) +
              (L2[(i+2)+SIZE*j] & 32);
     }
/*******************************
 * periodic boundary conditions *
 *******************************/
     for (i = 1; i < (SIZE-1); i += 2) {
        L1[i+SIZE*0] =
              (L2[i+SIZE*(0+1)] & 1)  +
              (L2[(i-1)+SIZE*(0+1)] & 2) +
              (L2[(i-1)+SIZE*0] & 4) +
              (L2[i+SIZE*(SIZE-1)] & 8) +
              (L2[(i+1)+SIZE*0] & 16) +
              (L2[(i+1)+SIZE*(0+1)] & 32);
        L1[(i+1)+SIZE*0] =
              (L2[(i+1)+SIZE*(0+1)] & 1) +
              (L2[i+SIZE*0] & 2) +
              (L2[i+SIZE*(SIZE-1)] & 4) +
              (L2[(i+1)+SIZE*(SIZE-1)] & 8) +
              (L2[(i+2)+SIZE*(SIZE-1)] & 16) +
              (L2[(i+2)+SIZE*0] & 32);
        L1[i+SIZE*(SIZE-1)] =
              (L2[i+SIZE*0] & 1)  +
              (L2[(i-1)+SIZE*0] & 2) +
              (L2[(i-1)+SIZE*(SIZE-1)] & 4) +
              (L2[i+SIZE*(SIZE-1-1)] & 8) +
              (L2[(i+1)+SIZE*(SIZE-1)] & 16) +
              (L2[(i+1)+SIZE*0] & 32);
        L1[(i+1)+SIZE*(SIZE-1)] =
              (L2[(i+1)+SIZE*0] & 1) +
              (L2[i+SIZE*(SIZE-1)] & 2) +
              (L2[i+SIZE*(SIZE-1-1)] & 4) +
              (L2[(i+1)+SIZE*(SIZE-1-1)] & 8) +
              (L2[(i+2)+SIZE*(SIZE-1-1)] & 16) +
              (L2[(i+2)+SIZE*(SIZE-1)] & 32);
     }
     for (j = 1; j < (SIZE-1); ++j) {
        L1[0+SIZE*j] =
              (L2[0+SIZE*(j+1)] & 1)  +
              (L2[(SIZE-1)+SIZE*j] & 2) +
              (L2[(SIZE-1)+SIZE*(j-1)] & 4) +
```

```
          (L2[0+SIZE*(j-1)] & 8) +
          (L2[(0+1)+SIZE*(j-1)] & 16) +
          (L2[(0+1)+SIZE*j] & 32);
       L1[SIZE-1+SIZE*j] =
          (L2[SIZE-1+SIZE*(j+1)] & 1) +
          (L2[(SIZE-1-1)+SIZE*(j+1)] & 2) +
          (L2[(SIZE-1-1)+SIZE*j] & 4) +
          (L2[SIZE-1+SIZE*(j-1)] & 8) +
          (L2[0+SIZE*j] & 16) +
          (L2[0+SIZE*(j+1)] & 32);
     }
     XPutImage(D,W,Gc,I1,0,0,0,0,SIZE,SIZE);
     }

}
```

## A4.10 FUNCTION FITTING

(10.1) *Another way to choose among models is to select the one that makes the weakest assumptions about the data; this is the purpose of* maximum entropy *methods. Assume that what is measured is a set of expectation values for functions $f_i$ of a random variable $x$,*

$$\langle f_i(x) \rangle = \int_{-\infty}^{\infty} p(x) f_i(x) \, dx \quad . \tag{A4.95}$$

(a) *Given these measurements, find the compatible normalized probability distribution $p(x)$ that maximizes the entropy*

$$S = -\int_{-\infty}^{\infty} p(x) \log p(x) \, dx \quad . \tag{A4.96}$$

Using the methods of Chapter 4, we want to make extremal a variational integral of the entropy with Lagrange multipliers added for the normalization and measurement conditions

$$\begin{aligned} 0 &= \delta \mathcal{I} \\ &= \delta \left[ -\int_{-\infty}^{\infty} p(x) \log p(x) \, dx + \lambda_1 \int_{-\infty}^{\infty} p(x) \, dx \right. \\ &\quad \left. + \sum_i \lambda_i \int_{-\infty}^{\infty} p(x) f_i(x) \, dx \right] \end{aligned}$$

$$\Rightarrow 0 = \frac{\partial}{\partial p}\left[-p\log p + \lambda_1 p + \sum_i \lambda_i p\, f_i(x)\right]$$

$$= -\log p - 1 + \lambda_1 + \sum_i f_i(x)$$

$$\Rightarrow p(x) = e^{-1+\lambda_1+\sum_i \lambda_i f_i(x)} \quad . \tag{A4.97}$$

The Lagrange multipliers are then found from the constraints.

(b) *What is the maximum entropy distribution if we know only the second moment*

$$\sigma^2 = \int_{-\infty}^{\infty} p(x)\, x^2\, dx \quad ? \tag{A4.98}$$

In this case $f(x) = x^2$, therefore

$$p(x) = e^{-(1+\lambda_1+\lambda_2 x^2)} \tag{A4.99}$$

(it will be convenient to switch the sign in the definition of the Lagrange multipliers). One constraint comes from normalization:

$$1 = \int_{-\infty}^{\infty} p(x)\, dx$$

$$= \left.\frac{\sqrt{\pi}\,\mathrm{erf}(x\sqrt{\lambda_2})\, e^{-(1+\lambda_1)}}{2\sqrt{\lambda_2}}\right|_{-\infty}^{\infty}$$

$$= \frac{\sqrt{\pi}\, e^{-(1+\lambda_1)}}{2\sqrt{\lambda_2}} + \frac{\sqrt{\pi}\, e^{-(1+\lambda_1)}}{2\sqrt{\lambda_2}}$$

$$= \frac{\sqrt{\pi}\, e^{-(1+\lambda_1)}}{\sqrt{\lambda_2}} \tag{A4.100}$$

and the other from the second moment

$$\sigma^2 = \int_{-\infty}^{\infty} p(x)\, x^2\, dx$$

$$= \left.\frac{\sqrt{\pi}\,\mathrm{erf}(x\sqrt{\lambda_2})e^{-(1+\lambda_1)}}{4\lambda_2^{3/2}} - \frac{xe^{-\lambda_2 x^2}\sqrt{\lambda_2}\, e^{-(1+\lambda_1)}}{2\lambda_2^{3/2}}\right|_{-\infty}^{\infty}$$

$$= \frac{\sqrt{\pi}\, e^{-(1+\lambda_1)}}{4\lambda_2^{3/2}} + \frac{\sqrt{\pi}\, e^{-(1+\lambda_1)}}{4\lambda_2^{3/2}}$$

$$= \frac{\sqrt{\pi}\, e^{-(1+\lambda_1)}}{2\lambda_2^{3/2}} \quad . \tag{A4.101}$$

These two equations in two unknowns are solved by

$$\lambda_1 = \log\sqrt{2\pi\sigma^2} - 1 \qquad \lambda_2 = \frac{1}{2\sigma^2} \tag{A4.102}$$

giving

$$p(x) = \frac{1}{\sqrt{2\pi\sigma^2}} e^{-x^2/(2\sigma^2)} \quad . \tag{A4.103}$$

This provides another justification for the choice of a Gaussian error model: given only the variance, it makes the weakest assumption about the underlying distribution. Similarly, if we know the autocorrelation function of a variable evaluated at a set of delays, the maximum entropy model is a linear delay process driven by Gaussian noise [Cover & Thomas, 1991]. Maximum entropy methods can be related to Bayesian estimation through the use of a "universal" prior, using the exponential of the entropy to measure the likelihood of a model.

(10.2) *Now consider the reverse situation. Let's say that we know that a data set $\{x_n\}_{n=1}^N$ was drawn from a Gaussian distribution with variance $\sigma^2$ and unknown mean $\mu$. Try to find an optimal estimator of the mean (one that is unbiased and has the smallest possible error in the estimate).*

The obvious guess is the average of the samples

$$\mu \approx f(x_1, \ldots, x_N) = \frac{1}{N} \sum_{n=1}^N x_n \quad . \tag{A4.104}$$

First, check to see if it's unbiased:

$$\begin{aligned}
\langle f \rangle &= \left\langle \frac{1}{N} \sum_{n=1}^N x_n \right\rangle \\
&= \frac{1}{N} \sum_{n=1}^N \langle x_n \rangle \\
&= \frac{1}{N} \sum_{n=1}^N \mu \\
&= \mu \quad .
\end{aligned} \tag{A4.105}$$

Not surprisingly, it is. Next, find the variance in the estimate of the mean:

$$\begin{aligned}
\langle (f - \mu)^2 \rangle &= \left\langle \left[ \frac{1}{N} \sum_{n=1}^N x_n - \mu \right]^2 \right\rangle \\
&= \left\langle \left[ \frac{1}{N} \sum_{n=1}^N (x_n - \mu) \right]^2 \right\rangle \\
&= \left\langle \frac{1}{N} \sum_{n=1}^N (x_n - \mu) \frac{1}{N} \sum_{n'=1}^N (x_{n'} - \mu) \right\rangle \\
&= \frac{1}{N^2} \sum_{n=1}^N \langle (x_n - \mu)^2 \rangle \\
&= \frac{\sigma^2}{N}
\end{aligned} \tag{A4.106}$$

(the last line follows because the errors are uncorrelated). Finally, calculate the

Fisher information of the mean of the distribution

$$
\begin{aligned}
J_N(\mu) &= N J(\mu) \\
&= N \left\langle \left[ \frac{\partial}{\partial \mu} \log \left( \frac{1}{\sqrt{2\pi\sigma^2}} e^{(x-\mu)^2/(2\sigma^2)} \right) \right]^2 \right\rangle \\
&= N \left\langle \left( \frac{x-\mu}{\sigma^2} \right)^2 \right\rangle \\
&= \frac{N}{\sigma^2} \quad .
\end{aligned}
\tag{A4.107}
$$

Comparing these, we see that

$$
\langle (f - \mu)^2 \rangle = \frac{1}{J_N(\mu)} \quad .
\tag{A4.108}
$$

Averaging the samples reaches the Cramér–Rao bound, and so it's not possible to do any better than that.

## A4.11 TRANSFORMS

(11.1) *Prove that the DFT is unitary.*

The $k, n$ element in the DFT matrix is

$$
\mathbf{M}_{kn} = \frac{e^{2\pi i k n/N}}{\sqrt{N}} \quad .
\tag{A4.109}
$$

Therefore the $k, n$ element in the product of the matrix with its adjoint is

$$
\begin{aligned}
(\mathbf{M}^\dagger \cdot \mathbf{M})_{kn} &= \sum_{j=0}^{N-1} \mathbf{M}^\dagger_{kj} \mathbf{M}_{jn} \\
&= \frac{1}{N} \sum_{j=0}^{N-1} e^{2\pi i j(k-n)/N} \quad .
\end{aligned}
\tag{A4.110}
$$

On the diagonal, $k = n$, and we see that

$$
(\mathbf{M}^\dagger \cdot \mathbf{M})_{kk} = \frac{1}{N} \sum_{j=0}^{N-1} e^0 = \frac{1}{N} \sum_{j=0}^{N-1} 1 = 1 \quad .
\tag{A4.111}
$$

For the off-diagonal terms we need to use the formula for a *geometric sum*

$$
\frac{1 - x^N}{1 - x} = 1 + x + x^2 + \cdots + x^{N-1}
\tag{A4.112}
$$

(which can be verified by multiplying both sides by $1 - x$). This lets us write equation (A4.110) as

$$\frac{1}{N} \sum_{j=0}^{N-1} \left[ e^{2\pi i(k-n)/N} \right]^j = \frac{1}{N} \frac{1 - e^{2\pi i(k-n)}}{e^{2\pi i(k-n)/N}}$$

$$= \frac{1}{N} \frac{1 - \cos[2\pi(k-n)] - i\sin[2\pi(k-n)]}{e^{2\pi i(k-n)/N}}$$

$$= 0 \ . \tag{A4.113}$$

Therefore

$$(\mathbf{M}^\dagger \cdot \mathbf{M})_{kn} = \delta_{kn} \tag{A4.114}$$

as needed.

(11.2) *Calculate the inverse wavelet transform, using Daubechies fourth-order coefficients, of a vector of length $2^{12}$, with a 1 in the 5th and 30th places and zeros elsewhere.*

The following Matlab program:

```
%
% invwave.m
% inverse wavelet transformation
% (c) Neil Gershenfeld   9/1/97
%

clear all

npts = 2^12;
norder = log(npts)/log(2) - 1;

c0 = (1+sqrt(3))/(4*sqrt(2));
c1 = (3+sqrt(3))/(4*sqrt(2));
c2 = (3-sqrt(3))/(4*sqrt(2));
c3 = (1-sqrt(3))/(4*sqrt(2));

w = zeros(1,npts);
w(5) = 1;
w(30) = 1;

for order = 1:norder
    xnew = w(1:(2^order));
    wnew = w((2^order + 1):(2^(order+1)));
    w(1:2:(2^(order+1))) = xnew;
    w(2:2:(2^(order+1))) = wnew;
    xnew(1) = c0*w(1) + c3*w(2) + c2*w(npts-1) + c1*w(npts);
    xnew(2) = c1*w(1) - c2*w(2) + c3*w(npts-1) - c0*w(npts);
    for i = 3:2:(2^(order+1))
        xnew(i) = c2*w(i-2) + c1*w(i-1) + c0*w(i) + c3*w(i+1);
        xnew(i+1) = c3*w(i-2) - c0*w(i-1) + c1*w(i) - c2*w(i+1);
```

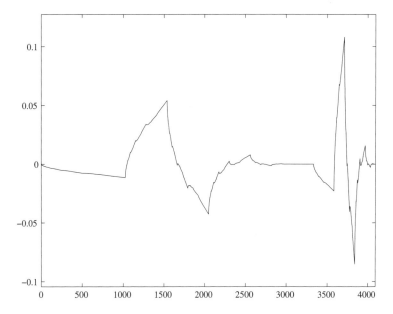

Figure A4.11. Inverse wavelet transform.

```
      end
   w(1:(2^(order+1))) = xnew;
   end
plot(w)
delta = 0.1*(max(w)-min(w));
axis([0 npts min(w)-delta max(w)+delta])
```

generates the output plotted in Figure A4.11, showing two of the wavelet basis functions in different positions and at different scales.

(11.3) *Consider a measurement of a three-component vector $\vec{x}$, with $x_1$ and $x_2$ being drawn independently from a Gaussian distribution with zero mean and unit variance, and $x_3 = x_1 + x_2$.*

(a) *Analytically calculate the covariance matrix of $\vec{x}$.*
$\langle x_1^2 \rangle = \langle x_2^2 \rangle = 1$, $\langle x_1 x_2 \rangle = 0$, $\langle (x_1 + x_2)^2 \rangle = \langle x_1^2 \rangle + \langle x_2^2 \rangle = 2$, therefore

$$\mathbf{C}_x = \begin{bmatrix} 1 & 0 & 1 \\ 0 & 1 & 1 \\ 1 & 1 & 2 \end{bmatrix} \quad . \tag{A4.115}$$

(b) *What are the eigenvalues?*
The characteristic equation $|\mathbf{C}_x - \lambda \mathbf{I}| = 0$ gives

$$(1 - \lambda)[(1 - \lambda)(2 - \lambda) - 1] - (1 - \lambda) = 0 \quad . \tag{A4.116}$$

This is solved by $\lambda = 1$, giving one root. Dividing by $1 - \lambda$ to remove it,

$$(1 - \lambda)(2 - \lambda) - 2 = 0 \quad . \tag{A4.117}$$

By inspection this is solved by $\lambda = 0$ and $\lambda = 3$, giving the others two roots.

(c) *Numerically verify these results by drawing a data set from the distribution and computing the covariance matrix and eigenvalues.*

(d) *Numerically find the eigenvectors of the covariance matrix, and use them to construct a transformation to a new set of variables $\vec{y}$ that have a diagonal covariance matrix with no zero eigenvalues. Verify this on the data set.*

The following output is produced by the Matlab program below:

...................................................................................................

```
> pca
covariance of x:
   1.002847 0.000656 1.003503
   0.000656 1.000499 1.001155
   1.003503 1.001155 2.004658
eigenvalues of covariance of x:
   1.001015 0.000000 0.000000
   0.000000 0.000000 0.000000
   0.000000 0.000000 3.006988
measured covariance of y:
   1.001015 0.000000
   0.000000 3.006988


%
% pca.m
% (c) Neil Gershenfeld   9/1/97
% PCA example
%
clear all
eps = 1e-10;
npts = 100000;
x(:,1) = randn(npts,1);
x(:,2) = randn(npts,1);
x(:,3) = x(:,1)+x(:,2);
Cx = cov(x);
fprintf('covariance of x:\n')
fprintf('   %f %f %f\n',Cx)
[M,Ex] = eig(Cx);
fprintf('eigenvalues of covariance of x:\n')
fprintf('   %f %f %f\n',Ex)
M = M';
index = (diag(Ex) > eps);
M = M(index,:);
y = (M*x')';
Cy = cov(y);
fprintf('measured covariance of y:\n')
fprintf('   %f %f\n',cov(y))
```

...................................................................................................

## A4.12 ARCHITECTURES

(12.1) *Find the first five* diagonal *Padé approximants* $[1/1], \ldots, [5/5]$ *to* $e^x$ *around the origin. Remember that the numerator and denominator can be multiplied by a constant to make the numbers as convenient as possible. Evaluate the approximations at* $x = 1$ *and compare to the correct value of* $e = 2.718281828459045$. *How is the error improving with the order?*

Here is the error, and the coefficients of the numerator and denominator polynomials:

```
N = 1, error = -0.28
    a:      2      1
    b:      2     -1
N = 2, error = 0.004
    a:     12      6      1
    b:     12     -6      1
N = 3, error = -2.8e-005
    a:    120     60     12      1
    b:    120    -60     12     -1
N = 4, error = 1.1e-007
    a:   1680    840    180     20      1
    b:   1680   -840    180    -20      1
N = 5, error = -2.8e-010
    a:  30240  15120   3360    420     30      1
    b:  30240 -15120   3360   -420     30     -1
```

The error is improving *faster* than exponentially in the order; for $e^x$ it in fact falls off as $N^{-1/2} x^{2N+1}/(2N)!$. The exclamation mark is certainly appropriate for an approximation that improves factorially. These calculations were done by the following Matlab program:

```
%
% pade.m
% find the diagonal Pade approximants to exp(x)
% (c) Neil Gershenfeld 9/1/97
%
clear all
x = 1;
for N = 1:5
   M = N;
   c = 1 ./ [1 cumprod(1:(N+M))]';
   C = [];
   for l = (N+1):(N+M)
      C = [C; c((1+l-1):-1:(1+l-M))'];
   end
   b = - inv(C) * c((1+N+1):(1+N+M));
   b = [1; b];
```

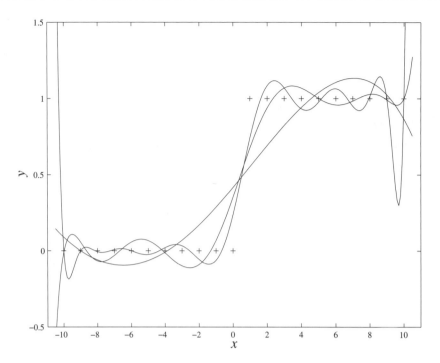

Figure A4.12. Polynomial fit.

```
    a(1) = c(1);
    for n = 1:N
        a(1+n,1) = c(1+n) + sum(b((1+1):(1+n)) ...
            .* c((1+n-1):-1:1));
    end
    y = sum(x.^(0:N) .* a') / sum(x.^(0:M) .* b');
    error = exp(x)-y;
    scale = min(abs(a));
    a = a / scale;
    b = b / scale;
    fprintf('N = %d, error = %.2g\n',N,error);
    fprintf('  a:'); fprintf('%6.0f ',a); fprintf('\n');
    fprintf('  b:'); fprintf('%6.0f ',b); fprintf('\n');
end
```

(12.2) *Take as a data set $x = \{-10, -9, \ldots, 9, 10\}$, and $y(x) = 0$ if $x \leq 0$ and $y(x) = 1$ if $x > 0$.*

  (a) *Fit the data with a polynomial with 5, 10, and 15 terms, using a pseudo-inverse of the Vandermonde matrix (such as Matlab's* pinv *function).*
  (b) *Fit the data with 5, 10, and 15 $r^3$ RBFs uniformly distributed between $x = -10$ and $x = 10$.*
  (c) *Using the coefficients found for these six fits, evaluate the total out-of-sample error at $x = \{-10.5, -9.5, \ldots, 9.5, 10.5\}$.*

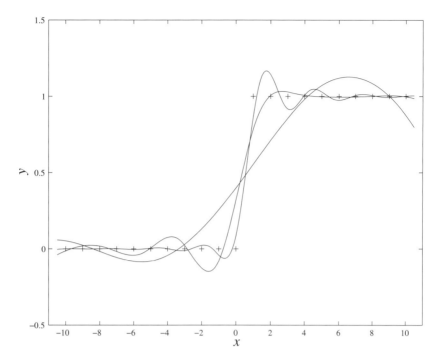

Figure A4.13. RBF fit.

(d) *Using a 10th order polynomial, fit the data with the curvature regularizer in equation (12.47), and plot the fits for $\lambda = 0, 0.1, .1, 1$ (this part is harder than the others).*

$$I = \sum_{n=1}^{N} [y_n - f(x_n)]^2 - \lambda \int_{x_{\min}}^{x_{\max}} \left[\frac{\partial^2 f}{\partial x^2}\right]^2 dx$$

$$= \sum_{n=1}^{N} \left[y_n - \sum_{m=0}^{M} a_m x_n^m\right]^2$$

$$\quad - \lambda \int_{x_{\min}}^{x_{\max}} \left[\sum_{m=2}^{M} a_m m(m-1) x^{m-2}\right]^2 dx$$

$$\frac{\partial I}{\partial a_l} = 0 = -\sum_{n=1}^{N} y_n x_n^l + \sum_{n=1}^{N} \sum_{m=0}^{M} a_m x_n^{m+l}$$

$$\quad - \lambda l(l-1) \sum_{m=2}^{M} a_m m(m-1) \frac{x_{\max}^{m+l-3} - x_{\min}^{m+l-3}}{m+l-3}$$

$$\equiv -\vec{A} + \mathbf{B} \cdot \vec{a} - \mathbf{C} \cdot \vec{a}$$

$$\Rightarrow \vec{a} = (\mathbf{B} - \mathbf{C})^{-1} \cdot \vec{A} \quad . \tag{A4.118}$$

Here is the Matlab program used to generate the Figures A4.12–A4.15:

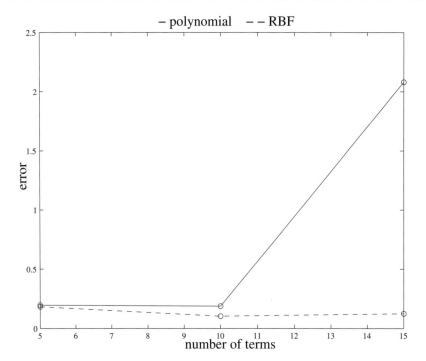

Figure A4.14. Out-of-sample error.

```
%
% fit.m
% (c) Neil Gershenfeld  9/1/97
% fitting examples
%
clear all
fitrange = 5:5:15;
xfit = (-10:10)';
yfit = (xfit > 0);
xcross = (-10.5:10.5)';
ycross = (xcross > 0);
xtest = (-10.5:.1:10.5)';
plot(xfit,yfit,'k+')
xlabel('x')
ylabel('y')
hold on
polyerror = [];
for nfit = fitrange
   A = [];
   for i = 1:nfit
      A = [A xfit.^(i-1)];
   end
   coeff = pinv(A)*yfit;
```

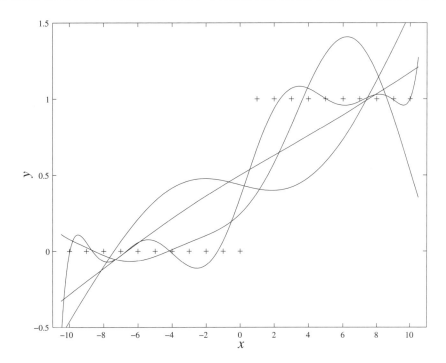

Figure A4.15. Regularized fit.

```
    A = [];
    for i = 1:nfit
       A = [A xcross.^(i-1)];
       end
    polyerror = [polyerror std(ycross - A * coeff)];
    A = [];
    for i = 1:nfit
       A = [A xtest.^(i-1)];
       end
    plot(xtest,A*coeff,'k-')
    end
axis([-11 11 -.5 1.5])
print -deps fitpoly.eps
hold off
figure
rbferror = [];
plot(xfit,yfit,'k+')
xlabel('x')
ylabel('y')
hold on
for nfit = fitrange
   A = [];
   centers = -10 + 20*(0:(nfit-1))/(nfit-1);
   for i = 1:nfit
```

```
        r2 = (xfit-centers(i)).^2;
        A = [A r2.^(3/2)];
      end
      coeff = A\yfit;
      coeff = pinv(A)*yfit;
      A = [];
      for i = 1:nfit
        r2 = (xcross-centers(i)).^2;
        A = [A r2.^(3/2)];
      end
      rbferror = [rbferror std(ycross - A * coeff)];
      A = [];
      for i = 1:nfit
        r2 = (xtest-centers(i)).^2;
        A = [A r2.^(3/2)];
      end
      plot(xtest,A*coeff,'k-')
    end
axis([-11 11 -.5 1.5])
print -deps fitrbf.eps
figure
hold off
plot(fitrange,polyerror,'k-',fitrange,rbferror,'k:',...
     fitrange,polyerror,'ko',fitrange,rbferror,'ko')
xlabel('Number of terms')
ylabel('Error')
title('- polynomial    -- RBF')
print -deps fitcross.eps
figure
M = 10;
plot(xfit,yfit,'k+')
xlabel('x')
ylabel('y')
hold on
lambda = [0 .01 .1 1];
for i = 1:length(lambda)
    A = [];
    for l = 0:(M-1)
        A(l+1,1) = sum(yfit .* (xfit/10).^l);
    end
    B = [];
    for l = 0:(M-1)
        for m = 0:(M-1)
            B(l+1,m+1) = sum((xfit/10).^(m+l));
        end
    end
    C = [];
```

```
        for l = 2:(M-1)
           for m = 2:(M-1)
              C(l+1,m+1) = lambda(i) * l*(l-1) * m*(m-1) * ...
                 (1^(m+l-3)-(-1)^(m+l-3))/(m+l-3);
           end
        end
        coeff = inv(B-C)*A;
        ypred = zeros(size(xtest));
        for m = 0:(M-1)
           ypred = ypred + coeff(m+1) * (xtest/10).^m;
        end
        plot(xtest,ypred,'k-')
        end
     axis([-11 11 -.5 1.5])
     print -deps fitreg.eps
     hold off
```

## A4.13 OPTIMIZATION AND SEARCH

(13.1) *Consider a 1D spin glass defined by a set of spins $S_1, S_2, \ldots, S_N$, where each spin can be either $+1$ or $-1$. The energy of the spins is defined by*

$$E = -\sum_{i=1}^{N} J_i S_i S_{i+1} \quad , \tag{A4.119}$$

*where the $J_i$'s are random variables drawn from a Gaussian distribution with zero mean and unit variance, and $S_{N+1} = S_1$ (periodic boundary conditions). Find a low-energy configuration of the spins by simulated annealing. The minimum possible energy is bounded by*

$$E_{min} = -\sum_{i=1}^{N} |J_i| \quad ; \tag{A4.120}$$

*compare your result to this. At each iteration flip a single randomly chosen spin, and if the energy increases by $\Delta E$ accept the move with a probability*

$$p = e^{-\beta \Delta E} \tag{A4.121}$$

*(always accept a move that decreases the energy). Take $\beta$ to be proportional to time, $\beta = \alpha t$ (where $t$ is the number of iterations), and repeat the problem for $\alpha = 0.1, 0.01,$ and $0.001$ for $N = 100$. Choose a single set of values for the $J$'s and the starting values for the spins and use these for each of the cooling rates.*

(13.2) *Now solve the same problem with a genetic algorithm (keep the same values for the $J$'s as the previous problem). Start with a population of 100 randomly drawn sets of spins. At each time step evaluate the energy of each member of*

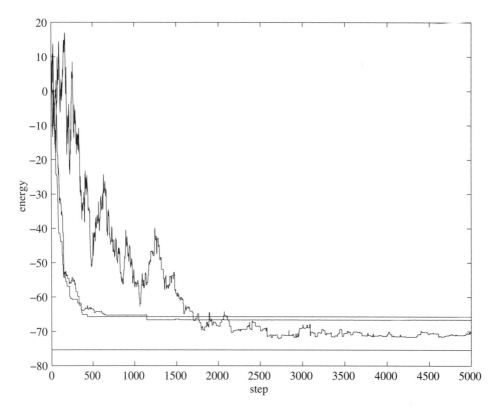

Figure A4.16. Simulated annealing solution for a spin glass.

*the population, and then assign it a probability for reproduction of*

$$p \propto e^{-\beta(E-E_{min})} \quad . \tag{A4.122}$$

*Generate 100 new strings by, for each string, choosing two of the strings from the previous population by drawing from this probability distribution, choosing a random crossover point, taking the bits to the left of the crossover point in the first string and the bits to the right in the second, and then mutating by randomly flipping a bit in the resulting string. Plot the minimum energy in the population as a function of time step for $\beta = 10, 1, 0.1, 0.01$.*

Both the GA and the simulated annealing solutions (Figures A4.16 and A4.17) show the trade-off between rapid early progress versus better eventual results. They also have the characteristic feature of progress occuring in jumps rather than uniformly, behavior that is not unlike the theory of *punctuated equilibrium* that evolution occurs in bursts [Gould & Eldredge, 1977].

The figures were generated by the following Matlab file:

```
%
% opt.m
% (c) Neil Gershenfeld 9/1/97
%
```

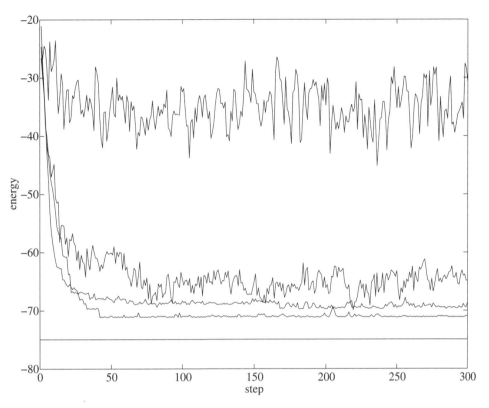

Figure A4.17. Genetic algorithm solution for a spin glass.

```
clear all
nbits = 100;
bonds = randn(1,nbits);
min_energy = -sum(abs(bonds));

%
% simulated annealing solution
%
nsteps = 5000;
nrates = 5;
rates = [0.001 0.01 0.1];
starting_spins = 2 * (rand(1,nbits) > 0.5) - 1;
for rate = 1:length(rates)
   spins = starting_spins;
   for step = 1:nsteps
      old_energy = - sum(bonds .* (spins .* ...
         [spins(2:nbits) spins(1)]));
      energy(step,rate) = old_energy;
      %
      % flip a random bit
      %
```

```
            bit = ceil(nbits * rand);
            spins(bit) = -spins(bit);
            new_energy = - sum(bonds .* (spins .* ...
                [spins(2:nbits) spins(1)]));
            delta_energy = new_energy - old_energy;
            %
            % check to see if we need to reject it
            %
            if ((delta_energy > 0) & ...
                 (rand(1) > exp(-delta_energy*rates(rate)*step)))
                spins(bit) = -spins(bit);
            end;
        end
    end
plot(energy,'b')
xlabel('step')
ylabel('energy')
hold on
plot([0 nsteps],[min_energy min_energy],'r')
hold off
figure

%
% genetic algorithms solution
%
nsteps = 300;
population = 100;
beta = [10 1 0.1 0.01];
for temp = 1:length(beta)
    spins = [];
    energy = [];
    newspins = [];
    for n = 1:population
        spins = [spins; (2 * (rand(1,nbits) > 0.5) - 1)];
    end
    for i = 1:nsteps
        %
        % evaluate fitness
        %
        for n = 1:population
            energy(i,n) = - sum(bonds .* (spins(n,:) .* ...
                [spins(n,(2:nbits)) spins(n,1)]));
        end
        prob = exp(-beta(temp)*(energy(i,:)-min_energy));
        prob = prob/sum(prob);
        prob = cumsum(prob);
        %
```

```
        % reproduce
        %
        for n = 1:population
            %
            % draw from distribution
            %
            index1 = 1+sum((rand) > prob);
            index2 = 1+sum((rand) > prob);
            %
            % crossover
            %
            crossover = ceil(rand*nbits);
            newspins(n,(1:crossover)) = ...
                spins(index1,(1:crossover));
            newspins(n,((crossover+1):nbits)) = ...
                spins(index2,((crossover+1):nbits));
            %
            % mutate
            %
            bit = ceil(nbits * rand);
            newspins(n,bit) = -newspins(n,bit);
        end
        spins = newspins;
    end
    plot(min(energy'),'b')
    hold on
  end
  xlabel('step')
  ylabel('energy')
  hold on
  plot([0 nsteps],[min_energy min_energy],'r')
  hold off
```

..................................................................................

## A4.14  CLUSTERING AND DENSITY ESTIMATION

(14.1) *Revisit the fitting problem in Chapter 12 with cluster-weighted modeling. Once again take as a data set $x = \{-10, -9, \ldots, 9, 10\}$, and $y(x) = 0$ if $x \leq 0$ and $y(x) = 1$ if $x > 0$. Take the simplest local model for the clusters, a constant. Plot the resulting forecast $\langle y|x \rangle$ and uncertainty $\langle \sigma_y^2|x \rangle$ for analyses with 1, 2, 3, and 4 clusters.*

Figures A4.18–A4.21 show the result, using the following Matlab program. Notice how there's no in-sample over-shoot or out-of-sample divergence because the clusters are limited to the behavior of the local models. Given the simplicity and small size of this example there are many other equally likely arrangements of the clusters.

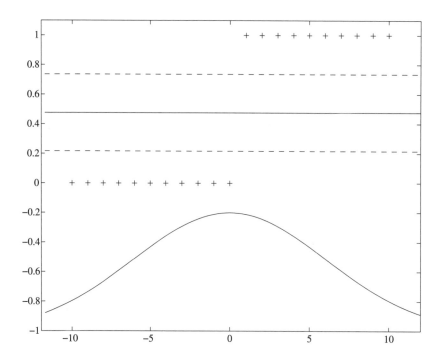

Figure A4.18. One cluster.

```
% cwm.m
% (c) Neil Gershenfeld  9/1/97
% 1D Cluster-Weighted Modeling example
clear all
x = (-10:10)';
y = (x > 0);
npts = length(x);
plot(x,y,'+')
xlabel('x')
ylabel('y')
nclusters = 3;
nplot = 100;
xplot = 24*(1:nplot)'/nplot - 12;
mux = 20*rand(1,nclusters) - 10;
muy = zeros(1,nclusters);
varx = ones(1,nclusters);
vary = ones(1,nclusters);
pc = 1/nclusters * ones(1,nclusters);
niterations = 50;
eps = .01;
for step = 1:niterations
   pplot = exp(-(kron(xplot,ones(1,nclusters)) ...
      - kron(ones(nplot,1),mux)).^2 ...
```

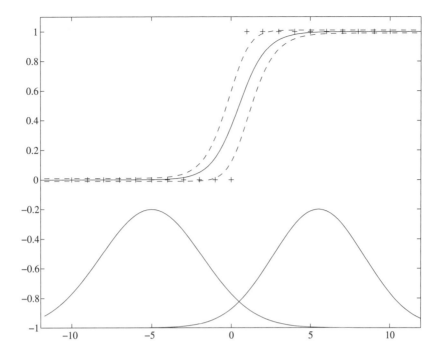

Figure A4.19. Two clusters.

```
            ./ (2*kron(ones(nplot,1),varx))) ...
            ./ sqrt(2*pi*kron(ones(nplot,1),varx)) ...
            .* kron(ones(nplot,1),pc);
    plot(xplot,pplot,'k');
    pause(0);
    px = exp(-(kron(x,ones(1,nclusters)) ...
            - kron(ones(npts,1),mux)).^2 ...
            ./ (2*kron(ones(npts,1),varx))) ...
            ./ sqrt(2*pi*kron(ones(npts,1),varx));
    py = exp(-(kron(y,ones(1,nclusters)) ...
            - kron(ones(npts,1),muy)).^2 ...
            ./ (2*kron(ones(npts,1),vary))) ...
            ./ sqrt(2*pi*kron(ones(npts,1),vary));
    p = px .* py .* kron(ones(npts,1),pc);
    pp = p ./ kron(sum(p,2),ones(1,nclusters));
    pc = sum(pp)/npts;
    yfit = sum(kron(ones(npts,1),muy) .* p,2) ...
            ./ sum(p,2);
    mux = sum(kron(x,ones(1,nclusters)) .* pp) ...
            ./ (npts*pc);
    varx = eps + sum((kron(x,ones(1,nclusters)) ...
            - kron(ones(npts,1),mux)).^2 .* pp) ...
            ./ (npts*pc);
    muy = sum(kron(y,ones(1,nclusters)) .* pp) ...
```

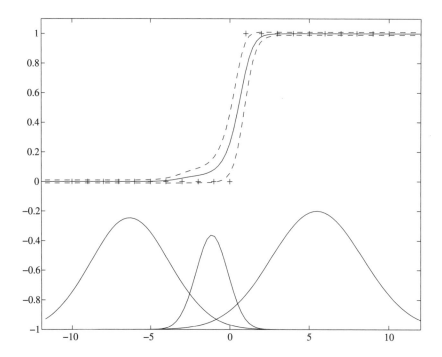

Figure A4.20. Three clusters.

```
      ./ (npts*pc);
   vary = eps + sum((kron(y,ones(1,nclusters)) ...
      - kron(ones(npts,1),muy)).^2 .* pp) ...
      ./ (npts*pc);
   end
pplot = exp(-(kron(xplot,ones(1,nclusters)) ...
   - kron(ones(nplot,1),mux)).^2 ...
   ./ (2*kron(ones(nplot,1),varx))) ...
   ./ sqrt(2*pi*kron(ones(nplot,1),varx)) ...
   .* kron(ones(nplot,1),pc);
yplot = sum(kron(ones(nplot,1),muy) .* pplot,2) ...
   ./ sum(pplot,2);
ystdplot = sum(kron(ones(nplot,1),(muy.^2+vary)) .* ...
   pplot,2) ./ sum(pplot,2) - yplot.^2;
plot(xplot,yplot,'k');
hold on
plot(xplot,yplot+ystdplot,'k--');
plot(xplot,yplot-ystdplot,'k--');
plot(x,y,'k+');
axis([-12 12 -1 1.1]);
plot(xplot,.8*pplot/max(max(pplot))-1,'k')
hold off
```

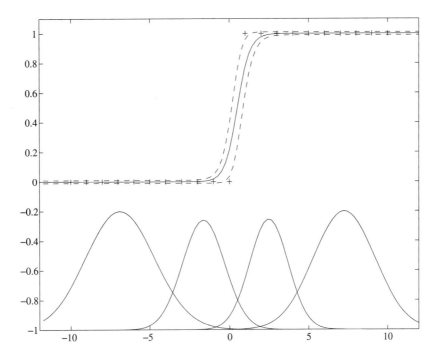

Figure A4.21. Four clusters.

## A4.15 FILTERING AND STATE ESTIMATION

(15.1) *Take as a test signal a periodically modulated sinusoid with noise added,*

$$y_n = \sin[0.1t + 4\sin(0.01t)] + \eta \equiv \sin(\theta_n) + \eta \quad , \tag{A4.123}$$

*where $\eta$ is a Gaussian noise process with $\sigma = 0.1$. Design a Kalman filter to estimate the noise-free signal. Use a two-component state vector $\vec{x}_n = (\theta_n, \theta_{n-1})$, and assume for the internal model a linear extrapolation $\theta_{n+1} = \theta_n + (\theta_n - \theta_{n-1})$. Take the system noise matrix $\mathbf{N}^x$ to be diagonal, and plot the predicted value of $y$ versus the measured value of $y$ if the standard deviation of the system noise is chosen to be $10^{-1}$, $10^{-3}$, and $10^{-5}$. Use the identity matrix for the initial error estimate.*

For this problem

$$\mathbf{A} = \begin{bmatrix} 2 & -1 \\ 1 & 0 \end{bmatrix} \quad \text{and} \quad \mathbf{B} = [\cos(\theta_n)\ 0] \quad . \tag{A4.124}$$

Figures A4.22–A4.24 show the output from the Matlab program below. When the internal noise in the state update is assumed to be small, the filter ignores the mismatch with changes in the signal. As the noise setting is increased the filter can do a good job of tracking the frequency. Finally, if the internal noise is set very large, the filter responds to the measurement noise as well as the underlying signal. This is an example of a *Phase-Locked Loop* (*PLL*) [Wolaver, 1991], used routinely to recover timing signals in radios and computers.

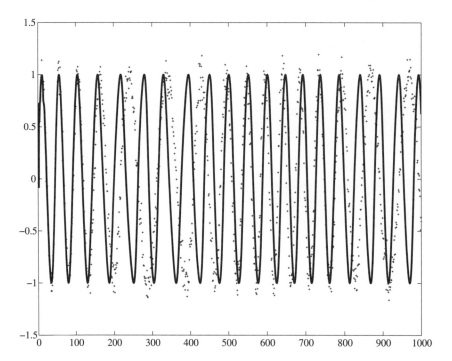

Figure A4.22. State noise standard deviation = $10^{-5}$.

```
%
% kalman.m
% Kalman filtering example
% (c) Neil Gershenfeld  9/1/97
%
clear all
nsteps = 1000;
y = sin(100*(1:nsteps)/nsteps ...
    + 4*(sin(10*(1:nsteps)/nsteps)));
E = [1 0 ; 0 1];
I = diag(ones(1,2));
stdx = 1e-5;
Nx = stdx^2*I;
stdy = 0.1;
Ny = stdy^2;
x = [.1 ; 0];
A = [2 -1 ; 1 0];
for i = 1:nsteps
   ypred = sin(x(1));
   ymeas = y(i) + stdy*randn;
   B = [cos(x(1)) 0];
   K = E*B'/(B*E*B'+Ny);
   x = x + K*(ymeas-ypred);
```

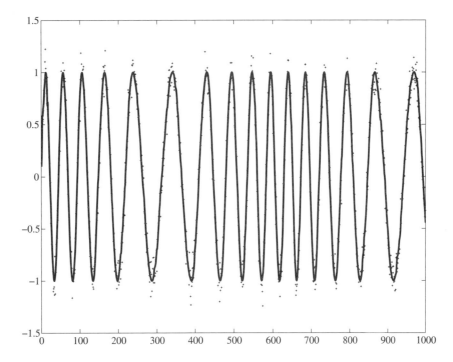

Figure A4.23. State noise standard deviation = $10^{-3}$.

```
      E = (I-K*B)*E;
      x = A*x;
      E = A*E*A'+Nx;
      meansplot(i) = ymeas;
      predplot(i) = ypred;
      end
plot(measplot,'k+','MarkerSize',2)
hold on
plot(predplot,'k','LineWidth',2)
hold off
```

## A4.16 LINEAR AND NONLINEAR TIME SERIES

(16.1) *Consider the Henon map*

$$x_{n+1} = y_n + 1 - ax_n^2 \qquad \text{(A4.125)}$$
$$y_{n+1} = bx_n$$

for $a = 1.4$ and $b = 0.3$.

(a) *Explore embedding by plotting*

1. $y_n$ versus $x_n$.
2. $x_n$ versus $n$.

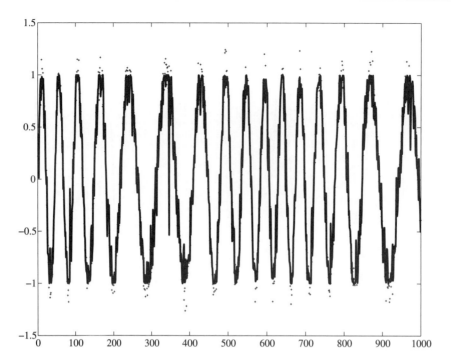

Figure A4.24. State noise standard deviation = $10^{-1}$.

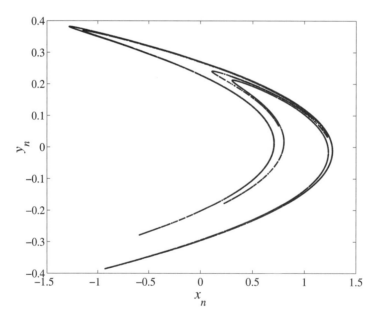

Figure A4.25.

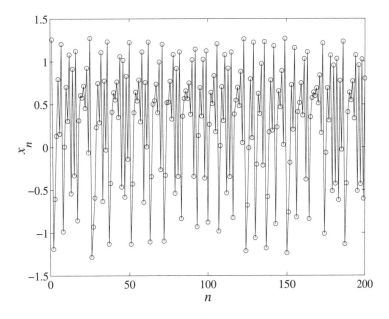

Figure A4.26.

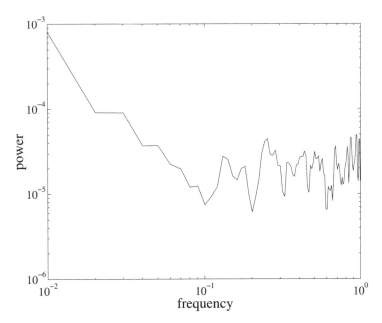

Figure A4.27.

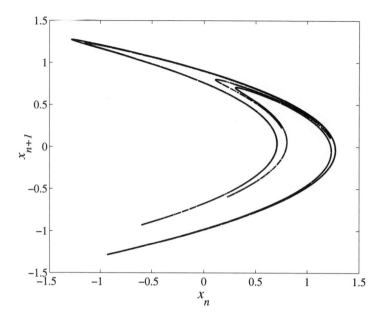

Figure A4.28.

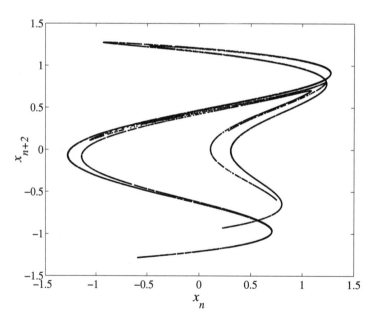

Figure A4.29.

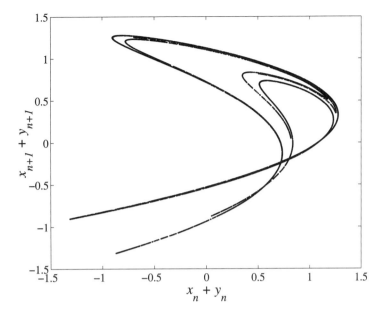

Figure A4.30.

3. The power spectrum of $x_n$ versus $n$.
4. $x_{n+1}$ versus $x_n$.
5. $x_{n+2}$ versus $x_n$.
6. $x_{n+1} + y_{n+1}$ versus $x_n + y_n$.

These are shown in Figures A4.25–A4.30.

(b) *Estimate as many of the nonlinear system properties as you can, such as the embedding dimension, attractor dimension, Lyapunov exponents, and source entropy.*

Numbers from the literature include the correlation dimension $\nu \approx 1.2$ [Grassberger & Procaccia, 1983b], which can be embedded in 2D with the $x_{n+1}$ versus $x_n$ series. The two exponents are $\lambda_1 \approx 0.6$ bits/step and $\lambda_2 \approx -2.3$ bits/step [Russell *et al.*, 1980]. Since there is just one positive exponent, the source entropy is equal to $\lambda_1$.

# Bibliography

[Abarbanel *et al.*, 1993] Abarbanel, H.D.I., Brown, R., Sidorowich, J.J., & Tsimring, L.S. (1993). The Analysis of Observed Chaotic Data in Physical Systems. *Reviews of Modern Physics*, **65**, 1331–92.

[Abelson *et al.*, 1997] Abelson, Harold, Mayer, Meinhard E., Sussman, Gerald J., & Wisdom, Jack. (1997). *A Computational Framework for Variational Mechanics*. Preprint.

[Abramowitz & Stegun, 1965] Abramowitz, Milton, & Stegun, Irene A. (1965). *Handbook of Mathematical Functions, with Formulas, Graphs, and Mathematical Tables*. New York, NY: Dover.

[Acton, 1990] Acton, Forman S. (1990). *Numerical Methods That Work*. Washington, DC: Mathematical Association of America.

[Aho *et al.*, 1974] Aho, Alfred V., Hopcroft, John E., & Ullman, Jeffrey D. (1974). *The Design and Analysis of Computer Algorithms*. Reading, MA: Addison-Wesley.

[Akaike, 1979] Akaike, H. (1979). A Bayesian Extension of the Minimum AIC Procedure of Autoregressive Model Fitting. *Biometrika*, **66**, 237–242.

[Ames, 1992] Ames, William F. (1992). *Numerical Methods for Partial Differential Equations*. 3rd edn. Boston, MA: Academic Press.

[Arfken & Weber, 1995] Arfken, George B., & Weber, Hans J. (1995). *Mathematical Methods For Physicists*. 4th edn. San Diego, CA: Academic Press.

[Auerbach *et al.*, 1992] Auerbach, D., Grebogi, C., Ott, E., & Yorke, J.A. (1992). Controlling Chaos in High dimensional Systems. *Phys. Rev. Lett.*, **69**, 3479–82.

[Baker & Graves-Morris, 1996] Baker, George A., & Graves-Morris, Peter. (1996). *Padé Approximants*. 2nd edn. New York, NY: Cambridge University Press.

[Balazs & Jennings, 1984] Balazs, N.L., & Jennings, B.K. (1984). Wigner's Function and Other Distribution Functions in Mock Phase Spaces. *Physics Reports*, **104**, 347–91.

[Balian, 1991] Balian, Roger. (1991). *From Microphysics to Macrophysics : Methods and Applications of Statistical Physics*. New York, NY: Springer–Verlag. Translated by D. ter Haar and J.F. Gregg, 2 volumes.

[Barron, 1993] Barron, Andrew R. (1993). Universal Approximation Bounds for Superpositions of a Sigmoidal Function. *IEEE Transactions on Information Theory*, **39**, 930–45.

[Batchelor, 1967] Batchelor, George Keith. (1967). *An Introduction to Fluid Dynamics*. Cambridge: Cambridge University Press.

[Bathe, 1996] Bathe, Klaus-Jürgen. (1996). *Finite Element Procedures*. Englewood Cliffs, NJ: Prentice-Hall.

[Bechtel & Abrahamsen, 1991] Bechtel, William, & Abrahamsen, Adele. (1991). *Connectionism and the Mind: an Introduction to Parallel Processing in Networks*. Cambridge, MA: B. Blackwell.

[Beck, 1990] Beck, C.(1990). Upper And Lower Bounds on the Renyi Dimensions and the Uniformity of Multifractals. *Physica D*, **41**, 67–78.

[Bell & Sejnowski, 1995] Bell, A.J., & Sejnowski, T.J. (1995). An Information-Maximization Approach to Blind Separation and Blind Deconvolution. *Neural Computation*, **7**, 1129–59.

[Bennett, 1988] Bennett, Charles H.(1988). Notes on the History of Reversible Computation. *IBM J. Res. Dev.*, **32**, 16–23.

[Bernardo & Smith, 1994] Bernardo, Jose M., & Smith, Adrian F.M. (1994). *Bayesian Theory*. New York, NY: Wiley.

[Besag *et al.*, 1995] Besag, J., Green, P.J., Higdon, D., & Mengersen, K. (1995). Bayesian Computation and Stochastic Systems. *Statistical Science*, **10**, 3–66.

[Bishop, 1996] Bishop, Christopher M. (1996). *Neural Networks for Pattern Recognition*. Oxford: Oxford University Press.

[Boer & Uhlenbeck, 1961] Boer, J. De, & Uhlenbeck, G. (eds). (1961). *Studies in Statistical Mechanics*. Vol. 1. New York, NY: Interscience Publishers.

[Box *et al.*, 1994] Box, George E.P., Jenkins, Gwilym .M., & Reinsel, Gregory C. (1994). *Time Series Analysis: Forecasting and Control*. Englewood Cliffs, NJ: Prentice Hall.

[Brenan *et al.*, 1996] Brenan, K.E., Campbell, S.L., & Petzold, L.R. (1996). *Numerical Solution of Initial-Value Problems in Differential-Algebraic Equations*. Philadelphia, PA: SIAM.

[Brown & Hwang, 1997] Brown, Robert Grover, & Hwang, Patrick Y.C.(1997). *Introduction to Random Signals and Applied Kalman Filtering*. 3rd edn. New York, NY: Wiley.

[Buntine, 1994] Buntine, W.L. (1994). Operations for Learning with Graphical Models. *Journal of Artificial Intelligence Research*, **2**, 159–225.

[Buntine, 1996] Buntine, W.L. (1996). A Guide to the Literature on Learning Probabilistic Networks from Data. *IEEE Transactions on Knowledge and Data Engineering*, **8**, 195–210.

[Buntine & Weigend, 1991] Buntine, Wray L., & Weigend, Andreas S. (1991). Bayesian Back-Propagation. *Complex Systems*, **5**, 603–43.

[Casdagli, 1992] Casdagli, Martin. (1992). A Dynamical Systems Approach to Modeling Input-Output Systems. *Pages 265–81 of:* Casdagli, M., & Eubank, S. (eds), *Nonlinear Modeling and Forecasting*. Santa Fe Institute Studies in the Sciences of Complexity. Redwood City, CA: Addison-Wesley.

[Catlin, 1989] Catlin, Donald E. (1989). *Estimation, Control, and the Discrete Kalman Filter*. Applied Mathematical Sciences, vol. 71. New York, NY: Springer Verlag.

[Chaitin, 1990] Chaitin, G.J. (1990). *Information, Randomness & Incompleteness*. 2nd edn. Series in Computer Science, vol. 8. Singapore: World-Scientific.

[Chaitin, 1994] Chaitin, G.J. (1994). Randomness & Complexity in Pure Mathematics. *International Journal of Bifurcation and Chaos*, **4**, 3–15.

[Cheeseman *et al.*, 1991] Cheeseman, P., Kanefsky, B., & Taylor, W. (1991). Where the Really Hard Problems Are. *Pages 331–37 of: Proc. of the 12th. International Joint Conference on A.I. (IJCAI-91)*. San Mateo, CA: Morgan Kaufmann.

[Chui *et al.*, 1994] Chui, Charles K., Montefusco, Laura, & Puccio, Luigia (eds). (1994). *Wavelets: Theory, Algorithms, and Applications*. San Diego, CA: Academic Press.

[Coddington & Levinson, 1984] Coddington, Earl A., & Levinson, Norman. (1984). *Theory of Ordinary Differential Equations*. Malabar, FL: R.E. Krieger.

[Collet & Eckmann, 1980] Collet, Pierre, & Eckmann, Jean-Pierre. (1980). *Iterated Maps on the Interval as Dynamical Systems*. Boston, MA: Birkhauser.

[Comon, 1994] Comon, Pierre. (1994). Independent Component Analysis: A New Concept? *Signal Processing*, **36**, 287–314.

[Cook *et al.*, 1989] Cook, Robert D., Malkus, David S., & Plesha, Michael E. (1989).

[Cooley & Tukey, 1965] Cooley, James W., & Tukey, John W. (1965). An Algorithm for the Machine Calculation of Complex Fourier Series. *Mathematics of Computation*, **19**, 297–301.

[Cover & Thomas, 1991] Cover, Thomas M., & Thomas, Joy A. (1991). *Elements of Information Theory*. New York, NY: Wiley.

[Cybenko, 1989] Cybenko, G.(1989). Approximation by Superpositions of a Sigmoidal Function. *Mathematics of Control, Signals, and Systems*, **2**, 303–14.

[Daubechies, 1988] Daubechies, I. (1988). Orthonormal Basis of Compactly Supported Wavelets. *Comm. Pure Applied Math.*, **41**, 909–96.

[Dawson et al., 1994] Dawson, S., Grebogi, C., Sauer, T., & Yorke, J.A. (1994). Obstructions to Shadowing When a Lyapunov Exponent Fluctuates about Zero. *Phys. Rev. Lett.*, **73**, 1927–30.

[Dempster et al., 1977] Dempster, A.P., Laird, N.M., & Rubin, D.B. (1977). Maximum Likelihood From Incomplete Data via the EM Algorithm. *J. R. Statist. Soc. B*, **39**, 1–38.

[Doolen et al., 1990] Doolen, Gary D. Frisch, Uriel, Hasslacher, Brosl, Orszag, Steven, & Wolfram, Stephen (eds). (1990). *Lattice Gas Methods for Partial Differential Equations*. Santa Fe Institute Studies in the Sciences of Complexity. Reading, MA: Addison-Wesley.

[Doyle et al., 1992] Doyle, John C., Francis, Bruce A., & Tannenbaum, Allen R. (1992). *Feedback Control Theory*. New York, NY: Macmillan.

[Dubrulle et al., 1991] Dubrulle, B., Frisch, U., Hénon, M., & Rivet, J.-P. (1991). Low Viscosity Lattice Gases. *Physica D*, **47**, 27–29.

[Duda & Hart, 1973] Duda, Richard O., & Hart, Peter E. (1973). *Pattern Classification and Scene Analysis*. New York, NY: Wiley.

[Durbin, 1960] Durbin, J. (1960). The Fitting of Time-Series Models. *Rev. Int. Inst. Statist.*, **28**, 233–43.

[Efron, 1983] Efron, Bradley. (1983). Estimating the Error Rate of a Prediction Rule: Improvements on Cross-Validation. *Journal of the American Statistical Association*, **78**, 316–31.

[Ekert & Jozsa, 1996] Ekert, Artur, & Jozsa, Richard. (1996). Quantum Computation and Shor's Factoring Algorithm. *Reviews of Modern Physics*, **68**(3), 733–753.

[Engl et al., 1996] Engl, Heinz W., Hanke, Martin, & Neubauer, Andreas. (1996). *Regularization of Inverse Problems*. Dordrecht: Kluwer Academic Publishers.

[Farmer & Sidorowich, 1987] Farmer, J.D., & Sidorowich, J.J. (1987). Predicting Chaotic Time Series. *Phys. Rev. Lett.*, **59**, 845–8.

[Feller, 1968] Feller, William. (1968). *An Introduction to Probability Theory and its Applications*. 3rd edn. New York, NY: Wiley.

[Foley et al., 1990] Foley, James D., van Dam, Andries, Feiner, Steven K., & Hughes, John F. (1990). *Computer Graphics: Principles and Practice*. 2nd edn. Reading, MA: Addison-Wesley.

[Forrest, 1993] Forrest, Stephanie. (1993). Genetic Algorithms: Principles of Natural Selection Applied to Computation. *Science*, **261**, 872–78.

[Fraser, 1989] Fraser, A.M. (1989). Information and Entropy in Strange Attractors. *IEEE Trans. Inf. Theory*, **35**, 245–62.

[Frederickson et al., 1983] Frederickson, P., Kaplan, J.L., Yorke, E.D., & Yorke, J.A. (1983). The Liapunov Dimension of Strange Attractors. *J. Diff. Eq.*, **49**, 185–207.

[Fredkin & Toffoli, 1982] Fredkin, Edward, & Toffoli, Tommaso. (1982). Conservative Logic. *Int. J. Theor. Phys.*, **21**, 219–53.

[Frisch et al., 1986] Frisch, E., Hasslacher, B., & Pomeau, Y. (1986). Lattice-Gas Automata for the Navier-Stokes Equation. *Physical Review Letters*, **56**, 1505–8.

[Frisch et al., 1987] Frisch, Uriel, d'Humiéres, Dominique, Hasslacher, Brosl, Lallemand, Pierre, Pomeau, Yves, & Rivet, Jean-Pierre. (1987). Lattice Gas Hydrodynamics in Two and Three Dimensions. *Complex Systems*, **1**, 649–707.

[Fukunaga, 1990] Fukunaga, Keinosuke. (1990). *Introduction to Statistical Pattern Recognition*. 2nd edn. Boston, MA: Academic Press.

[Gardiner, 1990] Gardiner, C.W. (1990). *Handbook of Stochastic Methods*. 2nd edn. New York, NY: Springer-Verlag.

[Gardner, 1970] Gardner, Martin. (1970). The Fantastic Combinations of John Conway's New Solitaire Game "Life". *Scientific American*, October, 120–3. Mathematical Games.

[Garey & Johnson, 1979] Garey, Michael R., & Johnson, David S. (1979). *Computers And Intractability: A Guide To The Theory Of NP-completeness*. San Francisco, CA: W.H. Freeman.

[Gear, 1971] Gear, C. William. (1971). *Numerical Initial Value Problems in Ordinary Differential Equations*. Englewood Cliffs, NJ: Prentice-Hall.

[Gershenfeld, 1989] Gershenfeld, Neil A. (1989). An Experimentalist's Introduction to the Observation of Dynamical Systems. *Pages 310–384 of:* Hao, Bai-Lin (ed), *Directions in Chaos*, vol. 2. Singapore: World Scientific.

[Gershenfeld, 1992] Gershenfeld, Neil A. (1992). Dimension Measurement on High-Dimensional Systems. *Physics D*, **55**, 135–54.

[Gershenfeld, 1993] Gershenfeld, Neil A. (1993). Information in Dynamics. *Pages 276–80 of:* Matzke, Doug (ed), *Proceedings of the Workshop on Physics of Computation*. Piscataway, NJ: IEEE Press.

[Gershenfeld, 1999a] Gershenfeld, Neil A. (1999a). *The Physics of Information Technology*. To be published.

[Gershenfeld, 1999b] Gershenfeld, Neil A. (1999b). *When Things Start To Think*. New York: Henry Holt.

[Gershenfeld & Chuang, 1997] Gershenfeld, Neil A., & Chuang, Isaac L. (1997). Bulk Spin Resonance Quantum Computation. *Science*, **275**, 350–6.

[Gershenfeld & Grinstein, 1995] Gershenfeld, Neil A., & Grinstein, Geoff. (1995). Entrainment and Communication with Dissipative Pseudorandom Dynamics. *Physical Review Letters*, **74**, 5024–7.

[Gershenfeld et al., 1983] Gershenfeld, Neil A., Schadler, Edward H., & Bilaniuk, Olexa M. (1983). APL and the Numerical Solution of High–Order Linear Differential Equations. *American Journal of Physics*, **51**, 743–6.

[Gershenfeld et al., 1998] Gershenfeld, Neil A., Schoner, Bernd, & Metois, Eric. (1999). Cluster-Weighted Modeling for Time Series Prediction and Characterization. *Nature*, **397**, 329–32.

[Gersho & Gray, 1992] Gersho, Allen, & Gray, Robert M. (1992). *Vector Quantization and Signal Compression*. Boston, MA: Kluwer Academic Publishers.

[Ghahramani & Hinton, 1998] Ghahramani, Zoubin, & Hinton, Geoffrey E. (1998). Hierarchical Nonlinear Factor Analysis and Topographic Maps. Jordan, M.I., Kearns, M.J., & Solla, S.A. (eds), *Advances in Neural Information Processing Systems*, vol. 10. Cambridge, MA: MIT Press.

[Giona et al., 1991] Giona, M., Lentini, F., & Cimagalli, V. (1991). Functional Reconstruction and Local Prediction of Chaotic Time Series. *Phys. Rev. A*, **44**, 3496–502.

[Girosi et al., 1995] Girosi, Frederico, Jones, Michael, & Poggio, Tomaso. (1995). Regularization Theory and Neural Networks Architectures. *Neural Computation*, **7**, 219–69.

[Goldstein, 1980] Goldstein, Herbert. (1980). *Classical Mechanics*. 2nd edn. Reading, MA: Addison-Wesley.

[Golub & Van Loan, 1996] Golub, G.H., & Van Loan, C.F. (1996). *Matrix Computations*. 3rd edn. Baltimore, MD: Johns Hopkins University Press.

[Gould & Eldredge, 1977] Gould, S.J., & Eldredge, N. (1977). Punctuated Equilibrium: The Tempo and Mode of Evolution Reconsidered. *Paleobiology*, **3**, 115–51.

[Goutte, 1997] Goutte, Cyril. (1997). Note on Free Lunches and Cross-Validation. *Neural Computation*, **9**, 1246–9.

[Grassberger & Procaccia, 1983a] Grassberger, P. & Procaccia, I.(1983a). Characterization of Strange Attractors. *Phys. Rev. Lett.*, **50**, 346–9.

[Grassberger & Procaccia, 1983b] Grassberger, P., & Procaccia, I. (1983b). Measuring the Strangeness of Strange Attractors. *Physica D*, **9**, 189–208.

[Guckenheimer & Holmes, 1983] Guckenheimer, John, & Holmes, Philip. (1983). *Nonlinear Oscillations, Dynamical Systems, and Bifurcations of Vector Fields*. New York, NY: Springer-Verlag.

[Guillemin & Pollack, 1974] Guillemin, Victor, & Pollack, Alan. (1974). *Differential Topology*. Englewood Cliffs, NJ: Prentice–Hall.

[Haar, 1910] Haar, A. (1910). Zur Theorie der Orthogonalen Funktionen-systeme. *Mathematics Analysis*, **69**, 331–71.

[Hair et al., 1998] Hair, Joseph, Black, William, Tatham, Ronald, & Anderson, Ralph. (1998). *Multivariate Data Analysis*. Englewood Cliffs, NJ: Prentice Hall.

[Hamilton, 1994] Hamilton, J. D. (1994). *Time Series Analysis*. Princeton, NJ: Princeton University Press.

[Hardy et al., 1976] Hardy, J., de Pazzis, O., & Pomeau, Y. (1976). Molecular Dynamics of a Classical Lattice Gas: Transport Properties and Time Correlation Functions. *Phys. Rev. A*, **13**, 1949–61.

[Hartigan, 1975] Hartigan, John A. (1975). *Clustering Algorithms*. New York, NY: Wiley.

[Hartman et al., 1990] Hartman, E.J., Keeler, J.D., & Kowalski, J.M. (1990). Layered Neural Networks with Gaussian Hidden Units As Universal Approximations. *Neural Computation*, **2**, 210–15.

[Hasslacher, 1987] Hasslacher, Brosl. (1987). Discrete Fluids. *Los Alamos Science*, 175–217.

[Hertz et al., 1991] Hertz, John A., Krogh, Anders S., & Palmer, Richard G. (1991). *Introduction to the Theory of Neural Computation*. Redwood City, CA: Addison-Wesley.

[Hildebrand, 1976] Hildebrand, Francis B. (1976). *Advanced Calculus for Applications*. 2nd edn. Englewood Cliffs, NJ: Prentice–Hall.

[Hill & Peterson, 1993] Hill, Fredrick J., & Peterson, Gerald R. (1993). *Computer Aided Logical Design with Emphasis on VLSI*. 4th edn. New York, NY: Wiley.

[Hlawatsch & Boudreaux-Bartels, 1992] Hlawatsch, F., & Boudreaux-Bartels, G.F. (1992). Linear and Quadratic Time-Frequency Signal Representations. *IEEE Signal Processing Magazine*, **9**, 21–67.

[Honerkamp, 1994] Honerkamp, Josef. (1994). *Stochastic Dynamical Systems: Concepts, Numerical Methods, Data Analysis*. New York, NY: VCH. Translated by Katja Lindenberg.

[Huang, 1987] Huang, Kerson. (1987). *Statistical Mechanics*. 2nd edn. New York, NY: Wiley.

[Hull, 1993] Hull, John. (1993). *Options, Futures, and Other Derivative Securities*. 2nd edn. Englewood Cliffs, NJ: Prentice Hall.

[Jordan & Jacobs, 1994] Jordan, M.I., & Jacobs, R.A. (1994). Hierarchical Mixtures of Experts and the EM Algorithm. *Neural Computation*, **6**, 181–214.

[Kamen, 1990] Kamen, Edward W. (1990). *Introduction to Signals and Systems*. 2nd edn. New York, NY: Macmillan.

[Kearns & Vazirani, 1994] Kearns, Michael J., & Vazirani, Umesh V. (1994). *An Introduction to Computational Learning Theory*. Cambridge, MA: MIT Press.

[Kempf & Frazier, 1997] Kempf, Renate, & Frazier, Chris (eds). (1997). *OpenGL*

*Reference Manual: The Official Reference Document to OpenGL, Version 1.1*. 2nd edn. Reading, MA: Addison Wesley.

[Kirk et al., 1993] Kirk, David B., Kerns, Douglas, Fleischer, Kurt, & Barr, Alan H. (1993). Analog VLSI Implementation of Multi-Dimensional Gradient Descent. *Advances in Neural Information Processing Systems*, vol. 5. San Mateo, CA: Morgan Kaufmann, pp. 789–96.

[Kirkpatrick, 1984] Kirkpatrick, S. (1984). Optimization by Simulated Annealing: Quantitative Studies. *J. Stat. Phys.*, **34**, 975–86.

[Kirkpatrick & Selman, 1994] Kirkpatrick, S., & Selman, B. (1994). Critical Behavior in the Satisfiability of Random Boolean Expressions. *Science*, **264**, 1297–301.

[Kirkpatrick et al., 1983] Kirkpatrick, S., Gelatt Jr., C.D., & Vecchi, M.P. (1983). Optimization by Simulated Annealing. *Science*, **220**, 671–80.

[Knuth, 1981] Knuth, Donald E. (1981). *Semi-Numerical Algorithms*. 2nd edn. The Art of Computer Programming, vol. 2. Reading, MA: Addison-Wesley.

[Kohonen, 1989] Kohonen, Teuvo. (1989). *Self-Organization and Associative Memory*. 3rd edn. New York, NY: Springer-Verlag.

[Kolmogorov, 1957] Kolmogorov, A.N. (1957). On The Representation of Continuous Functions of Several Variables by Superposition of Continuous Functions of One Variable and Addition. *Doklady Akameiia Nauk SSSR*, **114**, 953–6.

[Landauer, 1961] Landauer, Rolf. (1961). Irreversibility and Heat Generation in the Computing Process. *IBM J. Res. Dev.*, **5**, 183–91.

[Lawler & Wood, 1966] Lawler, E.W., & Wood, D.E. (1966). Branch-and-Bound Methods: a Survey. *Operations Research*, **14**, 699–719.

[Leff & Rex, 1990] Leff, Harvey S. & Rex, Andrew F. (eds). (1990). *Maxwell's Demon: Entropy, Information, Computing*. Princeton, NJ: Princeton University Press.

[Lei & Jordan, 1996] Lei, Xu, & Jordan, M.I. (1996). On Convergence Properties of the EM Algorithm for Gaussian Mixtures. *Neural Computation*, **8**, 129–51.

[Levinson, 1947] Levinson, N. (1947). The Wiener RMS (Root-Mean-Square) Error Criterion in Filter Design and Prediction. *Journal of Mathematical Physics*, **25**, 261–78.

[Lewis & Papadimitriou, 1981] Lewis, Harry R., & Papadimitriou, Christos H. (1981). *Elements of the Theory of Computation*. Englewood Cliffs, NJ: Prentice-Hall.

[Liu & Qian, 1995] Liu, Pei-Dong, & Qian, Min. (1995). *Smooth Ergodic Theory of Random Dynamical Systems*. Lecture Notes In Mathematics, vol. 1606. New York, NY: Springer-Verlag.

[Lloyd, 1993] Lloyd, Seth. (1993). A Potentially Realizable Quantum Computer. *Science*, **261**, 1569–1571.

[Lorenz, 1963] Lorenz, E. (1963). Deterministic Nonperiodic Flow. *J. Atmospheric Sci.*, **20**, 130–41.

[MacQueen, 1967] MacQueen, J. (1967). Some Methods for Classification and Analysis of Multivariate Observations. *Pages 281–97 of:* Cam, L. M. Le, & Neyman, J. (eds), *Proceedings of the Fifth Berkeley Symposium on Mathematical Statistics and Probability*, vol. 1. Berkeley: University of California Press.

[Macready & Wolpert, 1996] Macready, William G., & Wolpert, David H. (1996). What Makes an Optimization Problem Hard? *Complexity*, **1**, 40–6.

[Mandelbrot, 1983] Mandelbrot, Benoit B. (1983). *The Fractal Geometry of Nature*. New York, NY: W.H. Freeman.

[Marcienkiewicz, 1939] Marcienkiewicz, J. (1939). *Math. Z.*, **44**, 612.

[Marquardt, 1963] Marquardt, Donald W. (1963). An Algorithm for the Least-Squares Estimation of Nonlinear Parameters. *Journal of the Society for Industrial and Applied Mathematics*, **11**, 431–41.

[McLachlan & Basford, 1988] McLachlan, G.J., & Basford, K.E. (1988). *Mixture Models: Inference and Applications to Clustering*. New York, NY: Marcel Dekker.

[Merton, 1993] Merton, Robert K. (1993). *On The Shoulders Of Giants: A Shandean Postscript*. Chicago: University of Chicago Press.
[Metropolis et al., 1953] Metropolis, N., Rosenbluth, A.W., Rosenbluth, M.N., Teller, A.H., & Teller, E. (1953). Equation Of State Calculations by Fast Computing Machines. *Journal of Chemical Physics*, **21**, 1087–92.
[Mezard, 1987] Mezard, Marc. (1987). *Spin Glass Theory and Beyond*. Singapore: World Scientific.
[Miles, 1981] Miles, J.W. (1981). The Korteweg-de Vries Equation: A Historical Essay. *Journal of Fluid Mechanics*, **106**, 131–47.
[Minsky & Papert, 1988] Minsky, Marvin, & Papert, Seymour. (1988). *Perceptrons: An Introduction to Computational Geometry*. Expanded edn. Cambridge, MA: MIT Press.
[Nasrabadi & King, 1988] Nasrabadi, N.M., & King, R.A. (1988). Image Coding Using Vector Quantization: A review. *IEEE Trans. on Communications*, **COM-36**, 957–71.
[Nelder & Mead, 1965] Nelder, J.A., & Mead, R. (1965). A Simplex Method for Function Minimization. *Computer Journal*, **7**, 308–13.
[Newland, 1994] Newland, D.E. (1994). Wavelet Analysis of Vibration, Part I: Theory. *Journal of Vibration and Acoustics*, **116**, 409–16.
[Nye, 1992] Nye, Adrian (ed). (1992). *The Definitive Guides to the X Window System*. Sebastopol, CA: O'Reilly and Associates.
[Oppenheim & Schafer, 1989] Oppenheim, A.V., & Schafer, R.W. (1989). *Discrete-Time Signal Processing*. Englewood Cliffs, NJ: Prentice Hall.
[Packard et al., 1980] Packard, N.H., Crutchfield, J.P., Farmer, J.D., & Shaw, R.S. (1980). Geometry From a Time Series. *Phys. Rev. Lett.*, **45**, 712–16.
[Parker, 1977] Parker, R.L. (1977). Understanding Inverse Theory. *Ann. Rev. Earth Planet. Sci.*, **5**, 35–64.
[Parlitz et al., 1997] Parlitz, U., Junge, L., & Kocarev, L. (1997). Subharmonic Entrainment of Unstable Period Orbits and Generalized Synchronization. *Physical Review Letters*, **79**, 3158–61.
[Pearson, 1990] Pearson, Carl E. (1990). *Handbook of Applied Mathematics: Selected Results and Methods*. 2nd edn. New York, NY: Van Nostrand Reinhold.
[Pecora et al., 1997] Pecora, L.M., Carroll, T.L., Johnson, G.A., Mar, D.J., & Heagy, J.F. (1997). Fundamentals of Synchronization in Chaotic Systems, concepts, and applications. *Chaos*, **7**, 520–43.
[Powell, 1992] Powell, M.J.D.(1992). The Theory of Radial Basis Function Approximation in 1990. Pages pp. 105–210 of: Light, Will (ed), *Advances in Numerical Analysis*, vol. II. Oxford: Oxford University Press.
[Preparata & Shamos, 1985] Preparata, Franco P. & Shamos, Michael Ian. (1985). *Computational Geometry: An Introduction*. New York, NY: Springer-Verlag.
[Press et al., 1992] Press, William H., Teukolsky, Saul A., Vetterling, William T., & Flannery, Brian P. (1992). *Numerical Recipes in C: The Art of Scientific Computing*. 2nd edn. New York, NY: Cambridge University Press.
[Priestley, 1981] Priestley, Maurice B. (1981). *Spectral Analysis and Time Series*. New York, NY: Academic Press.
[Priestley, 1991] Priestley, Maurice B. (1991). *Non-Linear and Non-Stationary Time Series Analysis*. San Diego, CA: Academic Press.
[Prügel-Bennett & Shapiro, 1994] Prügel-Bennett, Adam, & Shapiro, Jonathan L. (1994). Analysis of Genetic Algorithms Using Statistical Mechanics. *Physical Review Letters*, **72**, 1305–9.
[Rabiner, 1989] Rabiner, Lawrence R.(1989). A Tutorial on Hidden Markov Models and Selected Applications in Speech Recognition. *Proceedings of the IEEE*, **77**, 257–86.

[Reichl, 1984] Reichl, L.E. (1984). *A Modern Course in Statistical Physics*. Austin, TX: University of Texas Press.
[Richards, 1988] Richards, Whitman (ed). (1988). *Natural Computation*. Cambridge, MA: MIT Press.
[Richardson & Green, 1997] Richardson, Sylvia, & Green, Peter J. (1997). On Bayesian Analysis of Mixtures with an Unknown Number of Components. *J. R. Statist. Soc. B*, **59**, 731–92.
[Rissanen, 1986] Rissanen, J. (1986). Stochastic Complexity and Modelling. *Ann. Stat.*, **14**, 1080–100.
[Rothman & Zaleski, 1994] Rothman, Daniel H., & Zaleski, Stéphane. (1994). Lattice-Gas Models of Phase Separation: Interfaces, Phase Transitions, and Multiphase Flow. *Review of Modern Physics*, **66**, 1417–79.
[Rothman & Zaleski, 1997] Rothman, Daniel H., & Zaleski, Stéphane. (1997). *Lattice-Gas Cellular Automata: Simple Models of Complex Hydrodynamics*. New York, NY: Cambridge University Press.
[Rumelhart *et al.*, 1996] Rumelhart, D. E., Durbin, R., Golden, R., & Chauvin, Y. (1996). Backpropagation: The Basic Theory. *Pages 533–66 of:* Smolensky, P., Mozer, M. C., & Rumelhart, D. E. (eds), *Mathematical Perspectives on Neural Networks*. Hillsdale, NJ: Lawrence Erlbaum Associates.
[Russell *et al.*, 1980] Russell, D.A., Hanson, J.D., & Ott, E. (1980). Dimension of Strange Attractors. *Physical Review Letters*, **45**, 1175–8.
[Saff & Snider, 1993] Saff, E.B., & Snider, A.D. (1993). *Fundamentals of Complex Analysis for Mathematics, Science, and Engineering*. 2nd edn. Englewood Cliffs, NJ: Prentice Hall.
[Sauer *et al.*, 1991] Sauer, T., Yorke, J.A., & Casdagli, M. (1991). Embedology. *J. Stat. Phys.*, **65**, 579–616.
[Sauer *et al.*, 1997] Sauer, T., Grebogi, C., & Yorke, J.A. (1997). How Long do Numerical Chaotic Solutions Remain Valid? *Phys. Rev. Lett.*, **79**, 59–62.
[Scheck, 1990] Scheck, Florian. (1990). *Mechanics : From Newton's Laws to Deterministic Chaos*. New York, NY: Springer-Verlag.
[Shaw, 1981] Shaw, R. (1981). Strange Attractors, Chaotic Behavior and Information Flow. *Z. Naturforsch. A*, **36A**, 80–112.
[Silverman, 1986] Silverman, B.W.(1986). *Density Estimation for Statistics and Data Analysis*. New York, NY: Chapman and Hall.
[Simmons, 1992] Simmons, G.J. (ed). (1992). *Contemporary Cryptology: The Science of Information Integrity*. Piscataway, NJ: IEEE Press.
[Simon *et al.*, 1994] Simon, M.K., Omura, J.K., Scholtz, R.A., & Levitt, B.K. (1994). *Spread Spectrum Communications Handbook*. New York, NY: McGraw-Hill.
[Sklar, 1988] Sklar, Bernard. (1988). *Digital Communications: Fundamentals and Applications*. Englewood Cliffs, NJ: Prentice Hall.
[Smyth *et al.*, 1997] Smyth, P., Heckerman, D., & Jordan, M.I. (1997). Probabilistic Independence Networks for Hidden Markov Probability Models. *Neural Computation*, **9**, 227–69.
[Stark *et al.*, 1997] Stark, J., Broomhead, D.S., Davies, M.E., & Huke, J. (1997). Takens Embedding Theorems for Forced and Stochastic Systems. *Nonlinear Analysis*, **30**, 5303–314.
[Stevens, 1990] Stevens, W. Richard. (1990). *UNIX Network Programming*. Englewood Cliffs, NJ: Prentice Hall.
[Stoer & Bulirsch, 1993] Stoer, J., & Bulirsch, R. (1993). *Introduction to Numerical Analysis*. 2nd edn. New York, NY: Springer-Verlag. Translated by R. Bartels, W. Gautschi, and C. Witzgall.
[Strang, 1986] Strang, Gilbert. (1986). *Introduction to Applied Mathematics*. Wellesley, MA: Wellesley-Cambridge Press.

[Strang, 1988] Strang, Gilbert. (1988). *Linear Algebra and its Applications.* 3rd edn. San Diego, CA: Harcourt, Brace, Jovanovich.

[Taft & Walden, 1990] Taft, Ed, & Walden, Jeff. (1990). *Postscript Language Reference Manual.* 2nd edn. Reading, MA: Addison-Wesley.

[Takens, 1981] Takens, Floris. (1981). Detecting Strange Attractors in Turbulence. *Pages 366–81 of:* Rand, D.A., & Young, L.S. (eds), *Dynamical Systems and Turbulence.* Lecture Notes in Mathematics, vol. 898. New York, NY: Springer-Verlag.

[Tanenbaum, 1988] Tanenbaum, Andrew S. (1988). *Computer Networks.* 2nd edn. Englewood Cliffs, NJ: Prentice-Hall.

[Taylor & Wheeler, 1992] Taylor, Edwin F., & Wheeler, John Archibald. (1992). *Spacetime Physics: Introduction to Special Relativity.* 2nd edn. New York, NY: W.H. Freeman.

[Temam, 1988] Temam, Roger. (1988). *Infinite-Dimensional Dynamical Systems in Mechanics and Physics.* Applied Mathematical Sciences, vol. 68. New York, NY: Springer-Verlag.

[Theiler *et al.*, 1992] Theiler, J., Eubank, S., Longtin, A., Galdrikian, B., & Farmer, J.D. (1992). Testing For Nonlinearity in Time Series: The Method of Surrogate Data. *Physica D*, **58**, 77–94.

[Therrien, 1989] Therrien, Charles W. (1989). *Decision, Estimation, and Classification: An Introduction to Pattern Recognition and Related Topics.* New York, NY: Wiley.

[Toffoli & Margolus, 1991] Toffoli, Tommaso, & Margolus, Norman. (1991). *Cellular Automata Machines: A New Environment for Modeling.* Cambridge, MA: MIT Press.

[Toffoli & Quick, 1997] Toffoli, Tommaso, & Quick, Jason. (1997). Three-Dimensional Rotations by Three Shears. *Graphical Models and Image Processing*, **59**, 89–95.

[Tong & Lim, 1980] Tong, H., & Lim, K.S. (1980). Threshold Autoregression, Limit Cycles and Cyclical Data. *J. Roy. Stat. Soc. B*, **42**, 245–92.

[Turing, 1936] Turing, A.M. (1936). On Computable Numbers, With An Application To The *Entscheidungsproblem. Proc. London Math. Soc.*, **42**, 230–65.

[Vapnik & Chervonenkis, 1971] Vapnik, V.N., & Chervonenkis, A.Y. (1971). On the Uniform Convergence of Relative Frequencies of Events To Their Probabilities. *Theory of Probability and its Applications*, **16**, 264–80.

[Weigend *et al.*, 1990] Weigend, A. S., Huberman, B. A., & Rumelhart, D. E. (1990). Predicting the Future: A Connectionist Approach. *International Journal of Neural Systems*, **1**, 193–209.

[Weigend *et al.*, 1996] Weigend, A. S., Zimmermann, H. G., & Neuneier, R. (1996). Clearning. *Pages 511–22 of:* Refenes, A.-P. N., Abu-Mostafa, Y., Moody, J., & Weigend, A. (eds), *Neural Networks in Financial Engineering (Proceedings of the Third International Conference on Neural Networks in the Capital Markets, NNCM-95).* Singapore: World Scientific.

[Weigend *et al.*, 1995] Weigend, A.S., Mangeas, M., & Srivastava, A.N. (1995). Nonlinear Gated Experts for Time Series: Discovering Regimes and Avoiding Overfitting. *International Journal of Neural Systems*, **6**, 373–99.

[Weigend & Gershenfeld, 1993] Weigend, Andreas S., & Gershenfeld, Neil A. (eds). (1993). *Time Series Prediction: Forecasting the Future and Understanding the Past.* Santa Fe Institute Studies in the Sciences of Complexity. Reading, MA: Addison–Wesley.

[Wernecke, 1994] Wernecke, Josie. (1994). *The Inventor Mentor: Programming Object-Oriented 3D Graphics With Open Inventor, Release 2.* Reading, MA: Addison-Wesley.

[Whitham, 1974] Whitham, Gerald B. (1974). *Linear and Nonlinear Waves*. New York, NY: Wiley-Interscience.

[Wolaver, 1991] Wolaver, Dan H. (1991). *Phase-Locked Loop Circuit Design*. Englewood Cliffs, NJ: Prentice Hall.

[Wolfram, 1986] Wolfram, Stephen. (1986). Cellular Automaton Fluids 1: Basic Theory. *Journal of Statistical Physics*, **45**, 471–526.

[Wolpert & Macready, 1995] Wolpert, David H., & Macready, William G. (1995). *No Free-Lunch Theorems for Search*. Santa Fe Institute working paper 95-02-010.

[Woo et al., 1997] Woo, Mason, Neider, Jackie, & Davis, Tom. (1997). *OpenGL Programming Guide: The Official Guide to Learning OpenGL, Version 1.1*. 2nd edn. Reading, MA: Addison Wesley.

[Wyld, 1976] Wyld, Henry W. (1976). *Mathematical Methods for Physics*. Reading, MA: W.A. Benjamin.

[Young & Gregory, 1988] Young, David M., & Gregory, Robert Todd. (1988). *A Survey of Numerical Mathematics*. New York, NY: Dover Publications. 2 volumes.

[Yule, 1927] Yule, G. U. (1927). On a Method of Investigating Periodicities in Disturbed Series with Special Reference to Wolfer's Sunspot Numbers. *Philosophical Transactions Royal Society London Ser. A*, **226**, 267–98.

[Zabusky, 1981] Zabusky, N.J. (1981). Computational Synergetics and Mathematical Innovation. *Journal of Computational Physics*, **43**, 195–249.

[Zwillinger, 1992] Zwillinger, Daniel. (1992). *Handbook of Differential Equations*. 2nd edn. New York, NY: Academic Press.

# Index

1/f noise, 214

Action, 38
Activation function, 150
Adams–Bashforth–Moulton method, 74
ADI, 84
Adjoint, 128
AFSR, 196
AIC, 207
Akaike information criteria, 207
Algorithmic information theory, 115
Aliasing, 130, 266
Alternating-direction implicit method, 84
Analog feedback shift register, 196
Analytical, 7
ANN, 138
Annealing, 162
Ansatz, 10
ANSI, 250
Applet, 245
AR, 205
Argmax, 202
ARIMA, 205
ARMA, 205
ARPANET, 250
Artificial neural networks, 138
Associative, 129
Asymptotic expansion, 18
Asynchronous connection, 237
Auto-regressive model, 205
Autocorrelation function, 51, 206
Autocovariance function, 51
Autonomous system, 212

Bézier curve, 142
Back propagation, 153
Backwards algorithm, 201
Backwards difference, 78
BASIC, 227
Baum–Welch algorithm, 201
Bayes' rule, 46, 116
Bayesian, 47, 116
Bayesian networks, 178
BBGKY hierarchy, 105
Bernstein polynomials, 142
Bessel's equation, 30
Bias/variance tradeoff, 113
BIC, 207
Bimodal distribution, 45
Bispectrum, 212

Blending functions, 141
Blessing of dimensionality, 167
Block entropy, 217
Boltzmann equation, 105
Boltzmann factor, 163
Bootstrap methods, 148
Boundary conditions, 10
Boundary-value problem, 10, 76
Box–Jenkins method, 207
Branch-and-bound algorithm, 156
Brownian motion, 53
Bubnov–Galerkin method, 95
Bulirsch–Stoer method, 75

C, 226
CA, 102
Calculator, 65
Catenary, 270
Cauchy–Schwarz inequality, 126
Cellular automata, 102
Central limit theorem, 49, 117
Chapman–Enskog expansion, 106
Chapman–Kolmogorov equation, 52
Characteristic equation, 10
Characteristic function, 48, 212
Characteristics, 26
Chebyshev polynomials, 145
Circulant matrix, 133
Classes, 245
Client, 235
Closed-form solution, 7
Cluster-weighted modeling, 178
Clustering, 175
Coarse-grained parallelism, 256
Collocation, 94
Colormap, 237
Commutative, 129
Compact support, 136, 172
Complete basis, 94, 144
Complete solution, 10
Conjugate gradient algorithm, 160
Connectionist, 150
Connectionless protocol, 251
Content-addressable memory, 185
Control points, 141
Control theory, 204
Corrector step, 74
Correlation dimension, 215
Cost function, 154
Courant–Friedrichs–Levy condition, 80

Covariance matrix, 136
Cramér–Rao bound, 125
Crank–Nicholson method, 84
Cross-validation, 147
Cumulant, 50
Cumulative distribution, 48, 172
Curse of dimensionality, 149

DAE, 67
Datagram, 250
Daubechies wavelets, 132
de Bruijn's identity, 126
Delta function, 51
Density estimation, 169
Detailed balance, 55
Device-independent, 230
DFT, 129
Diagonal approximation, 155, 309
Diagonalization, 13
Diffeomorphic, 210
Differential-algebraic equation, 67
Diffusion noise, 214
Dilation equation, 135
Dirichlet boundary conditions, 84
Discrete Fourier transform, 129
Discrete wavelet transformation, 135
Dispersion, 25
Distributive, 129
Divide-and-conquer, 130
Double buffering, 247
Downhill simplex method, 157
DSC, 231
DWT, 135

Early stopping, 153
Eispack, 228
Elliptic PDE, 27
EM, 177, 201
Embedding, 208, 210
Encapsulated PostScript, 231
Entrainment, 196
Entropy, 217
EPS, 231
Equipartition theorem, 55
Ergodic, 52, 212
Error model, 116
Euclidean norm, 129
Euler angles, 42
Euler equations for rigid bodies, 42
Euler's equation, 35
Euler's method, 68
Evidence, 116
Excel, 229
Expectation value, 45
Expectation-maximization, 177, 201
Extended Kalman filter, 195
Extrapolation, 139

Factor analysis, 137
Fast Fourier transform, 66, 130
Feed-forward network, 153
Fermat's principle, 38
FFT, 130
Filter bank, 132
Fine-grained parallelism, 256
Finite differences, 78

Finite elements, 76, 93
Finite Impulse Response, 205
FIR, 205
Fisher information, 125
Fixed boundary condition, 82
Fokker–Plank equation, 54
Fortran, 226
Forward algorithm, 200
Forward difference, 78
Fourier transform, 50
Fourier transform PDE method, 88
Fourier–Galerkin method, 95
Fourth-order Runge–Kutta method, 70
Fractal, 215
Frequentist, 47
Full rank, 119
Function approximation, 115

GA, 164
Galerkin method, 94
Game of life, 102
Gated experts, 178
Gauss elimination, 83
Gauss–Seidel method, 89
Gaussian distribution, 49
General solution, 10
Generalized coordinates, 38
Generalized dimension, 215
Generating partition, 219
Generic, 210
Genetic algorithms, 164
Geometric sum, 305
GLUT, 240
Gradient descent, 123
Gram–Schmidt orthogonalization, 144

Haar wavelets, 132
Halting problem, 115, 231
Haptic interfaces, 225
Hard clustering, 184
Harmonic wavelets, 136
Hat functions, 95
Hausdorff dimension, 215
Helmholtz's equation, 28
Hermite interpolation, 98
Hermite polynomials, 145
Hessian, 123
Heun method, 76, 276
Hidden Markov model, 186, 197
Hidden units, 150
HMM, 197
Homogeneous differential equation, 9
Homogeneous fractal, 216
HTML, 246
Hyper-parameters, 147
Hyperbolic functions, 37
Hyperbolic PDE, 27

ICA, 137
IID, 48, 117
IIR, 205
IMSL, 227
In-sample error, 147
Independent components analysis, 137
Independent variables, 47
Index of refraction, 38

Inertia tensor, 41
Infinite impulse response, 205
Information, 217
Information dimension, 215
Inhomogeneous differential equation, 9
Initial-value problem, 10
Innovations, 205
Integral of the motion, 39
Internet, 250
Internetworking, 250
Interpolation, 139
Inverse problems, 154
IP, 250
ISO, 250

Jacobi's method, 89
Jacobian, 48
Java, 227, 244

k-D tree, 170
k-means, 184
Kalman filter, 186
Kalman gain matrix, 192
Karhunen–Loéve transform, 137
Kernel density estimation, 175
KLT, 137
Knots, 141

Lagrange interpolation, 97
Lagrange multiplier, 37
Lagrangian, 38
Laguerre polynomials, 145
Langevin equation, 55
Laplace transform, 13
Laplace's equation, 27, 84
Lattice gas, 102
Law of large numbers, 49
Lax method, 80
Leapfrog method, 81
Least squares error model, 117, 118
Least squares weighting, 94
Legendre's equation, 32
Levenberg–Marquardt method, 124
Levinson–Durbin recursion, 207
Lexicographic sort, 169
LFSR, 59
Likelihood, 116
Line minimization, 159
Linear congruential random numbers, 57
Linear feedback shift register, 59
Linear least squares, 118
Linpack, 228
LISP, 227
Log-likelihood, 117
Lorenz set, 95, 208
Lossy compression, 137
Lyapunov exponents, 216

MA, 205
Macsyma, 228
MAP, 117
Maple, 228
Marginalize, 46
Markov chain, 51
Markov process, 51
Master equation, 55

Matched filter, 186, 187
Mathcad, 230
Mathematica, 228
Matlab, 228
Maximal LFSR, 59
Maximum entropy, 127, 302
Maximum likelihood, 46, 117
Maximum *a posteriori*, 117
MDL, 115
Mean value, 45
Mesh generation, 98
Meta-modeling, 2
Methods, 245
Mflop, 257
Midpoint method, 70
Minimum description length, 2, 115
Mixture models, 175
Mixture of experts, 178
ML, 117
MLP, 138
Mod map, 207
Model errors, 113, 147
Molecular dynamics, 102
Moment generating function, 49
Moment of inertia, 41
Momentum, 161
Monte-Carlo sampling, 116
Moving average model, 205
MPI, 254
Multigrid methods, 91
Multilayer perceptrons, 138
Multistationary, 153
Mutual information, 218

NAG, 227
NAR, 208
Narrow-sense stationarity, 51
Natural splines, 141
Navier–Stokes equation, 107
Nelder–Mead method, 157
Netlib, 227
Neumann boundary conditions, 84
Neural networks, 150
Newton's method, 123
No free lunch theorem, 113
Noether's theorem, 39
Noise reduction, 186
Nonlinear least squares, 122
Nonparametric fitting, 115
Nonuniform B-splines, 141
Nonuniform rational B-splines, 142
Normal distriibution, 49
Normal modes, 13
Nullspace, 119
Numerical dissipation, 81
Numerical quadrature, 73
Numerical Recipes, 227
NURBS, 142
Nyquist frequency, 130

$\mathcal{O}$, 18
Object oriented programming, 245
ODE, 9
One-sided $z$-transform, 19
One-sided Laplace transform, 13
Open Inventor, 244

OpenGL, 240
Operator splitting, 84
Order of differential equation, 9
Ordinary differential equation, 7, 9
Orthogonal functions, 94, 143
Orthogonal matrix, 119, 128
Orthogonal transformation, 128
Orthogonal vectors, 128
Orthonormal functions, 143
Orthonormal vectors, 128
OSI, 250
Out-of-sample error, 147
Outer product, 136
Overfitting, 147

Packet, 250
Padé approximant, 139
Parabolic PDE, 27
Parametric fitting, 115
Partial differential equation, 7, 9
Partial fraction expansion, 14, 15
Particular solution, 10
Pattern recognition, 184
PCA, 137
PDE, 9
PDF, 231
Penalty function, 154
Perceptrons, 152
Periodic boundary condition, 82
Pesin's identity, 219
Phase-locked loop, PLL, 323
Point predictions, 221
Poisson's equation, 27, 84
Poles, 15, 206
Polynomial expansion, 139
Port, 251
Portable Document Format, 231
Positive definite matrix, 177
PostScript, 230
Powell's method, 159
Power spectral density, 50
Prediction, 186
Predictor step, 73
Predictor-corrector method, 72
Principal axes, 41
Principal components analysis, 137
Printer's point, 231
Prior, 47, 116, 154
Probability density, 169
Probability distribution, 7
Pseudo-inverse, 119
Punctuated equilibrium, 316

$Q$, quality factor, 22, 261
Quadrature mirror filter, 133

Radial basis functions, 146
Radial correlation function, 214
Random variable, 44
Range, 119
Rank, 119
Rational function, 15
Rayleigh–Ritz method, 93, 100
RBF, 146
Realization of stochastic process, 45
Recurrent network, 153

Recursion relation, 59
Redundancy, 218
Regularization, 154
Relaxation technique, 89
Representation, 128
Residual, 94, 120
Richardson extrapolation, 75
Rigid body motion, 40

Schwarz' inequality, 187
Score, 125
Second moment, 46
Second-order Runge–Kutta method, 70
Separation of variables, 27
Server, 235
Shape functions, 97
Sherman–Morrison formula, 87
Shooting method, 76
Short-time Fourier transform, 131
Side-lobes, 130
Signal separation, 186
SIMD, 107
Simplex, 157
Simulated annealing, 163
Sinc function, 130
Singular value decomposition, 119
Singular values, 119
Smoothing, 186
Snell's law, 270
Socket, 251
Soft clustering, 184
Solitons, 65
Sonification, 225
SOR, 90
Source entropy, 219
SPEC, 257
Spectral radius, 89
Spin glass, 167
Splines, 98, 141
Spread spectrum, 196
Spreadsheet, 229
Standard deviation, 46
State-space reconstruction, 208
Stationary process, 51
Steepest descent, 123
STFT, 131
Stiff differential equation, 75
Stochastic processes, 7, 50
Stochasticity, 44
Strange attractor, 217
Strict stationarity, 51
Successive over-relaxation method, 90
Support, 45
Surrogate data, 214
SVD, 119

TAR, 208
TCP, 251
Threads, 246
Time series analysis, 204
TLA, 137
Transfer function, 17, 206
Transpose, 128
Transversality, 211
Trellis, 199
Trending, 205

Tridiagonal matrix, 83
Two-sided Laplace transform, 13

UDP, 251
Uncorrelated variables, 47
Underfitting, 147
Uniform Resource Locator, 2
Unitary, 128
Unreliable protocol, 251
Unsupervised learning, 175
Upper-diagonal matrix, 83
URL, 2

Vandermonde matrix, 121
Vapnik–Chervonenkis dimension, 148
Variable subset selection, 137
Variance, 46
Variational methods, 7
VC, 148
Vector quantization, 185
Visualization, 225
Viterbi algorithm, 202
Volterra series, 208
von Neumann stability analysis, 80

Voronoi tesselation, 184
VRML, 244

Wavelets, 131
Weak stationarity, 51
Weighted residuals, 93
Whitney embedding theorem, 211
Wide-sense stationarity, 51
Wiener filter, 186, 187
Wiener process, 54
Wiener–Hopf equation, 189
Wiener–Khinchin theorem, 51
Wigner function, 136
Windowing, 130
Wold decomposition, 207
World Wide Web, 2

X Windows, 234

Yule–Walker equation, 207

z-transform, 19
Zeros, 15, 206